Lecture Notes in Computer Science 3311

Commenced Publication in 1973
Founding and Former Series Editors:
Gerhard Goos, Juris Hartmanis, and Jan van Leeuwen

Vincent Roca   Franck Rousseau (Eds.)

Interactive Multimedia and Next Generation Networks

Second International Workshop
on Multimedia Interactive Protocols and Systems, MIPS 2004
Grenoble, France, November 16-19, 2004
Proceedings

 Springer

Volume Editors

Vincent Roca
INRIA Rhône-Alpes
655 av. de l'Europe, Montbonnot, 38334 ST ISMIER cedex, France
E-mail: vincent.roca@inrialpes.fr

Franck Rousseau
LSR-IMAG
681 rue de la Passerelle - BP. 72, 38402 Saint Martin d'Hères, France
E-mail: Franck.Rousseau@imag.fr

Library of Congress Control Number: 2004115079

CR Subject Classification (1998): H.5.1, C.2, H.4, H.5, H.3, D.2

ISSN 0302-9743
ISBN 3-540-23928-6 Springer Berlin Heidelberg New York

Springer is a part of Springer Science+Business Media

springeronline.com

© Springer-Verlag Berlin Heidelberg 2004
Printed in Germany

Typesetting: Camera-ready by author, data conversion by PTP-Berlin, Protago-TeX-Production GmbH
Printed on acid-free paper   SPIN: 11354420   06/3142   5 4 3 2 1 0

Preface

Interactive Distributed Multimedia Systems (IDMS) and Protocols for Multimedia Systems (PROMS) have been two successful series of international events bringing together researchers, developers and practitioners from academia and industry in all areas of multimedia systems. These two workshops successfully merged in 2003 and now constitute the MIPS workshop.

After the outstanding MIPS 2003 workshop, organized in Naples, Italy, by Giorgio Ventre and Roberto Canonico, from the University of Naples Federico II, MIPS 2004 moved to Grenoble, France. Following the great tradition, MIPS 2004 was intended to contribute to scientific, strategic and practical advances in the area of distributed multimedia applications, protocols and intelligent management tools, with emphasis on their provision over novel network architectures. This is undoubtedly a rather broad area, which is confirmed by the large range of topics that were addressed in the submitted (and accepted) papers.

This year the Call for Papers attracted 74 submissions, essentially from Europe and Asia, plus a few contributions from North America, the Middle East, and Africa, for a total of 20 countries. We would like to express our warmest gratitude to all the authors, without whom organizing this event would have been impossible!

Thanks to the outstanding work of the Program Committee and the additional reviewers, 20 full-sized papers and 5 additional short papers were finally accepted, which was not an easy task. Like MIPS 2003, MIPS 2004 remained a highly selective event (33% acceptance ratio, including the short papers) which is the best warrant of a good program quality. Additionally, all accepted papers were carefully shepherded by some members of the Program Committee, in order to warrant a high quality to the papers included in this proceedings. We want to warmly acknowledge the hard and never sufficiently rewarded work done by the Program Committee, the additional reviewers, and the shepherds. Thanks to all of you!

The 25 selected papers were organized into 9 single-track sessions: VoIP and audio transport, video encoding (I and II), multi-source multimedia, multicasting and broadcasting, scheduling schemes, content management, multimedia services and, last but not least, security. This rich program was nonetheless coherent, and many (if not most) papers were written with transmission and networking aspects in mind, which remains, historically, an important field of interest of many contributors and members of the Steering and Program Committees.

The MIPS 2004 workshop featured two half-day, outstanding tutorials. The first one was given by Prof. Vera Goebel and Thomas Plageman, from Oslo University, and was entitled "Data Stream Management Systems (DSMS): Concepts, Prototypes, and Applications." The fundamental difference with a classical database system is the data stream model. Instead of processing a query over a persistent set of data that is stored in advance on disk, queries are performed in

DSMSs over a data stream, with data elements that arrive dynamically and are only available for a limited time period.

The second tutorial was given by Rod Walsh, Toni Paila and Harsh Meta, three leading industrial experts very active in several international standardization organizations, and working at Nokia Corporation, Finland. Their tutorial, entitled "Advances in Mass Media Delivery to Mobiles," gave a state of the art of the current standardization efforts surrounding the area of Mass Media Delivery for mobile devices, in particular at the IETF, 3GPP and DVB. We would like to thank these five tutorial authors for their work and believe that these two tutorials perfectly complemented MIPS technical program.

Finally we would like to express our gratitude to all organizations and companies that supported MIPS 2004 in one way or another: first of all, the INRIA institute that accepted all the financial risks, and managed many technical details thanks to the inestimable help of Daniele Herzog. We would also like to thank the IMAG institute as well as the INPG for their technical and financial support. France Télécom R&D and STMicroelectronics also had a major impact on this event, both from a financial and human aspect. Finally we would like to thank ACM for the confidence they expressed in the quality of this workshop.

We hope all participants really appreciated this workshop and found it useful.

November 2004   Vincent Roca and Franck Rousseau

Organization

MIPS 2004 was organized by INRIA (Institut National de Recherches en Informatique et en Automatique), IMAG (Informatique et Mathematiques Appliquées de Grenoble), and INPG (Institut National Polytechniques de Grenoble), in cooperation with the ACM-SIGCOMM and ACM-SIGMM Special Interest Groups.

Program Chairs

Vincent Roca   (INRIA Rhône-Alpes, France)
Franck Rousseau (ENSIMAG, LSR-IMAG, France)

Steering Committee

Fernando Boavida   (University of Coimbra, Portugal)
Roberto Canonico   (University of Naples Federico II, Italy)
Laurent Mathy   (Lancaster University, United Kingdom)
Edmundo Monteiro   (University of Coimbra, Portugal)
Zdzislaw Papir   (AGH University of Technology, Poland)
Hans Scholten   (Twente University, The Netherlands)
Marten van Sinderen (Twente University, The Netherlands)
Giorgio Ventre   (University of Naples Federico II, Italy)

Program Committee

Arturo Azcorra   (Carlos III University, Madrid, Spain)
Fernando Boavida   (University of Coimbra, Portugal)
Olivier Bonaventure (CSE Dept. UCL, Belgium)
Torsten Braun   (University of Berne, Switzerland)
Andrew Campbell   (Columbia University, USA)
Roberto Canonico   (University of Naples Federico II, Italy)
Augusto Casaca   (INESC, Portugal)
Jon Crowcroft   (University of Cambridge, United Kingdom)
Jose de Rezende   (Brazil)
Michel Diaz   (LAAS-CNRS, France)
Andrzej Duda   (ENSIMAG, LSR-IMAG, France)
Wolfgang Effelsberg (University of Mannheim, Germany)
Frank Eliassen   (University of Oslo, Norway)
Serge Fdida   (LIP6, France)

Additional Reviewers

Salvatore d'Antonio	(CINI Consortium, Naples, Italy)
Maurizio d'Arienzo	(University of Naples Federico II, Italy)
Marylin Arndt	(France Télécom R&D, France)
Guillaume Auriol	(LAAS-CNRS, France)
Simon Balon	(University of Liège, Belgium)
Julien Bourgeois	(University of Franche-Comté, France)
Claude Castelluccia	(INRIA and Univ. of California/Irvine, France)
Christophe Deleuze	(ASSEDIC, France)
Donato Emma	(University of Naples Federico II, Italy)
Ernesto Exposito	(National ICT, Australia)
David Garduno	(LAAS-CNRS, France)
Martin Karsten	(Waterloo University, Canada)
Andreas Kleinschmitt	(TU Braunschweig, Germany)
Zefir Kurtisi	(TU Braunschweig, Germany)
Nicolas Larrieu	(LAAS-CNRS, France)
Gerald Maguire	(KTH, Sweden)
Ian Marsh	(SICS, Sweden)
Adrian Munteanu	(University of Brussels (VUB), Belgium)
Antonio Pescape	(CINI Consortium, Napoli, Italy)
Hans Ole Rafaelsen	(University of Tromso, Norway)
Jean-Luc Richier	(LSR-IMAG, France)
Julien Ridoux	(LIP6, France)

Local Organizing Committee

Daniele Herzog	(INRIA Rhône-Alpes, France)
Vincent Roca	(INRIA Rhône-Alpes, France)
Franck Rousseau	(ENSIMAG, LSR-IMAG, France)
Christoph Neumann	(INRIA Rhône-Alpes, France)

Sponsoring Institutions

INRIA (Institut National de Recherches en Informatique et en Automatique)
IMAG (Informatique et Mathematiques Appliquées de Grenoble)
INPG (Institut National Polytechniques de Grenoble)
France Télécom R&D
STMicroelectronics
ACM-SIGCOMM and ACM-SIGMM Special Interest Groups

Table of Contents

Multicasting and Broadcasting

Scheduling Schemes

Video Encoding II

Content Management

Multimedia Services

Security

Author Index 287

Minimising *Perceived* Latency in Audio-Conferencing Systems over Application-Level Multicast

Nick Blundell and Laurent Mathy

Lancaster University, UK
{n.blundell, laurent}@comp.lancs.ac.uk

Abstract. In this paper, we propose a scalable and dynamically-adapting application-level multicast (ALM) routing protocol, designed specifically for audio-conferencing systems over the Internet.

Currently proposed ALM protocols try to optimise delay for the whole group of participating nodes during construction and maintenance of an overlay network which, when using standard packet flooding, can result in a number of the participants experiencing unacceptably-high latencies, unsuitable for real-time audio communication; whereas we propose to dynamically prioritise routing for those participants who are currently *in* conversation (i.e. those who require the lowest latencies in order to react to conversational cues) and allow higher latencies for participants who simply listen to the conversation without taking an active part in it in that particular moment in time.

Thus, we aim to provide low *perceived* latency for *all* of the audio-conference participants without any support from the underlying network.

1 Introduction

As a result of improvements in computer hardware, research into voice-over-IP (VoIP) technologies, and increased network bandwidth available to the home, point-to-point audio communication can more-or-less be achieved over the Internet, provided that network delay and packet loss do not exceed their tolerable thresholds (see section 2.1).

Group audio communication, on the other hand, has proven more challenging in the way of deployment, scalability, and of communication-channel quality: network-level multicast was proposed over a decade ago [6] as a solution for efficient, large-scale group communication over the Internet, but wide-scale deployment of the service has since been hampered due to various technical and administrative issues that surround it [4]. In response to the lack of a group-communication service in the network, various application-level techniques have been proposed [7].

In application-level multicast (ALM), many-to-many communication is achieved through using overlay trees in one of three ways: (1) a sender floods

V. Roca and F. Rousseau (Eds.): MIPS 2004, LNCS 3311, pp. 1–12, 2004.

data to their subtree through their children and also sends data to the tree root, who, in turn, floods the data to the rest of the group through its children; (2) a sender floods data to the group through their children and through their tree parent such that data flow is bi-directional on tree links [2]; or (3) several trees are built, rooted at each source, allowing data to be flooded to the whole group by each source as does a single source in one-to-many communication [4].

ALM has been proposed as a solution for audio conferencing [4], however, as an ALM group grows in size there is, inevitably, an increased imbalance in the degrees of latency (end-to-end delay) experienced by different, communicating node pairs within the group such that, and with regard to studies into user tolerance of latency in audio systems [9], a significant number of participant pairs will begin to experience difficulty in communicating with each other due to excessive latency in the audio channel.

In this paper, we consider a dynamic ALM-routing approach over standard flooding as a way to minimise the perceived latency experienced by audio-conference participants, drawing from patterns in conversation and from a user's perception of audio-channel quality.

The remainder of this paper is organised as follows: firstly, in section 2, we describe related work which has influenced our design rationale; next, in section 3, we present, in a preliminary study, our own observations of turn-taking in actual samples of conversation, before, in section 4, describing the proposed application-level network audio-conferencing routing protocol (ALNAC). In section 5, we present, through simulation, the effects of the proposed routing protocol on the group and the underlying network. Finally, in section 6 we give concluding remarks on the paper and describe future directions of the presented work.

2   Related Work

In this section, we describe two areas of particular relevance to the proposed work, namely: Internet packet-audio transmission and conversation analysis — the study of conversation.

2.1   Internet Packet-Audio Transmission

In packet-switched networks, audio transmissions are typically subjected to several latency components: sampling, packetisation, pre-processing (silence-suppression and compression), network transmission, network propagation, un-compression, and finally, playout buffering; with network-propagation delay being typically the least-predictable and most-dominant component for audio transmission over the Internet.

An abundance of studies into user tolerance of round-trip latency in audio-communication systems has been conducted and generally agrees upon the following levels of tolerance: excellent, 0–300 ms; good, 300–600 ms; poor, 600–700 ms; and quality becomes unacceptable for round-trip latencies in excess of 700 ms [9].

As latency increases, it is miss-interpreted by the user as extended pause in speech, causing confusion when they fail to get immediate responses from the other user(s); this, in turn, results in their eventual loss of synchronisation with the conversation.

2.2   Conversation Analysis

Conversation analysis is the study of verbal communication between people, with an emphasis on how that communication is structured and on how it is affected by social or cultural settings [10].

An area of conversation analysis of particular relevance to this work is the study of turn taking: the basic form of organisation in conversation. In conversation, people naturally organise their spoken contributions (utterances) into turns, where each person silently waits, listening to the current speaker, for their turn to speak [10]. It is also worth noting that overlapping speech occurs rarely in conversation, since one person must remain silent to effectively listen to what another person is saying.

A person will typically wait for a duration of time after the current speaker becomes silent before taking their turn to speak: typically, the pause is one second for Anglo-Saxon English speakers [11].

A large part of verbal communication, among any number of participants, consists of turns that are somehow related to each other, known as adjacency pairs [10].

This organisation of conversation turns into adjacency pairs naturally leads to a degree of localisation, where, over a given interval of time, a small proportion of the participants present exchange turns with one another. Figure 1 gives an example of everyday conversation, illustrating localisation of interest through nested adjacency pairs. This localisation property forms the basis of our work.

```
Neil: Would you like to go out tonight, Jane? (question)
   Jane: Where to? (response and question)
      Neil: The cinema. (response)
         Jane: What film is on tonight? (response and question)
            Neil: "Big", with Tom Hanks. (response)
   Jane: Sounds good, would you like that, Issac? (re-routed to another person)
      Issac: Yes. (response)
   Jane: Yes, I would like that, Neil, if Issac is coming too. (response)
Neil: Right then, lets get ready! (non–adjacency-pair)
```

Fig. 1.   An example of nested adjacency pairs, leading to localisation of interest in natural conversation.

Table 1. Accuracy of Next-Speaker Prediction when Considering a Backlog of n Distinct Speakers.

Back-Log Size	1	2	3	4	5	6	7
Accurate Predictions (%)	58	73	82	88	92	94	95

3   Preliminary Study of Next-Speaker Predictability

This section presents a preliminary study of patterns in samples of actual conversation, helping to support our design rationale in the following section.

To examine the extent of turn localisation (see section 2.2) that occurs in conversation, we performed a simple analysis of two audio-conference trace files and two public-meeting transcripts.

The trace files, which contained time stamps of talk spurts produced by participants, were logged from a locally developed audio-conference system using multiple-unicast transmission over a LAN. One audio-conference session included ten players of an online game which lasted for a duration of twenty minutes, and the other session was an informal discussion among eighteen people which lasted forty minutes.

The public meetings whose transcripts we analysed included 38 [5] and 42 [1] speaking participants respectively.

By stepping through the participant turns in each of the trace files and transcript files, we calculated the probability that the next speaker was amongst the set of n previous, distinct speakers, for various values of n. These probabilities, which represent the accuracy with which the next speaker can be predicted, are presented in table 1.

These results show two interesting properties that suggest turn localisation can be exploited in audio-conferencing systems: firstly, in more than half the cases, it is the previous speaker who answers the current speaker; secondly, the improvement in prediction accuracy quickly diminishes as more previous speakers are considered.

4   Dynamic Overlay-Routing Protocol Design

The proposed design for the application-level network audio-conferencing routing protocol (ALNAC) is derived from our understanding both of a user's perception of audio-channel quality (see section 2.1), and of turn-taking patterns in conversation (see section 2.2).

From this understanding, we hypothesize that, if a pair of overlay nodes are joined with minimal overlay distance when their respective participants are most sensitive to high latency (when engaged in conversation with each other) but are allowed to become further apart, with respect to overlay distance, when they are least sensitive to latency (when passively listening), then their perception will remain that of minimal network latency.

An ALM routing strategy that adapts in such a way will effectively reduce participants' sensitivity to scaling of the overlay network provided that, (1) the unicast distance between participant nodes is not excessively high, and (2) latency adaptation of the adaptive-routing protocol is sufficiently responsive to their changing state of interaction.

Thus, the general strategy of ALNAC is to enable direct delivery (i.e. in unicast trip time) of audio packets from a transmitting node to the nodes hosting participants who will most likely take up the next conversation turn, whilst ensuring overlay scalability, potentially to a large number of audio-conference participants.

Our analysis of next-speaker prediction (see section 3) shows that the participants most likely to speak next are those who have already spoken in the recent past.

4.1 Adaptive Routing of Audio Packets

We describe ALNAC in terms of a shared-tree overlay network and choose, for the basic-protocol operation, to adopt uni-directional tree routing (i.e. where a transmitting node floods data packets through its children and supplies the rest of the tree via the tree root), since uni-directional routing ensures lower maximum hop counts on delivery paths than does bi-directional routing. It then follows, from describing shared-tree routing, that ALNAC can also be applied to per-source trees by considering the transmitting node to be the tree root.

An ALNAC node will obtain a set of child nodes and a parent node from the (non-specific) overlay-tree construction protocol used to build the shared tree, and, where available, to optimise routing, will make use of inter-node distance information collected and exposed by the overlay-construction protocol.

On transmitting an audio packet, an ALNAC node will send a copy of the packet to the tree's root node, but, rather than simply flooding the packet to its children — as performed in standard tree flooding — the transmitting node will choose to directly send the packet to a set of target nodes which will include a number of nodes hosting recently-speaking participants and a number (zero or more) of the transmitting node's children. Section 4.2 describes the process of selecting target nodes.

Note that, to avoid inter–talk-spurt jitter, we choose to update the target-node set between and not during talk spurts of the transmitting node: for example, if a participant is speaking, and at that same time begins to receive packets from a new speaker — not currently in the target-node set — the original speaker will defer inclusion of the newly-detected speaker in the target-node set until the start of the original speaker's next talk spurt.

Typically, resource constraints — network or application — impose a limit (a.k.a out-degree) on the number of nodes to whom an overlay node may simultaneously send a data packet. Therefore, by including recent speakers in its target-node set, the current speaker, of constrained out-degree, must deprive some, or all, of its children from receiving audio packets directly from itself;

we therefore propose that a node may temporarily delegate responsibility for supplying its deprived children to some target nodes.

To ensure that ALNAC is sufficiently responsive to the changing set of recent speakers, whilst ensuring consistency in tree routing state, the ALNAC header of transmitted audio packets may contain a delegation chain. The delegation chain is simply a list of delegated-node addresses, composed in such a way that it can be split up into smaller delegation chains for efficient re-delegation of delegated nodes when only a subset of the nodes may be supplied (see section 4.3).

On receiving an audio packet containing a delegation chain, a node will try to supply as many (deprived) nodes from the delegation chain as its out-degree limit will allow — splitting the chain as necessary — and may also deprive some of its own children in a similar manner to a speaking node when supplying recently-speaking participant nodes. Note that, in order to minimise the number of delegations per delegate, the receiving node will deprive no more than half of its own children to supply nodes in the incoming delegation chain.

To avoid loops and duplications in the overlay network, each audio-packet header will contain the address of nodes that were supplied on non-tree links by the transmitting node. Thus, the parent of a supplied node will choose simply not to forward an audio packet to a child that is indicated, in the header, to have been already supplied.

Figure 2 illustrates adaptive routing of ALNAC, where node S, with an out-degree limit of five, transmits an audio packet directly to three recently-speaking participant nodes, A_1, A_2, and A_3, and to the tree-root node, R. S also sends the audio packet to one of its children, D, nominating D as the most suitable delegate for S's three other children that have been deprived of receiving the packet directly from S. Consequently, D deprives two of its own children, allowing it to supply directly two of its siblings, and the third one, indirectly, through a 'chained' re-delegation (see section 4.3).

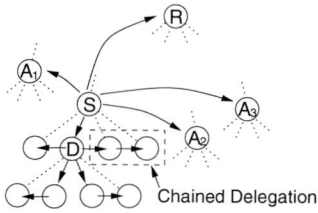

Fig. 2. Adaptive routing of ALNAC to three recently-speaking participant nodes.

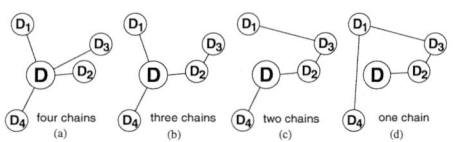

Fig. 3. Construction of an efficient delegation chain through all of the delegated nodes, D_1, D_2, D_3, and D_4, assigned to the delegate node, D.

4.2   Target-Node Selection

The number of selected target nodes is ultimately constrained by the out-degree of the transmitting node. There is, therefore, a trade-off between the number of recently-speaking participants who can be directly sent the audio frame and the extent of tree-route disruption caused by child-node deprivation and consequential delegation.

We propose a simple optimisation control for this trade-off by enforcing that a minimum number, n, of the transmitting node's children are always included in the target-node set; where the value for n can be chosen between zero and the transmitting-node's maximum fanout (out-degree-1) — at which value, routing will become standard tree flooding.

Where target-node selection allows the inclusion of a subset of children, the central-most children will be selected as approximations to the most suitable delegates.

4.3   Delegation

A transmitting node will nominate a subset of its target-node set as delegate nodes: if distance information is available among a subset of the target nodes, they will be nominated as delegates; otherwise, where no distance information is available, all of the target nodes will be nominated as delegates in favor of a balanced number of delegations per delegate.

Once a transmitting node has determined its sets of target nodes, delegate nodes, and deprived children, the transmitting node will attempt to assign each deprived child to its closest nominated delegate node, whilst ensuring a balanced distribution of delegated nodes per delegate node. If no distance information is available, the assignment will be made arbitrarily, based on the difference between network addresses.

As an optimisation to the delegation process, where inter-node distance information is available between a delegate node and each of its assigned delegated nodes, a transmitting node will give each delegate sufficient information in its delegation chain to allow efficient re-delegation should a delegate be assigned more nodes than it is able to supply.

The transmitting node will calculate delegation chains for a delegate node using the following algorithm:

- start with optimal chains (paths) from the delegate to each of its assigned delegated nodes (see figure 3(a))
- merge the two chains that produce the shortest path, producing $n-1$ chains (see figure 3(b))
- store the address of the first node in the chain that was joined to the appended chain, such that the chain can later be separated again at that point.
- continue, in the same manner, to merge the chains and to store the address of the joining node, until a single chain remains that spans all of the delegated nodes (see figure 3(c)–(d))

The result of this process is that we have a single, ordered chain from the delegate node through each delegated node. Thus, when a delegate node receives this information in the audio packet, it is able to break the chain into as many chain fragments as it is able to directly serve, efficiently re-delegating any remaining nodes. The reader should note that target nodes receive different delegation chains, comprising the deprived children assigned to them by the transmitting node.

4.4   Extended Protocol Operation

We have described, for the basic protocol operation, how we can avoid duplication in the overlay network when adopting a uni-directional routing approach, however, since a transmitting node may send a data packet directly to any node in the group — a recently-speaking participant node — there is an opportunity to achieve more uniform overall overlay latency by allowing a receiving node to forward a packet up to their parent.

We therefore define a protocol parameter to toggle a receiving node's ability to forward a packet up the tree if the packet is not from, or indicated (in the packet header) to have been already supplied to, its parent node. The effectiveness of such a bi-directional routing technique for reducing overall latency has been shown through random packet jumping in the context of resilient multicast over ALM [3].

5   Simulation Experiments

In this section, we analyse, through simulation, the performance of the ALNAC routing protocol in comparison to non-adaptive overlay-tree flooding typically used in ALM group communications.

5.1   Performance Measurements

Since ALNAC is designed to adapt latency in the audio-communication channel to the conversational patterns of audio-conference participants we choose to measure delay of received audio packets against a participant node's elapsed time since audio-packet transmission.

In addition to examining adaptation, we examine the impact that such adaptation has on the overlay network as a whole and on the underlying network by observing group delay characteristics and network stress (the extent of packet replication on network links).

5.2   ALNAC Protocol Parameters

In the proposed ALNAC design we left the following two parameters open for experimentation: the minimum number of children (MC) that a transmitting node is forced to include in the target-node set; and a flag (UP), enabling a node to forward a received packet up the tree if its parent has not been already supplied.

5.3 Simulation Set-Up

We simulated the ALM protocols using a locally-developed, packet-level network simulator, which implemented shortest-path routing over a 600-node, Internet-like GT-ITM[13]–generated core-network topology.

To simulate the presence of audio-conference client nodes in different network domains, we connected, with one-millisecond links, a further 100 nodes to random nodes at the core-network's edge; upon these client nodes, we ran the various ALM routing protocols.

As a benchmark for the best-attainable delay for all client nodes, akin to the delay attainable through network-level multicast, we implemented a naïve multiple-unicast protocol client.

For an overlay-tree construction protocol we implemented the Tree Building Control Protocol [8] (TBCP): a scalable, tree-first ALM protocol for building efficient, cost-constrained overlay trees. Note that, to further optimise TBCP for latency, we used the TBCP score function proposed in [12]. In the experiments, we fixed the maximum fanout of the overlay tree at five in consideration of turn-prediction backlog sizes (see section 3).

Over the overlay tree we ran both standard uni-directional flooding, representative of the non-adaptive routing typically used in ALM, and ALNAC using various parameter values.

To test adaptation of ALNAC under realistic conditions we instructed simulation clients to reproduce traffic from packet-trace files which were generated by a simple, unicast-based audio conference with 18 speaking participants that ran on the university LAN; also, for testing the impact of ALNAC whilst heavily exercised, we synthesised audio-conference traffic to produce random client interaction (i.e. periodically a client node, picked at random, was instructed to transmit an audio packet).

Thus, each simulation ran as follows: initially, 100 ALM clients were instructed to join the overlay network over a period of ten minutes in a random, uniformly-distributed fashion; then, the clients proceeded to transmit traffic in their designated patterns.

5.4 Simulation Results

The results in this section, presented for each protocol, are averaged over ten simulations using different client-node topology placement seeds.

Figures 4(a) and 4(b) show the average and maximum delay experienced by client nodes against elapsed time since they last spoke.

As expected, given the adaptive nature of ALNAC, figure 4(a) shows, clearly, a reduction in delay for recently-speaking nodes over that experienced when using standard tree flooding. We see, with one exception, a decreased delay when ALNAC enforced fewer child targets on nodes, since more of the recently-speaking nodes could be directly supplied; the exception occurred when no child targets were enforced (MC = 0) and when upward routing was disabled (UP

= 0), since nodes were not being forced to use any efficiently-constructed tree routes when transmitting packets.

We see from figure 4(b) how ALNAC maintained unicast delay for a minimum elapsed time-since-transmission of 7 seconds for all client nodes that had recently spoken; this time increased to 11 seconds when fewer child targets were enforced by ALNAC since more recently-speaking nodes could be accommodated by transmitting nodes. The best maximum delay achieved for recently-speaking nodes occurred when no child targets were enforced and upward routing was enabled since audio packets were being broadly disseminated over the tree by transmitting nodes.

Figures 5(a) and 5(b) show the cumulative distribution functions (CDFs) of the average and maximum delay experienced by client nodes. The simulation traffic was synthesised so as to fully exercise the ALNAC protocol with frequent and random client interaction.

We see from figure 5(a) that, in general, ALNAC improved average delay of the group over standard flooding; this occurred as a result of both packet flow being no longer bound only to overlay-tree paths and leaf nodes of the tree becoming more useful for disseminating packets to the group; the improvement was even greater when bi-directional routing was enabled.

Interestingly, the enforcement of child targets on transmitting nodes had little effect on the average latency of the group; this is an effect of dilution, since such an enforcement has no effect on the leaf nodes (with no children), which made up the majority of nodes in the tree.

In figure 5(b), we see that the maximum delay experienced by nodes when using ALNAC was slightly worse than that experienced when using standard flooding. Maximum delay increased slightly with fewer enforced target children, but the combination of both broad packet dissemination, by transmitting nodes, and efficient delegations that made use of local inter-node distance information, lessened the impact of ALNAC on the group as a whole.

Table 5.4 shows the maximum stress placed on network links by the routing protocols in simulations using synthesised, high-interaction audio-conference traffic. In general, the network incurred higher stress from ALNAC than from standard flooding; this was the result of overlay-tree circumvention by ALNAC when routing to recently-speaking nodes, since the tree-construction protocol (TBCP), by design, tries to minimise network stress through clustering overlay nodes that are topologically close together.

In summary, ALNAC effectively reduced delay for recently-speaking nodes whilst minimising the impact on the whole group, but did so at the cost of slightly increased stress on network links.

6   Conclusions

In this paper, we have justified the case, and presented a design, for an ALM routing protocol that can dynamically adapt audio-packet routing to changing patterns in application usage.

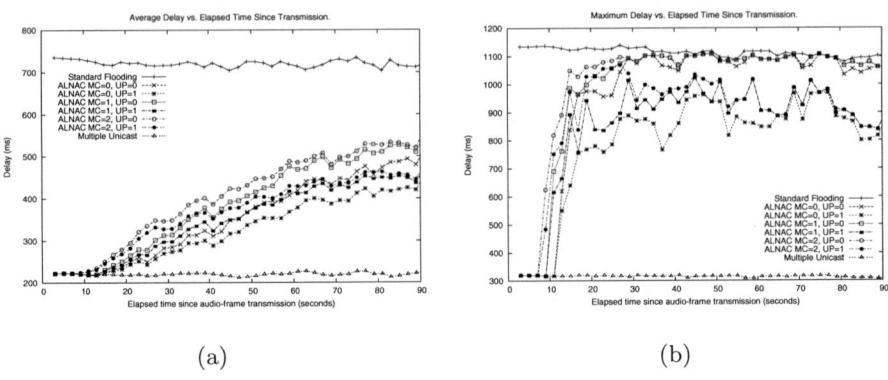

(a)   (b)

Fig. 4. Maximum and average delay experienced by recently-speaking participant nodes.

Table 2. Maximum network stress of routing protocols.

Routing Protocol	Maximum Stress
Standard Flooding	11
ALNAC (MC=0,UP=0)	15
ALNAC (MC=0,UP=1)	16
ALNAC (MC=1,UP=0)	15
ALNAC (MC=1,UP=1)	16
ALNAC (MC=2,UP=0)	15
ALNAC (MC=2,UP=1)	15

The approach we have taken essentially fuses the benefits — latency and scalability — of multiple-unicast and ALM group-communication techniques through consideration of patterns observed in conversation and through an understanding of a user's perception of audio-channel quality.

As further work, we plan to implement ALNAC within an ALM proxy client, creating a group-communication service for existing audio-conferencing applications, with which we will conduct subjective user trials. Such trials will help us to better understand the limitations of ALNAC: for example, as the size of a group increases, at what point does competition for conversation turns become so unfair for those participants who have not spoken in the recent past that they are unable to effectively communicate in the group.

Another interesting study would be to see how, say, a tree-first protocol using ALNAC would compare, subjectively, to a mesh-first protocol using standard flooding, where mesh-first protocols produce more-optimal trees though generally require more control overhead than the former.

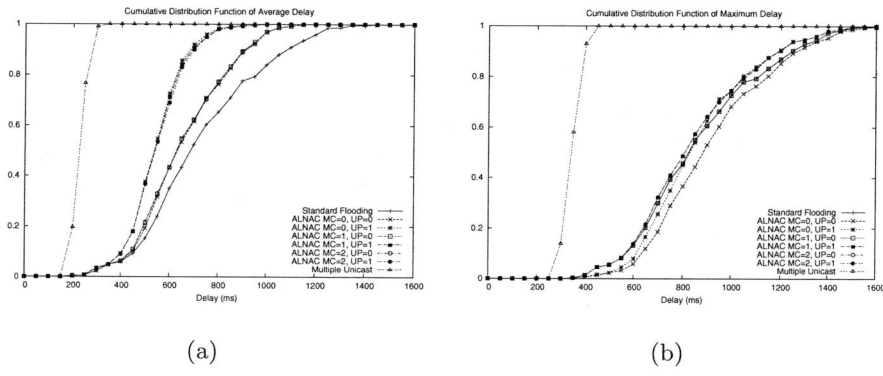

(a)   (b)

Fig. 5. Cumulative distribution of delay experienced by participant nodes.

References

1. Athelstan. Transcript of a Forty-Two–Member Meeting in the Corpus of Spoken Professional American English (CSPAE). http://www.athel.com/sample.html.
2. S. Banerjee, B. Bhattacharjee, and C. Kommareddy. Scalable Application Layer Multicast. In *ACM SIGCOMM*, Aug 2002.
3. Suman Banerjee, Seungjoon Lee, Bobby Bhattacharjee, and Aravind Srinivasan. Resilient Multicast using Overlays. In *Proceedings ACM Sigmetrics 2003, San Diego, CA.*, June 2003.
4. Y-H. Chu, S. Rao, and H. Zhang. A Case for End System Multicast. In *ACM SIGMETRICS*, pages 1–12, Santa Clare, CA, USA, June 2000.
5. Competition Commission. Lloyds TSB / Abbey National Merger Inquiry Open-Meeting Transcript. http://www.competition-commission.org.uk/inquiries/completed/2001/lloyd%s/lloydstran.htm.
6. S. Deering and D. Cheriton. Multicast Routing in Datagrams Internetworks and Extended LANs. *ACM Trans. Comp. Syst.*, 8:85–110, May 1990.
7. A. El-Sayed, V. Roca, and L. Mathy. A Survey of Proposals for an Alternative Group Communication Service. *IEEE Network*, 17(1):46–51, Jan/Feb 2003.
8. L. Mathy, R. Canonico, and D. Hutchison. An Overlay Tree Building Control Protocol. In *Proc. of Intl. workshop on Networked Group Communication (NGC)*, pages 76–87, Nov 2001.
9. Princy C. Mehta and Sanjay Udani. Overview of VoIP, Technical Report MS-CIS-01-31. Technical report, University of Pennsylvania, February 2001.
10. H. Sacks. *Lectures on Conversation*. Blackwell, Oxford, UK, 1992.
11. Ulrich Schmitz. Eloquent silence. *Linguistik-Server Essen (LINSE)*, 1994.
12. Su-Wei Tan and Gill Waters. Building Low Delay Application Layer Multicast Trees. In *Proceeding of 4th Annual PostGraduate Symposium (PGNet 2003)*, pages 27–32. John Moore University, Liverpool, UK, June 2003.
13. E. Zegura, K. Calvert, and S. Bhattacharjee. How to Model an Internetwork. In *IEEE Infocom*, pages 40–52, Mar 1996.

Background Noise Influence on VoIP Traffic Profile

Rafael Estepa, Juan Vozmediano, and Antonio Estepa

Área de Ingeniería Telemática Universidad de Sevilla*
Camino de los Descubrimientos s/n, E-41092 Sevilla
Tel.: +34 95448 7384
{rafa,jvt,aestepa}@trajano.us.es

Abstract. Modern audio codecs used in VoIP can improve the listening quality by transmitting the main characteristics of the background noise signal during the silence periods. This traffic has been traditionally neglected in the codec mean bit-rate estimation. Nevertheless, when considering an IP environment, the packet overhead increases significantly the required mean transmission bit-rate. Hence, the transmission of the background noise signal can result into either a poor network resource dimensioning in network planning or in the violation of the SLA traffic specifications in a DiffServ scenario.

This paper presents a study on the influence of the background noise signal in the mean transmission bit rate required by conversations in IP networks. A new traffic pattern generation model is presented, for which an analytical expression for the mean bit rate is derived. This model is parametrized for the G.729B and the GSM AMR codecs. Experimental results show that this new model significantly enhances the current mean bit rate estimation. The traffic profile of aggregated audio traffic is also addressed, obtaining results which improve the current ON-OFF aggregated traffic models.

1   Introduction and Previous Work

Current audio codecs can improve the speech quality by reproducing the talker's background noise. This feature is supported by a special frame type called Silence Insert Descriptor (SID), generated at the speaker's side, which describe the main characteristics of the background noise. The SID frames are generated during voice inactivity periods, and their coding scheme differs from that of the active voice frames (ACT frames). SID frames are generated by the codec's DTX algorithm according to changes in the background noise energy and to some specific rules depending on the codec's implementation [1,2].

Human speech has been traditionally modeled as a sequence of alternate talk and silence periods whose duration is exponentially distributed in the so-called

---

* The work leading to this article has been partly supported by CICYT and the EU under contract number TIC2003-04784-C02-02.

V. Roca and F. Rousseau (Eds.): MIPS 2004, LNCS 3311, pp. 13–24, 2004.

ON-OFF model [4]. The traditional ON-OFF model does not consider the effect of the SID frames in the traffic pattern generated by voice sources. This have been commonly accepted since the effect of short-size SID frames in the codec bit rate is reduced (typically less than 2 %) [1].

The encoded speech is then packetized into packets before transmission. Each such packet includes the headers at the various protocol layers (namely, RTP, UDP, and IP header as well as the data link layer headers) and the payload, comprising a sequence of N consecutive codec frames (we will name this factor N_{fpp}). The packet overhead significantly increases the mean bit rate, specially for low values of N_{fpp}. Thus, the presence of SID frames can severely affect the traditional ON-OFF traffic mean bit rate estimation [3].

We propose a new model for the frame generation pattern of one voice source in which the generation of SID frames is included. An analytical expression for the mean bit rate at the input of the IP network is deduced for this model as a function of the packetization factor N_{fpp}. Experimental validation will let us compare the mean bit rate obtained from both models with trace measurements for the G.729B and AMR codecs. This model is also extended to the aggregated traffic of voice sources, obtaining better results than with the traditional aggregation ON-OFF models.

The remainder of this paper is organized as follows: we next review the previous works about voice source traffic modeling, section 2 proposes a new voice source which differs from the traditional ON-OFF model in taking into account the traffic generated by the emission of the background noise. Section 3 addresses the traffic pattern of the aggregated voice sources which follow the new model. Section 4 is devoted to a experimental validation of the proposed model and finally, section 5 concludes the paper.

1.1   One Voice Source Modeling

Human speech traffic generated by codecs equipped with the silence suppression feature has been traditionally modeled with the ON-OFF model [4]. The voice frames are continuously generated during ON periods while no frames are transmitted during OFF periods. Although better fits can be found in [5], the duration of both periods is considered to follow an exponential distribution. The mean ON periods duration is T_{ON}, while mean OFF periods duration is T_{OFF}, where $1/T_{ON}$ and $1/T_{OFF}$ are the states transition rate. According to [6], an ON-OFF source can be defined by the triplet: T, T_{ON} and ρ, where ρ accounts for the portion of the time during which voice frames are being sent (ON state). This can be calculated from the balance equation of the Markovian process as $\rho = T_{ON}/(T_{ON} + T_{OFF})$.

According to the ON-OFF model, the mean bit rate due to the frame generation process is:

$$R = \rho \cdot \frac{L_{ACT}}{T} \tag{1}$$

where L_{ACT} is the voice frames size in bits.

In VoIP environments, the aforementioned codec frames are carried in IP packets toward their destination. In the packetization process it is necessary to set the number of codec frames that will be included in each IP packet (N_{fpp}). The size of the packet overhead (H) can be several times greater than one voice frame and, consequently, it significantly influences the conversation bit rate in a IP network during the ON periods as follows:

The network mean bit rate is then [3]:

$$R = \rho \cdot R_{ON} = \rho \cdot (\frac{L_{ACT}}{T} + \frac{H}{N_{fpp}T})$$ (2)

Since during the OFF periods no frames are generated, the SID frames are not considered in the ON-OFF model. In [1] it is shown that the SID frames influence in the codec mean bit rate (due only to the codec frames) is always minor than 1.5%, but also shows that after the packetization process, the packets generated during OFF periods due to SID frames, increase significantly the mean bit rate in the IP network.

1.2   Aggregation of Voice Sources

The analysis of systems involving Markov modulated traffic is a well known subject, e.g., [7,8,9,6]. Previous works usually consider the traffic profile resulting from the aggregation of N independent ON-OFF sources, each of which continuously generates packets during the ON state. The packets generated are fed to a common queue from which a server removes them for transmission over a communication link.

In such a system the probability of packet loss can be given as a function of the queue buffer size. In [6] the probability of packet loss for N sources (each one characterized by the triplet: $\rho_i, T_{ON}^i, R_{ON}^i$) is derived based on the results from [10]. For ATM links (also applies to IP) it is shown that the cell loss probability in the large buffer region (i.e. burst level region or fluid model region) depends strongly on the mean offered load: $N \cdot \rho \cdot R_{ON}/C$ [9,11] where N is the number of multiplexed sources, C is the output link capacity and it is asummed that the sources are homegeneous, with identical peak rate and ρ value.

The asymptotic behaviour of the buffer overflow probability is useful for systems with large buffer sizes and small loss probabilities, where the solutions take the form $c_0 \cdot exp(z_0x)$ and both z_0 and c_0 are strongly related to the mean offered load as shown in either [11] or [12].

In [13] an experimental validation of the aggregation of the ON-OFF traffic model through a token-bucket filter multiplexor is performed. There it is shown that the ON-OFF model accurately estimates the aggregated traffic profile, specially when the number of conversations is significant. The differences between the traces and the ON-OFF aggregation model still exists, and it is mainly due to the non-exponential nature of the ON and OFF periods in real conversations. The methodology followed and the use of the token-bucket filter allows the performance comparison of several codecs in a DiffServ traffic conditioning environment.

To our knowledge, there is no previous work aimed either to determine the influence of the SID frames in the aggregated traffic profile or to validate the ON-OFF multiplexing model when the codecs generate SID frames.

2   One Voice Source: ON-SID Model

Since the ON-OFF model is not valid to capture the traffic generated during the silence periods, we present a simplified analysis of our previous model [14] which overcomes this problem.

2.1   The ON-SID Model

The proposed model is based on the following frame generation pattern: during the ON periods, the ACT frames are generated with deterministic interarrival time T (as in the ON-OFF model), and during the voice inactivity periods SID frames are generated randomly.

For mathematical tractability we assume that, in the discrete time space $t_i = i \cdot T$, the codecs continuously generate frames that can be either type: ACT, SID or NoTXN. The latter corresponds to a zero-length frame used to model the instants where no frames are generated. The sizes of the previous frame types are listed in [1] for the G.729,G723.1. and GSM AMR codecs.

We define the random variable X as the inter-arrival discrete time between the SID frames during a voice inactivity period. Its probability density function $P_i = Prob\{X = i\}$ can be experimentally determined for different background noise environments.

This new model is determined by the parameters ρ, T_{ON}, T , from the ON-OFF model [6], and the additional parameter P_i.

2.2   Mean Bit Rate Rate of the ON-SID Model

The overall mean rate can be decomposed into the sum of the contribution of the packets sent during the active voice periods (R_{ON}) and voice inactivity periods (R_{SID}) separately. Thus:

$$R = \rho \cdot R_{ON} + (1 - \rho) \cdot R_{SID} \tag{3}$$

where ρ is the conversation activity factor, which is also the probability of being in an ON period.

Rate During ON periods. We assume that during the ON periods the packets payload is N_{fpp} ACT frames, thus packets are generated every $N_{fpp} \cdot T$. This is the figure traditionally considered in the ON-OFF model, as shown in eq. 2.

Rate during OFF periods. In order to deduce an analytical expression for R_{OFF}, recall that the discrete random variable X is the inter-arrival number of periods between two consecutive SID frames generated during an OFF period. Let P_i be the probability that X takes the value i ($Prob\{X = i\}$).

The average bit-rate during the OFF periods can be decomposed into two parts accounting for the contribution of the SID frames (R_{OFF}^{SID}) generated, and for the packet overhead (R_{OFF}^{H}).

The first term can be obtained as:

$$R_{OFF}^{SID} = \frac{L_{SID}}{T} \cdot \tau_1 + \frac{L_{SID}}{2 \cdot T} \cdot \tau_2 + ... = \frac{L_{SID}}{T} \cdot \sum_{i=1}^{\infty} \frac{1}{i} \cdot \tau_i \qquad (4)$$

where τ_i is defined as the portion of time during which the SID frame inter-arrival time is i time intervals ($i \cdot T$). Applying that:

$$\tau_i = \frac{i \cdot P_i}{E[X]} \qquad (5)$$

the expression of R_{OFF}^{SID} can be expressed as:

$$R_{OFF}^{SID} = \frac{L_{SID}}{T \cdot E[X]} \qquad (6)$$

The influence of the packet overhead in the bit-rate depends on the packet inter-arrival time as well as on the number of SID frames carried into each packet. Thus, there are two possible cases:

I : $X \geq N_{fpp}$. In this case each packet will carry one and only one SID frame, and the inter-arrival time for packets will be the equal to the inter-arrival time of the SID frames. Then, replacing L_{SID} by H in eq. 6:

$$R_{OFF}^{H} = \frac{H}{T \cdot E[X]} \quad , \quad X \geq N_{fpp}$$

II : $X < N_{fpp}$. In this case two or more SID frames will fit into each packet, reducing the overhead influence in the bit-rate as the inter-arrival time between packets increases proportionally to the number of SID frames which are carried in each packet. This number can be obtained rounding up to the integer of $\frac{N_{FPP}}{i}$:

$$R_{OFF}^{H} = \frac{H}{T} \cdot \sum_{i=1}^{\infty} \frac{\tau_i}{\lceil \frac{N_{FPP}}{i} \rceil \cdot i} = \frac{H}{T \cdot E[X]} \cdot \sum_{i=1}^{\infty} \frac{P_i}{\lceil \frac{N_{FPP}}{i} \rceil} \quad , \quad X < N_{fpp}$$

The latter expression is also valid for both cases, as the ceil function equals to one in case of having $i \geq N_{fpp}$.

Adding up the contributions of SID frames and packet overhead during the OFF periods it can be stated that:

$$R_{OFF} = R_{OFF}^{SID} + R_{OFF}^{H} = \frac{1}{T \cdot E[X]} \cdot (L_{SID} + H \cdot \sum_{i=1}^{\infty} \frac{P_i}{\lceil \frac{N_{FPP}}{i} \rceil}) \qquad (7)$$

Eq. 7 gives the bit-rate required by a single conversation during its silence periods due to the generation of SID frames, in a IP environment.

3   Aggregated ON-SID Traffic

The aggregated model for ON-SID would correspond to the aggregation of N two-state VBR sources, having each state its respective associated rate. Nevertheless, for simplicity, we propose the use of the ON-OFF model for the aggregation of ON-SID sources. Since the traffic generation pattern of the ON-OFF model restricts the transmission to the ON periods, we propose to adjust the ON-OFF model activity rate factor ρ, enlarging the ON periods to reach the same mean bit rate as the ON-SID model would have.

This will result in a new set of parameters $(\rho', T'_{ON}, T'_{OFF})$. Parameter T does not change since it only depends on the codec. The correctness of this model, called Modified ON-OFF model from now on, will be validated experimentally in section 4.

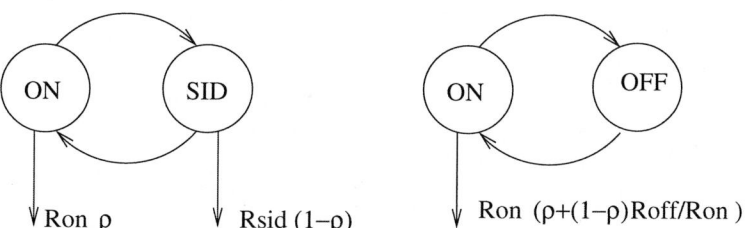

Fig. 1. ON-SID and Modified ON-OFF models

As shown in figure 1, the overall mean bit-rate required by a conversation can be viewed as the sum of the contributions of the ON (R_{ON}) and OFF (R_{OFF}) periods. Thus, it is possible to define the new activity factor as:

$$\rho' = \rho + (1 - \rho) \cdot \frac{R_{OFF}}{R_{ON}} \qquad (8)$$

Replacing in eq. 8 the values of R_{ON}(eq. 2) and R_{OFF} (eq. 7) results in:

$$\rho' = \rho \cdot (1 + \frac{1-\rho}{\rho} \cdot \frac{1}{E[X]} \cdot \frac{L_{SID} + H \cdot \sum_{i=0}^{\infty} \left\lceil \frac{P_i}{\frac{N_{fpp}}{i}} \right\rceil}{L_{ACT} + \frac{H}{N_{fpp}}}) \qquad (9)$$

The parameters T'_{ON} and T'_{OFF} can be derived from ρ' taking into account the following relations:

$$\rho' = \frac{T'_{ON}}{T'_{ON} + T'_{OFF}} \qquad (10)$$

$$T'_{ON} + T'_{OFF} = T_{ON} + T_{OFF} \qquad (11)$$

The first one readily comes from the ON-OFF model definition. The second one is straightforward since the duration of the conversation does not change.

Using the new values for one source (ρ', T'_{ON}, T) or (T'_{ON}, T'_{OFF}, T), the analytical expressions for multiple ON-OFF sources can be used for the aggregated traffic profile resulting from codecs which generate SID frames. Next section experimentally validates the Modified ON-OFF model and its goodness to capture the aggregate traffic profile of ON-SID sources.

4   Validation

This section describes the experimental validation of the ON-SID model by comparing the traffic profile obtained using the ON-OFF model to those obtained from real traffic.

4.1   Experiment Set-Up

Source material. The test-bed and source speech material used is described in [1]. Both ends of a set of 10 conversations 15 minutes long were recorded from an ISDN line in a low-noise office environment (SNR $> 20dB$). The raw sample audio files were encoded with the G.729B, G.723.1 and GSM AMR codecs, giving three sets of encoded trace files. The output of the codecs was processed to obtain the sequences of generated frame types. These sequences (ftype files) were analyzed to determine the experimental distribution of P_i as well as other model parameters (ρ, T_{ON}, T). The values obtained for these parameters can be found in [1]. The percentage of SID frames obtained for the codecs G.729B, GSM AMR and G.723.1 was 7.69%, 7.58% and 3.59%, respectively. The activity rate ρ was 0.4559, 0.4717 and 0.4697, respectively. The mean duration of the ON periods (T_{ON}) was 336, 1026 and 1490 ms, while the OFF periods (T_{OFF}) mean duration was 420, 1171 and 1722 ms respectively.

One source measurements. To measure the mean bit rate at the network edge, the packetization process over the ftype files was repeated for different values of N_{fpp}, taking a header size which included the IP, UDP and RTP protocols [1].

Multiplexing N voice sources. An event-driven network simulator [15] was used for multiplexing the ftype files corresponding to N speech sources. The system defined was similar to the one used in [13]: a number of voice sources connected to a statistical token-bucket-based multiplexer. The parameters used for the token-bucket filter of the multiplexer were: the token filling rate $r = R \cdot N \cdot R_{ON}$ while the bucket size is $b = B \cdot N \ (H + N_{fpp} \cdot L_{ACT})$. The factors R and B represent the complementary of the multiplexing gain and the normalized buffer size for any multiplexing factor N respectively. Packets exceeding the token-bucket bounds (out-of-profile packets) are discarded, thus the probability of out-of-profile is equivalent to packet loss probability, which is the measured value.

The multiplexer was fed by either trace files from the real conversation samples or by synthetic ON-OFF traffic sources. Ten different experiments three minutes long were held for each tested model, for a confidence level of 90%.

4.2   Results for One Source

Table 1 shows the mean bit-rate of one voice source as a function of the packetization coefficient. In the table it is shown the trace values as well as ON-OFF model mean bit-rate estimation using both, ON-OFF (OO) (eq. 2) and ON-SID (OS) (eq. 3) mean bit rate estimation values.

Table 1. One conversation mean bit-rate

N_{fpp}	G.729B-OS	G.729B-trace	G.729B-OO	AMR-OS	AMR-trace	AMR-OO
1	20.711	20.814	18.236	11.042	11.135	7.272
2	13.417	13.341	10.941	7.269	7.305	4.756
3	10.985	10.832	8.510	6.011	6.025	3.917
4	9.133	8.938	7.294	5.322	5.359	3.498
5	8.297	8.086	6.564	4.945	4.947	3.246
6	7.739	7.518	6.078	4.693	4.674	3.079
7	7.125	7.005	5.731	4.494	4.479	2.959
8	6.822	6.664	5.470	4.359	4.331	2.869
9	6.550	6.400	5.268	3.748	3.772	2.799
10	6.254	6.160	5.106	3.654	3.681	2.743

---

[1] The G.723.1 will not be considered in the rest of this paper since the percentage of SID frames is small and the frame period is long compared with the other two codecs, which makes this codec less sensitive than the other two to SID frames.

The error obtained using the traditional ON-OFF model is approximately a 20% for the codec G.729B, and rises up to a 35% for the GSM AMR (MR4.75), while in the ON-SID Model the error is always less than 3% as can be observed from figure 2. The envelope of the error function is explained though the values of P_i showed in [1].

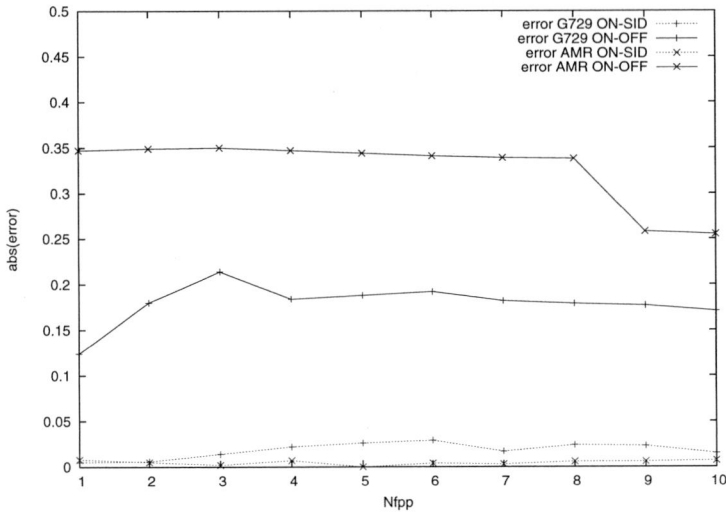

Fig. 2. One voice source mean bit-rate estimation error

4.3   Results for Aggregated Traffic

In this subsection we present the results obtained when multiplexing N conversations using codecs which generate SID frames. We compare the P_0 obtained with both ON-OFF and Modified ON-OFF models, versus the P_0 measured from the traces as described in previous section.

Figure 3 shows the estimated P_0 and confidence intervals (unnoticeable for the modified ON-OFF model) obtained for $R = 0.55$, $N = 20$ and $N_{fpp} = 2$ for the G.729B codec, when using the ON-OFF model with either the traditional parameters value and the modified values proposed. Similar results were obtained for the GSM AMR codec. The results show that the SID frames effect can severely affect the prediction obtained with the traditional ON-OFF aggregation model. The Modified ON-OFF model greatly improves the fitting. The value of the activity factor increases with the generation of SID frames, resulting in higher values of P_0. This could even break the multiplexing stability condition ($\rho < R$) when not taken into account.

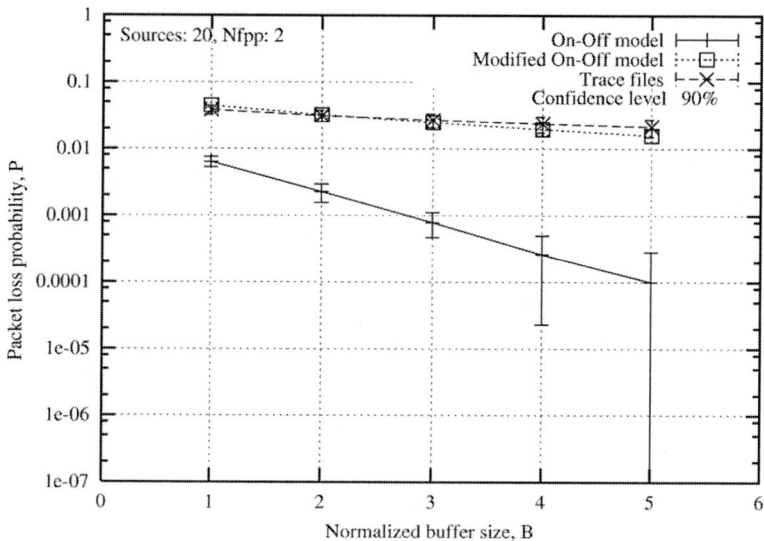

Fig. 3. Estimated average and confidence intervals for Packet Loss Probability (N=20, R=0.55, G.729B codec)

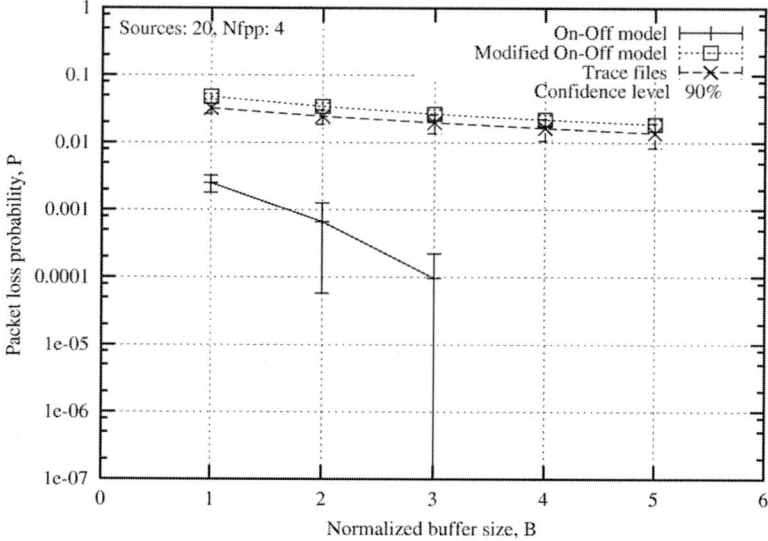

Fig. 4. Estimated average and confidence intervals for Packet Loss Probability (N=20, R=0.55, G.729B codec)

Figure 4 shows P_0 obtained for the same conditions but rising N_{fpp} up to 4. The reason why the Modified ON-OFF model incurs into higher values of P_0 than traces can be inferred from the behavior of ρ' with respect to N_{fpp} (which can be also observed in the ON-OFF error function in figure 2).

The value of the token-bucket filling rate changes according to the ON-OFF model dependency with N_{fpp} but the traces dependency with N_{fpp} is different and is better approximated by the ON-SID model. Nevertheless, the Modified ON-OFF model, although exhibit the same dependency with N_{fpp} than the traces, it does not transmit during the OFF periods, making the traffic burstier and increasing the effect of N_{fpp} in the aggregated traffic. That explains the progresive separation of the traces and the modified ON-OFF model when increasing N_{fpp}.

As stated in the introduction, the cell or packet loss probability for the aggregated voice traffic has an asymptotic behavior following $c_0 \cdot exp(z_0 x)$. This calculus is not trivial and is strongly dependent of the mean offered load and the output link capacity (i.e. the rate ρ/R for ON-OFF model, or ρ'/R for our Modified ON-OFF model). This explains that ρ' dependency of N_{fpp} is translated into P_0 dependency of N_{fpp}, a feature that can not be captured by the traditional ON-OFF model.

5   Conclusions

The traditionally accepted ON-OFF model ignores the SID frames, which causes significant errors in the estimated bandwidth requirements of a voice source in an IP network. The proposed ON-SID model allows to use an analytical expression for the mean bit rate at the input of the IP network. The experimental validation of this new model shows an improvement of the mean bit rate estimation obtained with the ON-OFF model up to a 32%. This new model allows dimensioning with more precision the network resources in VoIP environments.

The Modified ON-OFF model improves the estimation of the mean offered load offered to a statistical multiplexor when the audio codecs generate SID frames. It allows the establishment of more accurate SLA in a DiffServ environment and a better performance in the network resources dimensioning.

References

1. Estepa, A., Estepa, R., Vozmediano, J. "Paquetization and Silence Influence on VoIP traffic profile". Lecture Notes on Computer Science 2899 - Multimedia Interactive Protocols and Systems MIPS 2003. Naples Nov. 2003
2. Benyassine, A., Shlomot, E., Su, H.-Y., Massaloux, D., Lamblin, C., Petit, J.-P. "ITU-T Recommendation G.729 Annex B: a silence compression scheme for use with G.729 optimized for V.70 digital simultaneous voice and data applications". IEEE Communications Magazine, Vol. 35 Issue 9, Sep. 1997.
3. Goode, B. "Voice over Internet protocol (VoIP)". Proceedings of the IEEE, Vol. 90 Issue 9, Sep. 2002.

4. Sriram, K., Whitt, W. "Characterizing superposition arrival processes in packet multiplexers for voice and data". IEEE JSAC, Vol. SAC-4(6):833-846. Sep. 1986.
5. Bellalta, B., Oliver, M., Rincon, D. "Performance of the GPRS RLC/MAC protocols with VoIP traffic". Personal Indoor and Mobile Radio Communications, 2002. IEEE international symposium on, Vol. 1, 15-18. Sep. 2002.
6. Guerin, R., Ahmadi, H., Naghshineh, M. "Equivalent capacity and its application to bandwidth allocation in high-speed networks". IEEE JSAC., vol. 9, no. 7, pp. 968-981, Sep. 1991.
7. Anick, D., Mitra, D., Shondi, M. "Stochastic theory of data-handling system with multiple sources". Bell Sys. Tech. Journal 61(8): 1871-1894. Oct. 1982.
8. Daigle, J., Langford, J. "Models for Analysis of Packet Voice Communications Systems". IEEE Journal on Selected Areas in Communications. Vol SAC-4. No 6, Sep. 1986.
9. Baiocchi, A., Melazzi, N.B., Listante, M., Roveri, A., Winkler, R. "Loss performance Analysis of an ATM Multiplexer Loaded with High-Speed ON-OFF Sources.".IEEE Journal on Selected Areas in Communications, Vol 9: 388-393, Apr. 1991.
10. Mitra, D. "Stochastic theory of produced and consumers coupled by a buffer". Adv. Appl. Prob, 20: 646-676, 1988.
11. Pitts, J., Schormans, J. "Introduction to IP and ATM design and Performance". Ed. John Wiley & sons. 2nd edition. 2000.
12. Qian, K., McDonald, D. "An approximation method for complete solutions of Markov modulated Fluid Models". Queueing Systems 30: 365-384, 1988.
13. Jiang, W., Schulzrinne, H. "Analysis of On-Off patterns in VoIP and their effect on Voice Traffic Aggregation.". In The 9th IEEE International Conference on Computer Communication Networks, 2000.
14. Estepa A., Estepa R., Vozmediano J. "A new approach for VoIP traffic characterization". IEEE Communication Letters. Accepted, publication pending.
15. Vozmediano J. "AitSim: Discrete-event based network simulator". http://trajano.us.es/clases/lsim/aitsim.

Adaptive VoIP Transmission over Heterogeneous Wired/Wireless Networks

Abdelbasset Trad[1], Qiang Ni[1], and Hossam Afifi[2]

[1] INRIA, Planete Project
2004 Route des Lucioles, BP-93
06902 Sophia-Antipolis, France
{atrad, qni}@sophia.inria.fr
[2] INT-INRIA
9 rue Charles Fourier
91011 Evry, France
{Hossam.Afifi}@int-evry.fr

Abstract. In this paper, we present an adaptive architecture for the transport of VoIP traffic over heterogeneous wired/wireless Internet environments. This architecture uses a VoIP gateway associated with an 802.11e QoS enhanced access point (QAP) to transcode voice flows before their transmissions over the wireless channel. The instantaneous bit rate is determined by a control mechanism based on the estimation of channel congestion state. Our mechanism dynamically adapts audio codec bit rate using a congestion avoidance technique so as to preserve acceptable levels of quality. A case study presenting the results relative to an adaptive system transmitting at bit rates typical of G.711 PCM (64 kbit/s) and G.726 ADPCM (40, 32, 24 and 16 kbit/s) speech coding standards illustrates the performance of the proposed framework. We perform extensive simulations to compare the performance between our adaptive audio rate control and TFRC mechanism. The results show that the proposed mechanism achieves better voice transmission performance, especially when the number of stations is fairly large.

1   Introduction

The Internet heterogeneity is increasing due to the fast deployment of wireless local area networks (WLANs). WLAN hold the promise of providing unprecedented mobility, flexibility and scalability than its wired counterpart. At the same time it seems inevitable that future telephony services will be based on IP-technology. There is a serious concern from the operators side as to offer at least the current "circuit switched" quality for future voice over IP (VoIP) communications. In order for this to become reality, a lot of issues related to VoIP transmission over heterogeneous wired/wireless networks must be solved. In WLAN environments bandwidth is scarce and channel conditions are varying and highly lossy. Even if a lot of voice codec can tolerate some small loss without severe degradation, voice traffic has unacceptable performance if long delays

V. Roca and F. Rousseau (Eds.): MIPS 2004, LNCS 3311, pp. 25–36, 2004.

are incurred. Moreover, the original IEEE 802.11 WLAN standard [1] has been mainly designed for data applications. Two different channel access mechanisms are specified in the 802.11 standard, namely, the contention-based DCF and the polling-based PCF access mechanisms. While DCF and PCF may provide satisfactory performance in delivering best-effort traffic, they lack the support for QoS requirements posed by real time traffic such as VoIP. These requirements make the DCF scheme an infeasible option to support QoS for VoIP traffic. Furthermore, apart from these limitations, a typical WLAN with 11Mbps bandwidth could only support a very limited VoIP connections in DCF mode. On the other hand, PCF mode, with a centralized controller, represent another promising alternative to providing QoS in WLAN [2]. However, studies on carrying VoIP over WLAN in PCF mode in [3] found that when the number of stations in a basic service set (BSS) is large, the polling overhead is high and results in excessive end-to-end delay and that VoIP still get poor performance under heavy load conditions. The medium access control (MAC) layer of the emerging IEEE 802.11e [4] standard tries to support QoS in 802.11 wireless networks using a new Hybrid Coordination Function (HCF) that provides stations with prioritized and parameterized QoS access to the wireless medium. The simple HCF scheduler proposed in 802.11e standard considers the QoS requirements of flows by allocating transmission time to stations based on their mean sending rate and mean frame size. In this work, we investigate the performance limitations in the case of a large number of VoIP flows transmitted over an IEEE 802.11e WLAN. We specifically address the problem of long distance VoIP transport over heterogeneous wired/wireless networks. In the studied case we consider VoIP traffic transmitted from a wired Internet part through a last-hop wireless link that represents the bandwidth bottleneck. All voice traffic needs to be routed through an 802.11e QAP (QoS-enhanced Access Point). The QAP becomes heavily loaded, especially when the number of active stations is fairly large and this results in different types of audio performance degradation (loss due to congestion, loss due to bit errors at the link layer and packet delay). We show through simulations the performance of VoIP according to the number of wireless stations in a BSS. We propose an architecture that is based on a VoIP gateway for interworking the wired and wireless networks. The gateway communicates with a QAP in order to adapt coding rate of voice flows according to the radio channel conditions. Simulations show that our adaptive audio rate control outperforms TFRC mechanism. The paper is organized as follows: Section 2 describes related work on rate and loss control for multimedia applications. Section 3 states the problem. In Section 4, we advance the proposed architecture and we explain our adaptation algorithm. We show simulation results in section 5. Finally, Section 6 concludes the paper.

2   Rate and Loss Control for Multimedia Applications

Rate control is an important issue for both wired and wireless multimedia applications using unresponsive transport protocols (i.e., UDP and RTP). A proper form of congestion control is needed in order for these applications to share congested bandwidth fairly with each other and with TCP-based applications. Many schemes were developed based on TCP-Friendly control mechanisms. These mechanisms can be classified into three main categories: equation-based mechanisms, window-based mechanisms and additive increase, multiplicative decrease (AIMD) mechanisms. Equation-based rate control [9][11] is a widely popular rate control scheme over wired networks, also known as TCP-Friendly Rate Control (TFRC). In this scheme, the sender uses an equation characterizing the allowed sending rate of a TCP connection as a function of the RTT and packet loss rate, and adjusts its sending rate according to those measured parameters. A key issue is than to choose a reliable characterization of TCP throughput. A formulation of the TCP response function was derived in [10], it states that the average throughput of a TCP connection is given by:

$$T(Bytes/sec) = \frac{s}{t_{RTT}\sqrt{\frac{2l}{3}} + t_{RTO}(3\sqrt{\frac{3l}{8}})l(1+32l^2)} \qquad (1)$$

Equation (1) roughly describes TCP's sending rate as a function of the frequency of loss indication l, round trip time t_{RTT} and packet size s. This equation reflects TCP's retransmit timeout behavior, as this dominates TCP throughput at higher loss rates. In the scheme proposed in [10] the receiver acknowledges each packet, and at fixed time intervals the sender estimates the packet loss rate experienced during the previous interval and updates the sending rate using equation (1). This scheme updates the sending rate at fixed time intervals, hence it is suitable for use with multimedia applications. Nevertheless it has the disadvantage of having a poor transient response at small time-scales [16]. In [11], Floyd et al. developed the TFRC protocol. TFRC estimates the recent loss event rate of a connection at the receiver. A loss event consists of one or more packets dropped within a single RTT. The algorithm used for calculating the loss event rate (average loss interval) offers a good responsiveness to changes in congestion while avoiding abrupt reductions of the sending rate in response to a single loss. To behave in a TCP-friendly manner, the sender adapts according to an equation that models TCP response function in steady-state. The main advantages of TFRC are: first it does not cause network instability, thus avoiding congestion collapse. Second, it is fair to TCP flows. Third, the TFRC's rate fluctuation is significantly lower than that of the standard TCP congestion control algorithm, making it more appropriate for real-time applications that require a smooth congestion control and a constant quality. Window-based mechanisms such as TEAR [20] maintain a congestion window to control the transmission of packets. TEAR shifts TCP emulation to the receiver and uses a sliding window to smooth sending rates. The main disadvantage of this type of mechanisms is the lack of flexibility related to the TCP window dynamics [16]. Unlike window-based mechanisms, AIMD mechanisms [18][19] are rate-based congestion control mechanisms that are not applied to a congestion window. The Rate Adaptation Protocol [18] implements an AIMD algorithm based on regular acknowledgments sent by the receiver. In [19], the authors propose an end-to-end rate adaptation scheme that adjusts the transmission rate of multimedia applications to the congestion level of the network. Based on the estimation of the loss rate and the RTT obtained from the regular information of RTCP [8] reports, the sender increases the transmission rate during underload periods and reduces this rate during congestion periods, while avoiding an aggressive adaptation behavior. Although TCP-friendly rate control mechanisms provide relatively smooth congestion control for real-time traffic, they are more appropriate for use over wired IP networks. For multimedia applications over wireless, packets can be corrupted by wireless channel errors at the physical layer and thus TFRC cannot distinguish between packet loss due to congestion and that due to bit errors. TFRC, designed to deal with congestion in wired networks, treats any loss as a sign of congestion. End-to-end statistics can be used to help detecting congestion when packet loss happens. For example, by examining trends in the one-way delay variation, loss could be interpreted as a sign of congestion if this delay is increasing, and as a sign of wireless channel error otherwise [14]. The scheme presented in [13] combines packet inter-arrival times and relative one-way delay to differentiate between congestion and wireless packet losses. This scheme is based on the observation that the one-way delay increases monotically when there is congestion and that the inter-arrival time is expected to increase if there is wireless channel packet loss. Loss differentiation algorithms can then be combined with TFRC to achieve a rate control over heterogeneous Internet environments. The second limitation of TFRC mechanisms is that they are originally designed for applications that use fixed packet size, and vary their sending rate in packets per second in response to congestion. Hence, they should not be used for applications that vary their packet size instead of their packet rate in response to congestion [12]. Varying the packet size during the time interval between two estimations of the sending rate distorts packet-based measurement of loss event. In some situations, using rate control alone does not solve the performance degradation. Such situations may be short-term transient congestion, congestion caused by others' traffic or residual bit errors caused by a noisy wireless link. Forward Error Correction (FEC) has been one of the main methods used to protect against packet loss over packet switched networks and also to improve the quality of noisy transmission wireless links. The amount of FEC information should be tuned according to the characteristics of the loss process in order not to increase bandwidth requirement (and hence the packet loss rate) when the channel is loss free and also not to increase the end-to-end delay since the destination typically has to wait longer to decode the original data packet [15]. The rate/error control advocated in [15] is based on an optimization problem. This approach lacks taking delay into consideration. In [16] an adaptive error control scheme for real-time audio over the Internet is developed. In this work the FEC scheme is selected according to its impact on the end-to-end delay using an utility function for assessing the perceived audio quality that consider the effect of the end-to-end delay. These error control schemes were designed to resolve audio packet loss over the wired IP networks; the packet loss process is different in wireless environments where loss may occur due to congestion or to residual bit errors at the link layer.

3   Problem Statement

The most sensitive case of multimedia traffic is VoIP. In particular the delay is most critical in VoIP applications. It is recognized that the end-to-end delay has a great impact on the perceived quality of interactive conversations with a threshold effect around $150ms$ [5]. For intra-continental calls, the packet delay is on the order of $30ms$ and for inter-continental calls the delay can be as large as $100ms$ [21]. While reducing the effect of a small jitter can be realized by a playout buffer introduced at the receiver, the avoidance of a high jitter/delay is much more complex. Especially retransmissions and contention-based medium access schemes are accountable for high delays and jitters. We consider the case of VoIP calls that are transmitted over heterogeneous wired/wireless networks, we assume that the wired Internet part is error-free and congestion-free and that the bandwidth bottleneck is the last-hop wireless link (Figure 1). In this case, all the voice traffic needs to be routed through the 802.11e QAP (the "bridge" between the wired and wireless networks). Hence, the QAP becomes heavily loaded, especially when the number of active stations is fairly large. Moreover, 802.11e EDCF mode grants different priorities to specific traffic classes (i.e., latency sensitive traffic) but not specific nodes. The VoIP packets may be queued at the QAP if it cannot gain TXOPs from the competition with other nodes, and will become a bottleneck in the network and this will result in additional delays. Three different types of degradation may occur in the last-hop wireless link: packet loss due to congestion, delay due to congestion and packet loss due to bit errors at the link layer. Although a lot of voice codec can tolerate some small loss without severe degradation, most of them operate under preset schemes for data and channel code rates making them vulnerable to the varying conditions on wired and wireless IP-based hops [17]. Some kind of adaptation is therefore needed to dynamically adapt the codec bit rate to the changing wireless network conditions so as to preserve acceptable levels of reliability and quality.

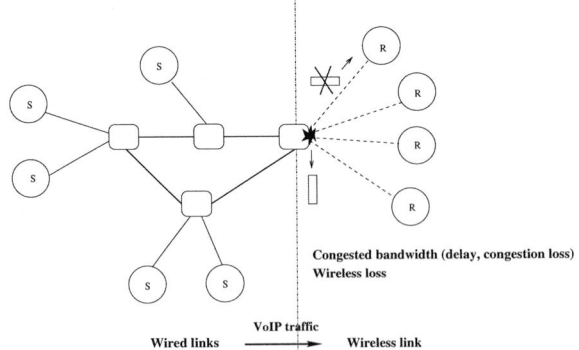

Fig. 1. Voice traffic transmitted from wired network through a last hop congested wireless link

4   Proposed Architecture and Rate Control Mechanism

4.1   VoIP Gateway Interworking Wired and Wireless Network

The proposed architecture (Figure 2) uses a VoIP gateway located at the edge
of the wired network and the wireless link, to transcode voice flows before
their transmissions over the wireless channel. The gateway is associated with
an 802.11e QAP. A QAP is required to support VoIP calls between wired and
wireless networks. In such a situation, the functionality of HC (Hybrid Coor-
dinator) is performed at the QAP. The QAP may gain high priority to access
the channel by piggybacking data packets on the QoS-Poll packets or the QoS
Ack packets, and thus speeds up dispatching packets from wired networks. The
instantaneous bit rate is determined by an adaptation algorithm (described in
section 4.2) based on the estimation of wireless channel congestion status. Congestion control information can be obtained through RTCP reports sent back to
sources via the HC.

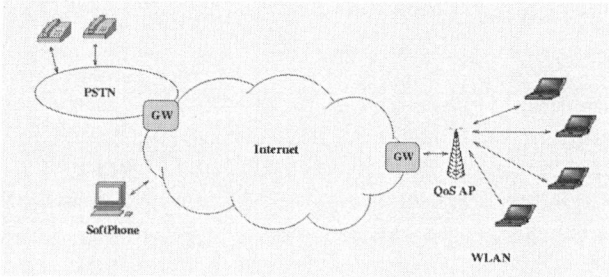

Fig. 2. VoIP gateway at the edge of wired and wireless network

4.2   Vegas-Like Audio Rate Control Mechanism

The proposed rate control mechanism is based on a TCP Vegas-like congestion
avoidance technique for the rate and loss control of VoIP flows over the WLAN.
The rate of the audio codec used for transcoding the voice flow at the VoIP
gateway is varied according to the RTT measured between the QAP and wireless
stations. In case of packet loss, a delay-based loss predictor is used to determine
the type of loss and apply the appropriate strategy depending on whether packet
losses are due to network congestion (transcode the voice flow using a lower audio
codec bit rate) or wireless link errors (increase robustness by adding FEC). The
source and channel adaptation algorithm residing at the gateway will converge to
the available bandwidth in the WLAN while attempting to optimize overall call
quality of several simultaneous voice communications. The input of the algorithm
is the estimation of current WLAN congestion state given by delay and loss

parameters. The basic idea of our Vegas-like audio rate control algorithm is to adapt the audio codec rate by varying audio packet size to avoid congestion, and this unlike TFRC mechanism that uses fixed size packets and varies the sending rate in packets/sec. The VoIP gateway keeps track of the BaseRTT defined as the minimum of all RTTs measured on the WLAN using RTCP receiver reports. When a receiver report related to the voice flow i is received at the gateway, the Expected Audio Data and the Actual Audio Data are calculated as:

$$ExpectedAudioData = R_i \times BaseRTT \tag{2}$$

$$ActualAudioData = R_i \times RTT_i \tag{3}$$

where R_i is the audio codec bit rate used for voice flow i RTT_i is the round-trip time between the gateway and the wireless station i estimated when the receiver report i is transmitted through the gateway. Actual Audio Data represents the amount of audio transmitted during RTT_i using codec rate R_i. The difference D is calculated as:

$$\begin{aligned} Diff &= ActualAudioData - ExpectedAudioData \\ &= R_i(RTT_i - BaseRTT) \end{aligned}$$

D is an estimation of the extra audio data the voice flow i has in transit, i.e. data that would not have been sent if the audio codec used for this voice flow exactly matched the available wireless channel bandwidth. The algorithm then compares the value of D to the α and β thresholds, these two thresholds are defined in terms of bytes. The farther away the actual audio data gets from the expected value, the more congestion there is in the WLAN, which implies that the sending codec rate should be reduced. This decrease is triggered by the β threshold. The α threshold triggers the increase of the audio codec rate in case

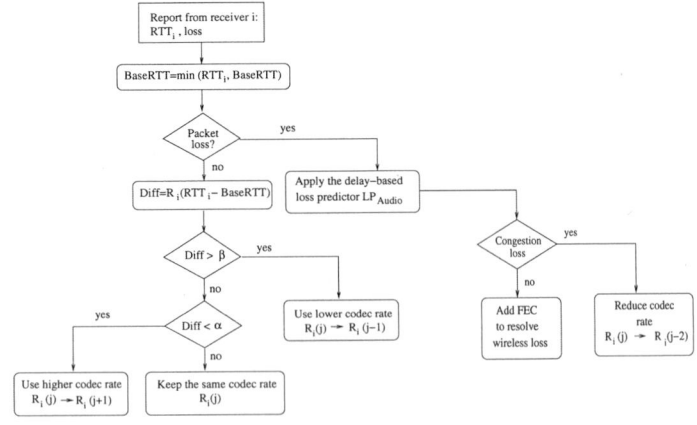

Fig. 3. Flow-chart for the Vegas-like audio rate control

the voice flow is not utilizing the available bandwidth. The goal is then to keep between α and β extra bytes transmitted over the wireless channel (Figure 3). In order to differentiate between congestion and wireless losses, we define the following delay-based loss predictor:

$$LP_{Audio} = \frac{Diff}{BaseRTT} = R_i\left(\frac{RTT_i}{BaseRTT} - 1\right) \tag{4}$$

LP_{Audio} would predict that next packet loss will be due to congestion when the Vegas-like audio rate control algorithm suggests that audio codec rate be decreased. If a loss occurs when the algorithm is recommending increasing codec rate, it may be reasonable to assume that the loss is due to transmission errors on the wireless channel and thus FEC information will be added in order to resolve this loss (Figure 3).

5   Simulation Experiments and Discussion

We provide $NS - 2$ simulation results obtained from downlink VoIP flows transmitted on a WLAN with CBR background traffic using the EDCF/HCF mode of operation. We consider a high-rate IEEE 802.11a WLAN with physical data rate of 36Mbps and an adaptive system in which sources can switch between five bit-rates, corresponding to widely used telephone speech coding standards (Table 1). The G.726 [7] codec makes a conversion of a 64 kbit/s pulse code modulation (PCM) channel to and from a 40, 32, 24 or 16 kbit/s channel. The conversion is applied to the PCM bit stream using an ADPCM (Adaptive Differential Pulse Code Modulation) transcoding technique. The relationship between the voice frequency signals and the PCM encoding/decoding laws is fully specified in Recommendation G.711 [6]. A variable bit-rate system operating at such bit-rates can be viewed as a system that always delivers "toll quality," but with different levels of complexity, delay and robustness. Codec rates and packet sizes used to simulate our mechanism are shown in Table 1. For TFRC flows, we use packets of $200bytes$ and for background flows we set packet size to $1500bytes$. The goal of the simulations is not a complete analysis of the considered system but, rather, an indication that interesting performance evaluation indices can be derived through the proposed approach. We consider only the adaptive codec

Table 1. Codec bit rate and packet size

Codec	Bit Rate (Kbit/s)	Payload Size (bytes)	Total Packet Size (with IP/UDP/RTP headers)
G.711	64	160	200
G.726	40	100	140
G.726	32	80	120
G.726	24	60	100
G.726	16	40	80

rate part of the Vegas-like control algorithm. In future work we will consider adding FEC information based on the Loss Predictor defined in (4). Simulations are carried out for the duration of 20 seconds and the presented results are averaged over 5 simulations. We set the EDCF VoIP flow priority to 6 and background flows priority to 1. The number of stations is increased from 2 to 24 (including the QAP). A VoIP and a background flow are transmitted from the QAP to each QSTA. In order to ensure more accurate responsiveness to the channel load, we use variable values for α and β parameters of our Vegas-like audio rate control algorithm that depend on an estimation of the number of audio packets transmitted by the voice flow i during the RTT_i :

$$\alpha = 30 \times \frac{RTT_i(ms)}{20ms}(Bytes)$$

$$\beta = 50 \times \frac{RTT_i(ms)}{20ms}(Bytes)$$

Figure 4 shows the average delay of VoIP traffic over the WLAN. Adaptive audio rate control presents good performance, as it keeps average delay below $4ms$ in both situations of small and large number of VoIP flows. With TFRC, VoIP average delay rises above $15ms$. The confirmation of these results is provided by the maximum voice packet delay depicted in Figure 5. Adaptive audio rate control is able to keep the maximum delay below $10ms$ when the number of VoIP flows is less than 10, and below $20ms$ in the case of more than 10 VoIP flows. For TFRC the maximum delay is about $25ms$. Reducing the audio packet delay by the value of about $10ms$ on the WLAN is important in order to cope with the before mentioned audio QoS requirements (one way delay is restricted to at most $150ms$) and since we have to consider the delay caused in the wide area network that must be traversed by an audio packet on its way to the destination in the WLAN. The adaptability of our control mechanism to the WLAN conditions ensures a reduced packet voice delay and this improves perceived voice quality. Figure 6 depicts the mean bandwidth of VoIP flows as the number of stations is increased. When the number of flows is below 8, the mean bandwidth of our adaptive VoIP mechanism is higher than that of TFRC. Our mechanism is less

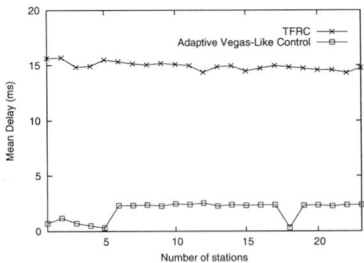

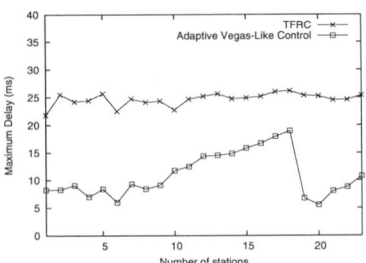

Fig. 4. Mean VoIP delay vs. the number of QSTAs

Fig. 5. Maximum VoIP delay vs. the number of QSTAs

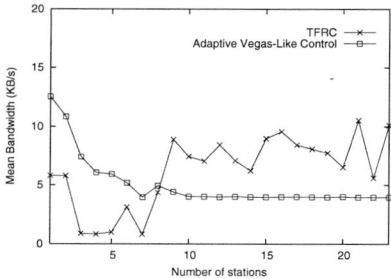

Fig. 6. Mean bandwidth of VoIP flows vs. the number of QSTAs

conservative than TFRC when the number of flows is reduced, and this avoids the under-utilization of network bandwidth and also avoids the voice transmission quality degradation in response to congestion. For more than 6 flows, the mean bandwidth used by our mechanism is steady (4 KB/s) and smoother than TFRC mean bandwidth. Maintaining low sending rate variation and avoiding abrupt rate changes will reduce the delay jitter and this will ameliorate the perceived voice quality. Besides, reducing the sending rate in case of high load network conditions will increase the WLAN capacity.

In order to evaluate the fairness between VoIP flows, we compute for each scheme the fairness index defined as:

$$FairnessIndex = \frac{(\sum_{i=1}^{n} T_i)^2}{n \times \sum_{i=1}^{n}(T_i)^2} \qquad (5)$$

where n is the number of flows using the same control scheme, and T_i is the throughput of flow i. The fairness index is equal to 1 if all T_i are equal (highest degree of fairness between flows). Figure 7 shows that our adaptive audio rate control achieves considerably better fairness than TFRC. With our mechanism,

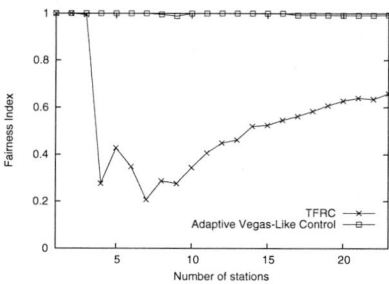

Fig. 7. Improved VoIP flow fairness with the adaptive Vegas-like codec rate control

fairness index is kept almost at 1 independently of the number of VoIP flows, however with TFRC, this index goes below 0.4 for a small number of VoIP flows (6 flows) and it is improved when the number of flows is increased (0.62 for 20 flows). This can be explained by the fact that our adaptive control uses delay information to avoid congestion and adapts the audio rate of each flow, however TFRC uses loss information for rate adaptation and this information is not so accurate for detecting congestion in WLANs.

6   Conclusion

In this paper we propose a novel adaptive architecture for the transport of VoIP traffic over heterogeneous wired/wireless Internet environments. This architecture supports adaptive VoIP coding on WLAN using a VoIP gateway located at the edge of the wired Internet and the wireless network. The adaptive coding mechanism is illustrated considering a specific control mechanism applied to variable bit-rate system operating at five VoIP coding bit rates (64, 40, 32, 24 and 16 Kbit/s). Simulation results show that our adaptive architecture responds constructively to network congestion and improves QoS support for VoIP in IEEE 802.11e networks. Using the 802.11e EDCF/HCF scheme, we reduce the transmission delay of VoIP traffic compared to current TFRC algorithm. Obtained results show that our adaptive rate control mechanism is fairer than TFRC especially when the number of VoIP flows is increased. The system capacity is also increased, since the sending rate is reduced in case of high-load network conditions.

Acknowledgments. This work has been supported by the french ministry of industry in the context of the national RNRT project VTHD++.

References

1. IEEE Std 802.11-1999, Part 11: *Wireless LAN MAC and Physical Layer specifications.* Reference number ISO/IEC 8802-11:1999(E).
2. M. Veeraraghavan, N. Cocker, and T. Moors. *Support of Voice Services in IEEE 802.11 Wireless LANs.* In Proceedings of INFOCOM 2001, Alaska, April 2001.
3. D.-Y. Chen, S. Garg, M. Kappes, and K. S. Trivedi. *Supporting VBR VoIP traffic with IEEE 802.11 WLAN in PCF mode.* In Proceedings of OPNETWork 2002, Washington D. C., August 2002.
4. IEEE 802.11e/D4.0. *Wireless MAC and Physical Layer specifications: MAC Enhancements for QoS.* November 2002.
5. ITU-T. *One-Way Transmission Time.* Recommendation G.114, February 1996.
6. ITU-T. *Pulse code modulation (PCM) of voice frequencies.* Recommendation G.711, November 1988.
7. ITU-T. *40, 32, 24, 16 kbit/s Adaptive Differential Pulse Code Modulation (AD-PCM).* Recommendation G.726, December 1990.
8. H. Schulzrinne, S. Casner, R. Frederick, and V. Jacobson. *RTP: A Transport Protocol for Real-Time Applications.* Request for Comments 1889, January 1996.

9. S. Floyd, and K. Fall. *Promoting the Use of End-to-End Congestion Control in the Internet*. IEEE/ACM Transactions on Networking, August 1999.

10. J. Padhye, V. Firoiu, D. Towsley, and J. Kurose. *Modeling TCP Throughput: A Simple Model and its Empirical Validation*. In Proceedings of ACM SIGCOMM'98, Vancouver, Canada, August/September 1998.

11. S. Floyd, M. Handley, J. Padhye, and J. Widmer. *Equation-Based Congestion Control for Unicast Applications*. In Proceedings of ACM SIGCOMM'2000, pp. 43-56, Stockholm, Sweden, August/September 2000.

12. M. Handley, S. Floyd, J. Padhye, and J. Widmer. *TCP Friendly Rate Control (TFRC): Protocol Specification*. Request for Comments 3448, January 2003.

13. S. Cen, P.C. Cosman, and G.M. Voelker. *End-to-end Differentiation of Congestion and Wireless Losses*. In Proceedings of Multimedia Computing and Networking (MMCN) conf. 2002 pp. 1-15, San Jose, CA, January 2002.

14. M. Chen and A. Zakhor. *Rate Control for Streaming Video over Wireless*. In Proceedings of INFOCOM'04, Hong Kong, March 2004.

15. J-C. Bolot, S. Fosse-Parisis, and D. Towsley. *Adaptive FEC-based Error Control for Internet Telephony*. In Proceedings of INFOCOM'99, New York, March 1999.

16. C. Boutremans and J. Y. Le Boudec. *Adaptive Delay Aware Error Control for Internet Telephony*. In Proceedings of 2nd IP-Telephony workshop, Columbia University, New York, April 2001.

17. J. Matta, C. Pépin, K. Lashkari, and R. Jain. *A Source and Channel Rate Adaptation Algorithm for AMR in VoIP Using the Emodel*. In Proceedings of NOSSDAV'03, San-Francisco, California, June 2003.

18. R. Rejaie, M. Handley, and D. Estrin. *RAP: An End-to-end Rate-based Congestion Control Mechanism for Realtime Streams in the Internet*. In Proceedings of IEEE INFOCOM'99, New York, March 1999.

19. D. Sisalem and H. Schulzrinne. *The Loss-Delay based Adjustment Algorithm: a TCP-friendly Adaptation Scheme*. In Proceedings of NOSSDAV'98, Cambridge, UK, July 1998.

20. I. Rhee, V. Ozdemir, and Y. Fi. *TEAR: TCP Emulation at Receivers – Flow Control for Multimedia Streaming*. NCSU Technical Report, April 2000.

21. M. Karam and F. Tobagi. *Analysis of the Delay and Jitter of Voice Traffic over the Internet*. In Proceedings of IEEE INFOCOM'01, Anchorage, AL, April 2001.

On Modeling MPEG Video at the Frame Level Using Self-Similar Processes

José C. López-Ardao, Pablo Argibay-Losada, and Raúl F. Rodriguez-Rubio

Telematics Engineering Department, University of Vigo (Spain)
ETSET Telecomunicacion, Campus Universitario
36200 Vigo (Spain)
jardao@det.uvigo.es

Abstract. MPEG video traffic is expected to represent most of the load in the future high-speed networks. Adequate traffic models for MPEG *Variable Bit-Rate* (VBR) video are thus important for network design, performance evaluation, admission control and resource allocation.

Many models for VBR video traffic have been proposed in the literature. However, while the GOP-level process has been widely analyzed in literature, and so the inter-GOP correlation, little effort has been devoted up to now to the frame-level processes, and so to the intra-GOP correlations, even though it is a fundamental characteristic of MPEG traffic and it might have an important impact on queueing performance.

In this work, we compare different solutions proposed in the literature to obtain MPEG frame-level processes, depending on the performance metric to study (loss rate, mean delay and jitter). Besides, we claim for the use of self-similar processes, and more concretely, Gaussian F-ARIMA$(1, d, 0)$ processes, to adequately capture the correlation structures involved in MPEG video.

1 Introduction

MPEG video traffic is expected to represent most of the load in the future high-speed networks. So, adequate traffic models for MPEG Variable Bit-Rate (VBR) video are important for network design, performance evaluation, admission control and resource allocation.

The output of a MPEG encoder consists of a deterministic periodic sequence of encoded frames where the period is called GOP (Group of Pictures). A GOP contains three different types of frames: one I-frame (encoded after removing the spatial redundancy within a still image) and several P- and B-frames (encoded adding motion compensation with respect to the previous I-frame).

VBR video traffic is characterized by a high burstiness and a strong positive correlation. These properties were found to be induced by the presence of a characteristic inherent to VBR video traffic, called self-similarity. This characteristic involves long-range correlations over arbitrarily long time-scales, a phenomenon usually referred to as Long-Range Dependence (LRD) [1]. LRD is mainly characterized by an Autocorrelation Function (ACF) with a hyperbolic decay.

V. Roca and F. Rousseau (Eds.): MIPS 2004, LNCS 3311, pp. 37–48, 2004.

Many models for VBR video traffic have been proposed in the literature. However, while the GOP-level process has been widely analyzed in literature [2,3,4], and so the inter-GOP correlation, little effort has been devoted up to now to the frame-level processes, and so to the correlations existing between the different MPEG frame-types (from now on, we will refer to it as intra-GOP correlations). These correlations are often neglected even though they represent a fundamental characteristic of MPEG traffic and they might have an important impact on queueing performance [4,5]. In fact, many works avoid to descend to the frame level and propose models only for the GOP process [6,7].

Different solutions to obtain MPEG frame-level processes have been proposed in the literature, but these have never been compared and analyzed jointly. So, by means of this comparison, this work intends to serve as a guideline to help researchers with designing MPEG video models and being in a good position to use the best model in terms of simplicity, computational cost and queueing performance, depending on the metric to study (loss rate, mean delay and jitter).

The paper is organized as follows: Section 2 starts classifying the different video models proposed in the literature and discussing their advantages and drawbacks. The problem of choosing the most suitable LRD process to adequately characterize MPEG video correlations is also addressed. In Sect. 3, the different solutions to obtain the MPEG frames within a GOP are presented. Section 4 shows the capabilities of these solutions for capturing empirical distributions and correlations. In Sect. 5 we show the main results related to queueing performance, and Section 6 analyzes issues related to the computational efficiency. Finally, Section 7 summarizes the main conclusions of this study.

2   VBR Video Modeling

Depending on how the relevant time-scales are captured, the VBR video models proposed in the literature (see [8] for a survey) can be classified into two types: hierarchical and non-hierarchical.

Hierarchical models typically consider two time-scales: scene or activity level (time-scale of minutes) and GOP/frame (time-scale of milliseconds/seconds) [9, 10,6]. Scene changes occur when the mean of the bit-rate process changes significantly as a result of a considerable change in picture content (camera cuts). Hierarchical models have several well-known drawbacks. On the one hand, the scene layer adds complexity to the model and, on the other hand, the model has an important subjective component in the scene detection process [11]

Non-hierarchical models consider only the GOP/frame level. For this reason, only those models based on stochastic processes that incorporate inherently the relevant time-scales are valid. In this field, the use of self-similar processes in video traffic modeling [2,3,5,11] has a clear advantage over the traditional use of Markovian or Autoregressive processes [7,9]. Self-similar processes are able to exhibit LRD over a wide range of time-scales, whereas the other types of processes ignore the correlations beyond a particular lag intending to keep the analytical tractability, and so it is said that they exhibit Short-Range Dependence

(SRD). For this reason, validity could be in doubt because modeling the LRD through SRD processes typically will require many parameters.

In other words, while the LRD present in MPEG video implies a hyperbolic decay in the ACF[1], SRD models give rise to a much more rapid exponential decay, and it is well-known that it is necessary a sum of infinite exponentials to match a hyperbolic function. In the case of MPEG, adequately modeling a frame-level process with a Markov chain could require a large number of states. However, SRD processes could be typically well-suited for being used in hierarchical models, an intermediate solution between SRD and LRD, as they only capture LRD for a finite range of time-scales.

On the other hand, the requirement to use few parameters is important because they must be estimated from the empirical data and each estimate incurs a certain amount of error. Besides, a video traffic model should be independent of any empirical data, and so it is very important that its parameters have physical meaning. For this reason, the rise of self-similar processes for video traffic modeling purposes has been very important due to their capability to exhibit LRD over all time-scales by making use of few parameters (parsimonious modeling).

In any case, the impact of the LRD on queueing performance depends much on the length of the busy period (the number of the relevant correlations that interact at the queue) and so this effect can be less important when the utilization is low or with small buffers [12,13,14].

However, though this discussion is not within the scope of this paper, it seems clear that, if it is feasible, the use of a simple, parsimonious and efficient model based on self-similar processes (or, in general, exhibiting LRD) will lead us to achieve better estimations of the system performance. Furthermore, the model will neither depend on particular and hypothetical assumptions about the load or the buffer sizes, nor on the subjectivity of the scene detection. In order to reinforce this statement, in [11] it is shown that the hyperbolic decay is the best match for the ACF of the different frame-types of many commonly used MPEG-encoded video sequences.

2.1   Choosing a LRD Process

We must choose a LRD process that is able to model the most important traffic characteristics parsimoniously (see [15] for a survey on LRD and self-similar processes). In our view, such a process should at least have a well-known marginal distribution and allow to set its mean and variance. For this reason, some LRD processes with unknown marginal distribution, like chaotic maps or the shifting level process are seldom utilized. Instead, self-similar processes like Fractional Gaussian Noise (FGN) and Gaussian Fractional-ARIMA processes, both with Gaussian marginal distribution, and $M/G/\infty$ processes, with Poissonian marginal distribution, are commonly used.

---

[1] $r(k) = k^{2H-2}$ $0.5 < H < 1$. H is the physical parameter that indicates the intensity of the LRD

Since the queuing performance of a network depends not only on the correlation structure, but also on the marginal distribution of the input processes [12, 14], it would be very useful to be able to modify the marginal distribution of the process without altering its correlation structure.

In this regard, FGN and Gaussian F-ARIMA(p, d, q) processes enjoy the attractive advantage that their marginal distribution can be changed by a sufficiently regular transformation[2] (parameters included) without modifying their long-range correlation structure [3].

FGN processes have only three parameters (μ, σ and H), and their correlation structure is determined by a single parameter, H, that only shapes their LRD structure, in such a way that their SRD structure remains fixed and can not be modified independently. So, FGN processes have a correlation structure too rigid to capture the wide range of low-lag correlation structures encountered in practice (VBR video, for example) [16]. This fact is a strong limitation of FGN because SRD can also have a significant impact on performance, involving the need to incorporate SRD into traffic models [17].

Instead, F-ARIMA(p, d, q) processes are much more attractive for traffic modeling purposes, since they are capable of modeling LRD (by means of the parameter d) and SRD (by means of the AR(p) and MA(q) components) independently. In fact, the simple addition of one AR(1) component to a F-ARIMA(0, d, 0) process creates much more flexible processes than FGN [16, 18].

Despite their flexibility, F-ARIMA processes have been little used in simulation studies, since the usual generation methods for FGN are significantly more efficient than those for F-ARIMA. However, this drawback can be easily overcome using the fast generation method for Gaussian F-ARIMA proposed in [15]. This method produces high-quality traces much more efficiently than the usual methods, even if compared with the best methods for FGN.

As for the choice of the type of LRD processes, we opt to use Gaussian F-ARIMA(1, d, 0) processes for efficiency (using the aforementioned method) and simplicity reasons (see [15] for a more extensive discussion). As it will be shown below, F-ARIMA(1, d, 0) processes permit to accurately capture the correlations present in the MPEG video traffic with only two parameters: the coefficient of the AR(1) component, used to match the SRD behavior, and the d parameter for LRD ($H = d + 0.5$). We have also verified that additional AR or MA components are not needed for an adequate fitting and add more parameters and complexity.

On the other hand, it is known that the marginal distribution transformation induces a slight decrease of the magnitude of the SRD. However, fortunately, the use of Gaussian F-ARIMA(1, d, 0) processes permits to easily overcome this inconvenience by adequately adjusting the AR(1) component.

With respect to the most suitable marginal distributions, an extensive study in [4] concludes that Lognormal or Gamma distributions are a useful approximation to the histograms of the I, P, and B frame sizes for a wide range of

---

[2] The distribution change will be done by $h(x) = F_Y^{-1}(F_X(x))$, where $F_X(x)$ is the original marginal distribution and $F_Y(x)$ is the target marginal distribution.

MPEG video sequences. Similar results are found for the GOP sizes. We will use Lognormal distributions for efficiency reasons.

3   On Obtaining MPEG Frames Within a GOP

It is important to note that all the sequences for the same type of MPEG frame ($P_i(n)$ and $B_i(n)$) have very close statistical properties and they can be considered independent of the place i within the GOP [11,19]. So, the target problem is to obtain the three frame-size processes, $I(n)$, $P(n)$ and $B(n)$.

An important work due to Lombardo et al. [5] shows the correlation existent among the different types of MPEG frames by obtaining the conditional distribution of P- and B-frames sizes, given an I-frame size. However, this model requires a huge computational cost and it is little flexible. We will focus on other more efficient and simpler solutions for obtaining MPEG frames within a GOP.

Next we briefly describe the different solutions proposed in the literature which will be analyzed in this paper. It is important to note that we are only interested in the solution taken to obtain three different processes, one for each MPEG frame-type, and not in other peculiarities involved into the model, such as the type of processes used or the marginal distributions.

Krunz et al. [10] propose a hierarchical model where the inter-GOP correlation is only taken into account for the I-frames, whereas P- and B-frames are modeled by two independent processes of i.i.d. random variables with Lognormal distribution. This solution does not consider intra-GOP correlation. We will refer to it as **solution KRU**.

Rose [20] uses a background process that matches the inter-GOP correlation to obtain a GOP-size process. The frames sizes are obtained by multiplying the GOP size by a scaling factor. The scaling factor for each frame-type is the quotient between the mean frame-size for that type and the mean GOP-size. With this simple method, P-frames (and B-frames) have the same size in every GOP, and so the frame-by-frame correlation is lost. However, it is the only simple solution where intra-GOP correlation is introduced directly since the size of the P- and B-frames depends on the GOP-size, that is, on the I-frame size. We will refer to this solution as **solution ROS**.

In [3] three different distribution transformations for each frame-type, $h_I(x)$, $h_P(x)$ and $h_B(x)$, are applied to a background self-similar process that matches the empirical ACF of the I-frame sizes (inter-GOP correlation). We will refer to it as **solution HUA**. Using the same background process for obtaining each type of frame induces a certain amount of correlation between the different frame-types, but it has the drawback of using the same correlation structure for all of them, often very different. We must also note that, like in **solution ROS**, P-frames (and B-frames) have the same size in every GOP.

Finally, Ansari et al. [11] propose to use three independent processes matching the ACFs of each frame-size process, $I(n)$, $P(n)$ and $B(n)$. This solution, referred to as **solution ANS**, does not introduce intra-GOP correlation either.

4   Modeling an Empirical MPEG Trace

For the sake of brevity, in this paper we present only the results for the empirical MPEG encoded data The Simpsons, although we have analyzed other different traces[3], to which the same considerations and conclusions apply. The trace under study consists of 40000 frames which is equivalent to approximately half an hour (at a rate of 25 frames per second). The GOP length is 12 (IBBPBBPBBPBB).

As we already said, in order to capture the correlations present in the MPEG video traffic, we opt for Gaussian F-ARIMA$(1, d, 0)$ processes, along with the fast generation method proposed in [15], to generate the background processes (GOP sizes for solution **ROS**; I-frame sizes for solutions **HUA** and **KRU**; and I, P and B-frame sizes for solution **ANS**). F-ARIMA processes are generated for matching the correlation structure (one-lag autocorrelation and Hurst parameter) of the target empirical process. Then, a Lognormal distribution transformation is applied. By means of simply adjusting the AR(1) component, it is easy to cancel the effect of this transformation on the SRD structure.

We have verified that, after the distribution transformation, the Probability Density Functions (PDFs) of all the solutions come close enough to the empirical distributions (for frames and GOPs).

In relation to the ACFs and the estimated autocorrelation parameters of the synthesized GOP/frame sizes (Hurst parameter, H, and the one-lag auto-correlation, r_1), shown in Table 1, we can observe that all the solutions except **KRU**, mainly for the SRD structure, achieve good matching for the ACF of the GOP-size process. This implies that inter-GOP short-range correlations are not adequately captured through I-frame correlations when P- and B-frames are modeled by two independent processes of i.i.d. random variables.

On the other hand, the ACFs of synthesized MPEG traces for all the solutions exhibit the well-known sub-periodic behavior, and regarding the correlation for each frame-type, we can make the following comments:

- The solution **KRU** fits well the I-frame sizes, as it was expected, but the simplicity of P- and B-frames modeling by means of two independent processes of i.i.d. random variables leads obviously to destroy the correlation structure.
- In the same way, the solution **HUA** fits well the I-frame sizes. However, while the LRD structure of the P- and B-frame sizes is captured well enough, the SRD magnitudes are much higher than those of the empirical trace. The reason lies in the fact that the correlation structure of these frames matches the I-frame sizes correlation, whose magnitudes are significantly higher.
- The use of averaging factors on the GOP sizes, proposed in the solution **ROS**, retains the underlying LRD structure of each frame-type, but induces higher magnitudes in the short-range correlations of the P- and B-frames, perhaps due to the repetition of frame-sizes within a GOP.

---

[3] The traces were courtesy of Dr. O. Rose [4]

Table 1. Estimated autocorrelation parameters of the synthesized GOP/frame sizes

	GOPs		I-frames		P-frames		B-frames	
	H	r_1	H	r_1	H	r_1	H	r_1
Simpsons trace	0.86	0.73	0.92	0.84	0.89	0.58	0.84	0.75
Solution **ANS**	0.89	0.68	0.92	0.83	0.90	0.59	0.84	0.76
Solution **ROS**	0.86	0.72	0.90	0.82	0.87	0.67	0.87	0.81
Solution **KRU**	0.81	0.17	0.92	0.83	0.51	0.01	0.50	0.01
Solution **HUA**	0.90	0.81	0.92	0.83	0.89	0.94	0.91	0.98

5   Queueing Performance

Although the validation of a source model should require an approximate distribution and autocorrelation fitting, the ultimate validity of a model is mainly determined by its performance predictions in comparison to empirical data.

A single-server FIFO queue with finite buffer size is commonly used in the literature to address this issue. In this paper, we also use this system, where the capacity of the server (bits per second), or its utilization factor, and the buffer size are system variables.

In order to analyze the aforementioned solutions for obtaining the different types of MPEG frames within a GOP, we have made exhaustive trace-driven simulation experiments. The video frames, whose sizes are object of modeling, arrive at the switch at a constant rate (25 frames per second in our case). When the buffer limit is reached, new arrivals are dropped.

Given that the "goodness" of a traffic model may depend on the metric under study, we analyze three key performance measures (statistics) of the system: loss probability, mean delay ($E[T_i]$) and jitter ($E[T_i - T_{i-1}]$). The mean values were estimated using 95% confidence intervals by means of a batch-means procedure, modified and improved to work with LRD data (see details in [21]).

We will represent the performance metrics versus the utilization factor for different values of the buffer size. Although many values were used in the simulation experiments, we show the results for some representative values. For utilization, a wide range was used: 0.4 (low load), 0.5, 0.6, 0.7, 0.8, 0.9 and 0.95 (high load). For buffer size, four values were taken: 200 kbits (a small buffer roughly equivalent to one GOP), 1 Mbits (a medium-size buffer equivalent to around five GOPs), 2.5 Mbits (a large buffer equivalent to about 12.5 GOPs) and 5 Mbits (a very large buffer, around 25 GOPs).

5.1   Loss Probability

Regarding the estimated loss probabilities, the results are very interesting (see Fig. 1). Curiously, the solution **ROS** achieves very good predictions, clearly better than the other solutions. The solution **ANS** always underestimates the loss rate, although it does not work badly for low and moderate utilization.

Nevertheless, the other two solutions fail. The solution **HUA** gives rise to a very high loss rates, indicating that the method for inducing the intra-GOP correlation is not suitable. Finally, the solution **KRU** hardly generates losses, confirming that modeling just the inter-GOP correlation, in this case through the I-frame sizes, even capturing all the marginal distributions, leads to a very poor model for loss-rate predictions.

These results indicate that the GOP-size process is dominant for loss performance. That is, loss rates will be better predicted by the solution that best matches the PDF and the ACF of the GOP-size process, and also takes into account the intra-GOP correlation, even by means of simple averaging factors. It does not seem necessary to devote efforts to capturing the correlation behaviors of the size of each frame-type.

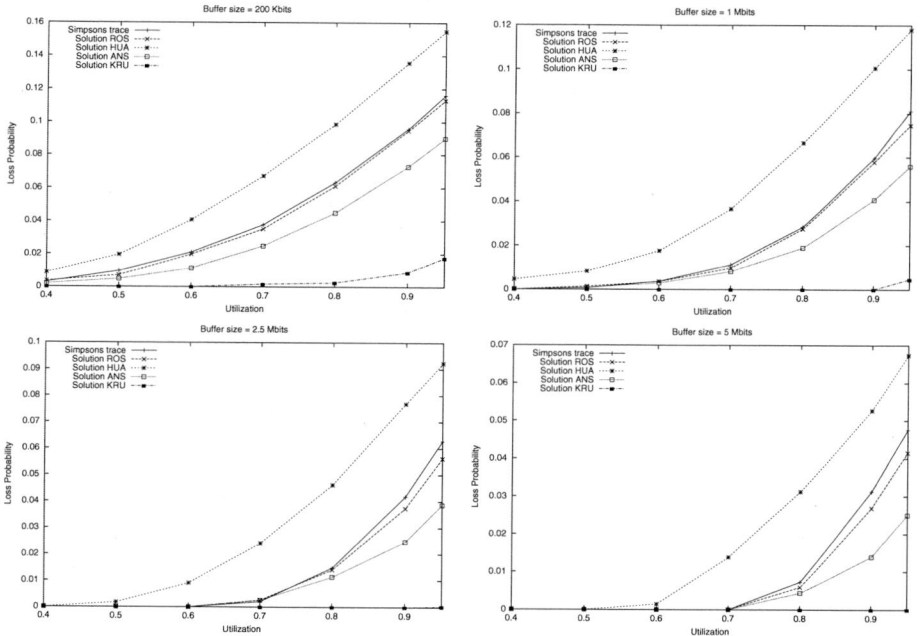

Fig. 1. Loss probability results

5.2   Mean Delay

In this case the results are slightly different (see Fig. 2). Now both the solution **ROS** and the solution **ANS** work well enough, although the solution **ROS** is still slightly better. Specifically, it is the best option for small and moderate buffer sizes. The predictions of both solutions get worse when the buffer size is increased, whereas the solution **HUA** improves as buffer size and/or utilization is increased. In fact, for infinite buffer the predictions are very accurate.

The results for the solution **KRU** are not shown because of its extremely poor predictions (very low mean delays) for the reasons cited in Sect. 4.

In short, when the queue size is not very large, a good approximation consists of modeling only the GOP-size process and introducing the intra-GOP correlation by simple averaging. Alternatively, adequately capturing the correlation structure of each type of frame also provides acceptable performance results. However, when queue size is very large, and so the correlation structure is much more relevant to performance, the results suggest that the GOP-size process loses relevance in favor of the intra-GOP correlation, and so another method for characterizing this type of correlation is necessary. Perhaps, that is the reason why the interdependence introduced by using the same background process in the solution **HUA** provides the best results for the largest queue sizes.

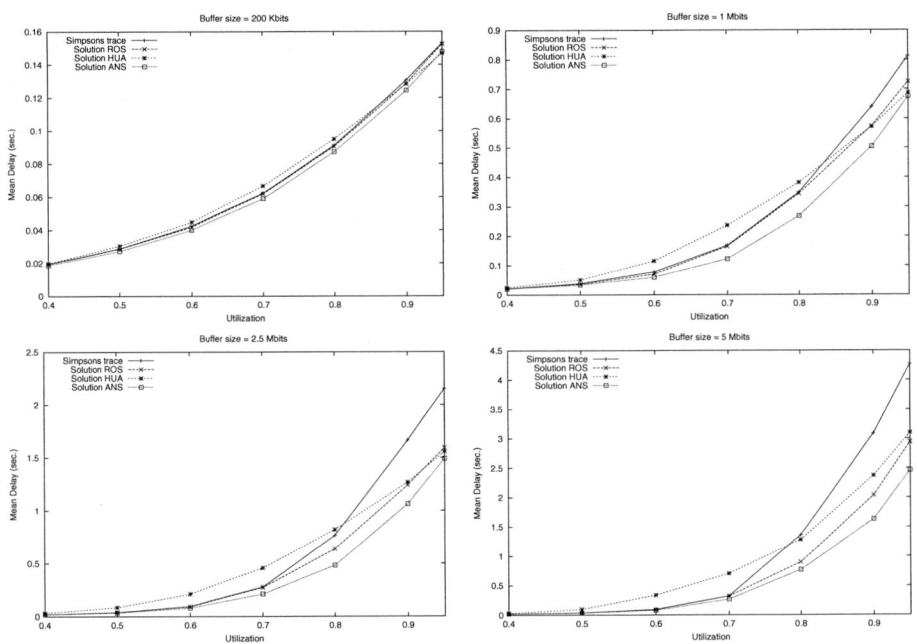

Fig. 2. Mean Delay results

5.3   Jitter

Jitter depends on the difference in size between consecutive frames. So, P-frames, and mainly B-frames, play a fundamental role owing to the relative number of these in comparison to I-frames. Properly characterizing the ACF and PDF of the B-frame sizes is still more important because they are sent consecutively. For this reason, the solution **ANS** is clearly the best.

Instead, the solutions **ROS** and **HUA** can never be valid because they ignore the real frame-by-frame correlation, introducing several consecutive frames of identical size in the same GOP, so leading to many zero-jitter samples. Simulation results (see Fig. 3) demonstrate that jitter is grossly underestimated.

On the other hand, the solution **KRU** is not valid either since it removes both the intra-GOP correlation and the autocorrelation of each frame-type, and so it is expected that jitter predictions are highly overestimated.

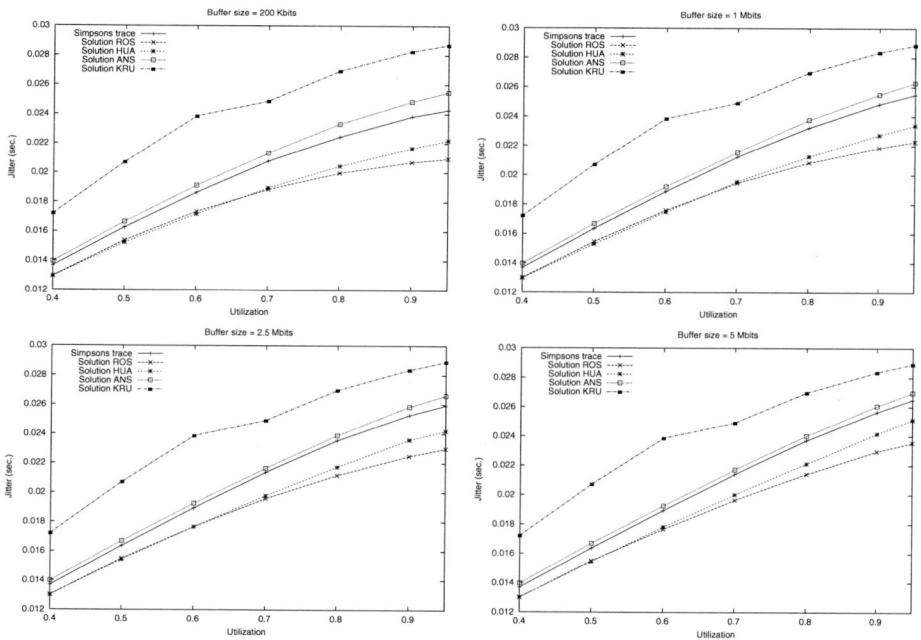

Fig. 3. Jitter results

6   Issues Related to Efficiency

Although the behaviour in relation to the queueing performance should play an important role in choosing the most suitable solution, another important decision factor is the computational efficiency due to the large number of long runs that must be used in an exhaustive simulation study in the context of LRD.

Let us assume that we want to generate a trace of N GOPs, that is, $12 * N$ frames. For large N, the computational cost of the **solution ROS** is roughly proportional to N. The solutions **HUA** and **KRU** entail the additional generation of $2 * N$ and $11 * N$ distribution transformations, respectively. Instead, the **solution ANS** requires the highest computational cost, roughly twelve times the computational cost of the **solution ROS**.

For example, the generation of 1 million GOPs (12 million MPEG frames) at a Pentium III 1GHz. (1GB RAM) requires 15 seconds for the **solution ROS**, 24 seconds for the **solution HUA**, 52 seconds for the **solution KRU** and 175 seconds for the **solution ANS**.

7   Conclusions

Many models for VBR video traffic have been proposed in the literature. However, while inter-GOP correlation has been widely analyzed, little effort has been devoted up to now to the obtaining of the different frame-types within a MPEG GOP for capturing the intra-GOP correlations.

We propose self-similar Gaussian F-ARIMA$(1, d, 0)$ processes to capture the LRD behavior of the MPEG traffic and analyze the impact on queueing performance of different solutions proposed in the literature for obtaining the different types of MPEG frames within a GOP. The results showed us that the suitability of a model depends much on the performance metric to estimate.

For loss rate, the results indicate that the GOP-size process is dominant, and so the best option is the one that best matches the PDF and, mainly, the ACF of the GOP-size process, but also taking into account the intra-GOP correlation, even by means of simple averaging factors applied to the GOP sizes, like it is proposed by the **solution ROS**. It does not seem necessary to devote efforts to capturing the correlation behaviors of the size of each frame-type.

For mean delay, when using small and moderate buffer sizes, the GOP-size process and the averaging factors proposed by the **solution ROS** are the best option again. However, when the queue size is very large (that is, for large buffer sizes and high utilization), and so the correlation structure dominates performance, the results suggest that the GOP-size process loses relevance in favor of the intra-GOP correlation, and so another method for characterizing this type of correlation might be necessary.

For jitter, adequately characterizing the PDF and, mainly, the ACF of the B-frames is very important because they are sent consecutively, and so the **solution ANS** is the best choice. Again, when the utilization and the buffer size are increased, the intra-GOP correlation gain relevance and, therefore, other solutions that take into account this correlation are advisable.

Finally, in the context of simulation experiments with LRD processes, the computational cost is usually critical. In this case, the **solution ROS** is clearly the best, but this issue could be a serious drawback for the **solution ANS**.

References

1. Beran, J., Sherman, R., Taqqu M.S., Willinger W.: Long-Range Dependence in Variable-Bit-Rate Video Traffic. IEEE Trans. on Comm. **43** (1995) 1566–1579
2. Garrett, M.W., Willinger, W.: Analysis, Modeling and Generation of Self-Similar VBR Video Traffic. In Proc. of ACM SIGCOMM '94, London, UK (1994) 269–280

3. Huang, C., Devetsikiotis, M., Lambadaris, I., Kayevol, A.R: Modeling and Simulation of Self-Similar Variable Bit Rate Compressed Video: A Unified Approach. In Proc. of ACM SIGCOMM '95, Cambridge, MA USA (1995) 114–125

4. Rose, O.: Statistical Properties of MPEG Video Traffic and their Impact on Traffic Modeling in ATM Systems. In Proc. of the 20th Annual Conference on Local Computer Networks, Minneapolis, MN (1995) 397–406

5. Lombardo A., Morabito G., Palazzo., Schembra, G.: MPEG Traffic Generation Matching Intra- and Inter-GoP Correlation. Simulation **47** (2001)

6. Jelenkovic, P.R., Lazar, A.A, Semret, N.: The Effect of Multiple Time Scales and Subexponentiality of MPEG Video Streams on Queueing Behavior. IEEE Journal on Selected Areas in Communications **43** (1995) 1566–1579

7. Frey, M., Nguyen-Quang, S.: A Gamma-Based Framework for Modeling Variable-Rate MPEG Video Sources: the GOP GBAR Model. IEEE/ACM Transactions on Networking **8** (2000) 710–719

8. Izquierdo, M.R., Reeves, D.S: A Survey of Statistical Source Models for Variable-Bit-Rate Compressed Video. Multimedia Systems **7** (1999) 199–213

9. Heyman D.P., Lakshman, T.V.: Source Models for VBR Broadcast-Video Traffic. IEEE/ACM Transactions on Networking **4** (1996) 40–48

10. Krunz, M., Tripathi, S.K.: On the Characterization of VBR MPEG Streams. In Proc. of ACM SIGMETRICS '97, Seattle, WA (1997) 192–202

11. Ansari N., Liu H., Shi Y.Q.: On Modeling MPEG Video Traffics. IEEE Transactions on Broadcasting **48** (2002) 337–347

12. Grossglauser, M, Bolot, J.-C.: On the Relevance of Long-Range Dependence in Network Traffic. In Proc. of ACM SIGCOMM '96, Stanford Univ., CA (1996) 15–24

13. Heyman, D.P., Lakshman, T.V.: What Are the Implications of Long-Range Dependence for VBR-Video Traffic Engineering?. IEEE/ACM Transactions on Networking **4** (1996) 301–317

14. Ryu B.K., Elwalid A.: The Importance of Long-Range Dependence of VBR Video Traffic in ATM Traffic Engineering: Myths and Realities. In Proc. of ACM SIGCOMM '96, Stanford University, CA (1996) 3–14

15. López-Ardao, J.C., Suárez-González A., López-García, C., Fernández-Veiga, M., Rodríguez-Rubio, R.: On the use of self-similar processes in network simulation. ACM Transactions on Modeling and Computer Simulation **10** (2000) 125–151

16. Leland, W. E., Taqqu, M. S., Willinger, W., Wilson, D. V.: On the self-similar nature of Ethernet traffic (extended version). IEEE/ACM Transactions on Networking **2** (1994) 1–15

17. Erramilli, A., Narayan, O., Willinger, W.: Experimental Queueing Analysis with Long–Range Dependent Packet Traffic. IEEE/ACM Transactions on Networking **4** (1996) 209–223

18. Hosking, J.R.M.: Fractional differencing. Biometrika **68** (1981) 165–176

19. Lombardo A., Morabito G., Palazzo., Schembra, G.: MPEG Traffic Generation Matching Intra- and Inter-GoP Correlation. Simulation **47** (2001)

20. Rose, O.: Simple and Efficient Models for Variable Bit Rate MPEG Video Traffic. Performance Evaluation **30** (1997) 69–85

21. Suárez-González A., López-Ardao, J.C., López-García, C., Fernández-Veiga, M., Sousa Vieira, E.: A Batch Means Procedure for Mean Value Estimation of Processes Exhibiting Long Range Dependence. In Proc. of WSC'02, San Diego (CA) (2002)

Dynamic Bit-Rate Allocation with Fuzzy Measures for MPEG Transcoding

TaeYong Kim[1] and Jong-Seung Park[2]

[1] Graduate School of Advanced Imaging Science,
Multimedia and Film, Chung-Ang University
HukSuk-dong 17, DongJak-gu, Seoul, 156-756, Republic of Korea
kimty@cau.ac.kr
[2] Dept. of Computer Science and Engineering, University of Incheon
177 Dohwa-dong, Nam-gu, Incheon City, 402-749, Republic of Korea
jong@incheon.ac.kr

Abstract. In this paper, we propose a dynamic bit-rate allocation algorithm for MPEG transcoding. The method consists of a bit-rate allocation with fuzzy measures and a least-distortion bit-rate reduction. Fuzzy measures are calculated by the code length, the discontinuity bluntness, and the neighborhood momentum in each DCT block. These measures are summed with weights and form a reduction fuzziness to indicate the degree of plausible reduction. Using the reduction fuzziness, each DCT block is filtered by the least-distortion reduction method to adjust the bit-rate for a target bandwidth. In the experiment, we show the results that the transcoded video quality by the method is better and the bandwidth is more regular than those of existing methods in both visually and quantitatively.

1 Introduction

With the explosive growth of the Internet and wireless access, there is an increasing demand for multimedia services over the network. Since in many video based applications original video files are encoded at high quality with high bit-rate, and heterogeneous networks such as ATM, TCP/IP, wireless and PSTN are interconnected, a network may not guarantee the required bandwidth quality for those services. Thus, it is essential to precisely control the bit-rate of incoming video streams to match the available bandwidth of outgoing network [11].

The standardized MPEG-2 scalability supports Spatial Scalability, SNR Scalability, and Temporal Scalability. The intention of scalable coding is to provide interoperability between different services and to flexibly support receivers with different display capabilities. Receivers that unable to reconstruct the full resolution video can decode subsets of the layered bit stream to display video at lower spatial or temporal resolution or with lower quality. However, to support the scalabilities, both encoder and decoder must have scalability functions, which are rare in current H/W systems. Dynamic bit-rate control can be achieved using the Scalable Coding schemes provided in current video coding standards

V. Roca and F. Rousseau (Eds.): MIPS 2004, LNCS 3311, pp. 49–60, 2004.

[4]. However, it can only provide up to three levels of discrete video quality because of the limit on the number of enhancement layers [7]. In many networked multimedia applications, a much finer scaling capability is desirable. Converting a previously compressed video stream to a lower bit-rate stream through transcoding can provide finer and more dynamic adjustment of the bit-rate.

The recent publications on video transcoding have mainly focused on low cost architecture or macro block-based bit allocation problems. In [1], various low complexity transcoding schemes were described with simple bit scaling methods for the rate allocation. In [13], the picture complexities based on the quantization scale or allocated bits by an encoder were used for the target bit-rate estimation. All these schemes have not considered the spatial information embedded in the compressed bit stream to reduce artifacts or to enhance visual perception. To achieve dynamic bit-rate adjustment without much degrading the visual quality, content-based bit-rate allocation is essential. We have proposed a content-based transcoding [8], which has a problem of estimating outgoing bit-rate.

In this paper, we propose a bit-rate allocation algorithm in the Discrete Cosine Transform (DCT) domain for MPEG transcoding. The method consists of a bit-rate allocation technique with fuzzy measures and a least-distortion bit-rate filtering method. Fuzzy measures are calculated by the code length, the discontinuity bluntness, and the neighborhood momentum in each DCT block. The fuzzy measure for discontinuity bluntness can preserve a dominant discontinuity and the measure for neighborhood momentum can decrease block artifacts. These measures are summed with weights and form a reduction fuzziness to indicate the degree of plausible reduction for each DCT block. Using the reduction fuzziness, each DCT block is filtered by the least-distortion reduction method to adjust the bit-rate for the target bandwidth. This DCT block-based technique can give more delicate and accurate control against macro block-based techniques. This scheme enhances the visual quality in the same bandwidth against constant bit allocation methods, and it is suitable for real-time applications by its fast processing time. In the experiment, we show the results that the transcoded video quality by the method is better and the bandwidth is more regular than those of existing methods.

2   MPEG Bit-Rate Control

The generic rate control problem can be separated into the following two steps: (1) Allocate target bits for each frame according to image complexities and buffer fullness for a given channel bit-rate (frame-layer rate control). (2) Derive the actual bits for each DCT block, and make the number of produced bits meet the target bits (block-layer bit allocation).

In the bit domain the run-length tuples are represented by variable length coded (VLC) codewords. The tuple decoding yields quantized coefficients in the coefficient domain. These coefficients are de-quantized to obtain the DCT coefficients. For the selective use of non-zero AC coefficients from the original into the reduced stream, three approaches are possible, such as full transcoding, coefficient selection, and codeword selection [9]. In full transcoding, from the DCT coefficients the encoding can be done as in a normal intra-frame encoder, namely, quantizing subject to a rate control, tuple encoding, and VLC encoding. In co-efficient selection, no re-quantization is needed, only tuple encoding and VLC encoding have to be carried out. Codeword selection is the lowest complexity approach in the bit domain. Selected VLC codewords from the original stream are simply copied into the reduced stream. Decoding or re-quantization is not necessary. In the method we start with the first AC coefficient and retain a certain number of consecutive non-zero AC coefficients for a DCT block.

In the block-layer bit allocation, the objective is to remove AC coefficients from the original stream such that a good quality picture results in the reduced stream. Generic bit-rate allocation problem is stated as follows:

Bit-rate allocation problem: Given a sequence of N DCT blocks in t-th frame with original R_t bits and a target bit budget B_t for the frame, determine a set of reducing bits $\Delta R_t = \{\Delta r_0, \Delta r_2, ..., \Delta r_{N-1}\}$ to each DCT block that minimizes a distortion measure of the frame and uses condition $R_t - \Delta R_t = R_t - \sum_{i=0}^{N-1} \Delta r_i \leq B_t$ for the target bits B_t.

To solve the optimal-distortion bit allocation problem, methods based on Lagrangian optimization [14], or Dynamic Programming [9] have been suggested. These methods typically perform an analysis of future video frames to measure their rate-distortion characteristics before applying a bit allocation. Though the rate-distortion techniques achieve better performance in video quality, these techniques increase the encoder complexity. If frame dependencies are taken into account, the complexity can become very high, thus some of these methods are not suitable for real time encoding[12]. In addition to the complexity, since these schemes have not considered the contents of blocks and filter blocks by uniform parameters, the visual quality of transcoded blocks is degraded without depending on their contents or importance [3].

3   Fuzzy Measures and Bit-Rate Allocation for DCT Block

The theoretical bit allocation methods based on information theory have some assumptions, such as the DCT coefficients are uncorrelated with a Gaussian distribution[5], and the coding bit-rate is approximately equal to the entropy. However, in the motion compensated P or B frames expressions of entropy and distortion for arbitrary distributions are generally unavailable, and there is a large mismatch between the theoretical entropy and the actual bit-rate [6].

In this section, in order to solve the DCT block-layer bit allocation problem, we suggest three fuzzy measures and a fuzzy bit allocation technique in the DCT domain. Since these measures are computed by using the DCT coefficients based on the features in a block, the bit allocation technique with these measures is faster and more intuitive than probability based methods.

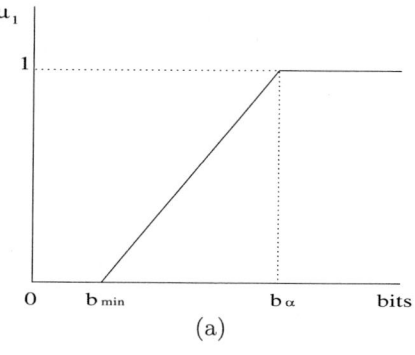 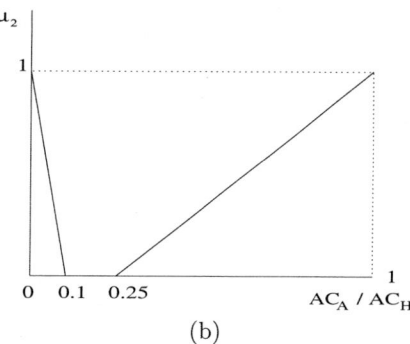

(a) (b)

Fig. 1. Fuzzy measures for a DCT block: (a) fuzzy measure for code length abundance, and (b) fuzzy measure for discontinuity feature bluntness.

3.1   Code Length Abundance

In the block-layer bit allocation, the code length of a DCT block is the critical restriction. If the allocated reducing bits (Δr_i) for the i-th DCT block is larger than the actual bit length, it is impossible to reduce the bits for the target budget. A DCT block with sufficient code length can be reduced to the amount of allocated bits to match the outgoing bandwidth. So, we first devise a fuzzy measure for code length abundance, which denotes the amount of possible reduction.

The code length abundance μ_1 is formulated as follows:

$$\mu_1 = \begin{cases} 0 & if \ b_i < b_{min} \\ 1 & if \ b_i > b_\alpha \\ \frac{b_i - b_{min}}{b_\alpha - b_{min}} & otherwise, \end{cases} \qquad (1)$$

where b_i, b_{min}, and b_α denote the number of bits for i-th DCT block, minimum bits for a DCT block of MPEG standard, and sufficient code length for the allocated target bits, respectively. b_{min} and b_α are constants for a video stream, and the variation of the measure with the number of bits is depicted in Figure 1 (a).

The measure μ_1 will be zero when the number of bits (b_i) is smaller than the minimum length (b_{min}). If b_i for a DCT block is larger than the estimated sufficient bits (b_α), then μ_1 will be one, which means that a DCT block has enough bits for reduction. b_α can be changed according to the compression standards (MPEG-1 or MPEG-2). Since the codeword length in a DCT block varies discretely and it is difficult to accurately describe the code length variation, we make μ_1 vary linearly with the code length in the range of b_{min} and b_α for the convenience of calculation.

3.2   Discontinuity Feature Bluntness

For the DCT block having a steep feature, such as an ideal step discontinuity, the high frequency coefficients in the block have to be retained to preserve the

Table 1. Comparison of $\frac{AC_H}{AC_A}$ ratio with discontinuity types and positions in a DCT block.

Discontinuity type	$P = 1, 7$	$P = 2, 6$	$P = 3, 5$	$P = 4$
Axis aligned	0.13	0.18	0.16	0.11
Diagonal	0.23	0.20	0.21	0.13

visual quality. For the block having a blunt discontinuity, coefficients can be removed without much deteriorating visual quality against the block with a steep feature. Thus, we suggest a heuristic technique that evaluates the bluntness of a discontinuity feature in a DCT block by using the ratio of summed AC coefficients. The summation of AC frequencies are formulated as follows:

$$AC_A = \sum_{u=0}^{7} \sum_{v=0}^{7} |F(u, v)| - F(0, 0),$$

$$AC_H = \sum_{v=6}^{7} \sum_{v=6}^{7} |F(u, v)|. \tag{2}$$

In the Eq.2 we decide the high frequencies as ACs of $u > 5$ or $v > 5$, which make the ratio discriminate a discontinuity from noises effectively. The DCT coefficients and position index of a discontinuity are depicted in Figure 2 (b).

If an ideal step discontinuity exists in a DCT block, the ratio between sum of all AC frequencies and that of high frequencies is calculated as shown in Table 1. From the table and the definition of DCT transformation, the ratio is independent of the discontinuity height, is slightly affected by the direction and the position, and is heavily affected by the shape of a discontinuity. By testing various shapes of features, if a blunt discontinuity exists in the center position of a block, the value is less than 0.05, and if line discontinuities or scattered noises exist, the values are larger than 0.5. Thus, we consider the shape of a discontinuity as the major factor to change the ratio and the discontinuity bluntness can be measured by the ratio. It indicates the allowable degree of reduction by preserving important features.

The fuzzy measure μ_2 for the discontinuity feature bluntness is calculated by following equation, and its variation is depicted in Figure 1 (b).

$$\mu_2 = \begin{cases} 1 - 10\frac{AC_H}{AC_A} & if \ 0.00 \leq \frac{AC_H}{AC_A} \leq 0.10 \\ 0 & if \ 0.10 < \frac{AC_H}{AC_A} \leq 0.25 \\ (\frac{AC_H}{AC_A} - 0.25)/0.75 \ if \ 0.25 < \frac{AC_H}{AC_A} \leq 1.00. \end{cases} \tag{3}$$

In the case of chrominance block, since the coefficients in the block are not related with the discontinuity feature, μ_2 for the chrominance block is estimated by averaging those of the luminance blocks within the same macro block.

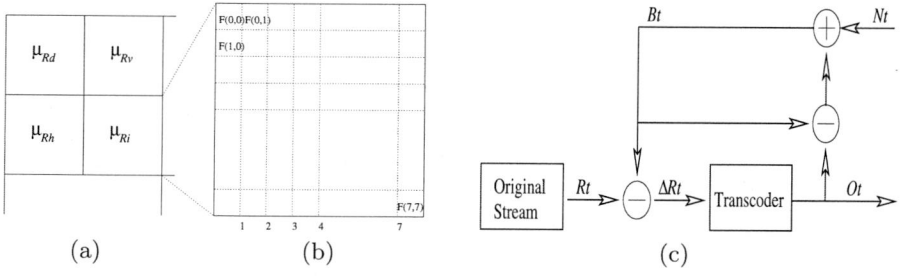

Fig. 2. Neighborhood labeling for block momentum in a frame and the bit-rate control sequence: (a) d, v, and h denote the neighborhood blocks for the i-th block as diagonal, vertical, and horizontal block, respectively, (b) DCT coefficient labeling scheme and discontinuity positions, and (c) bit-rate control sequence for the t-th frame.

3.3   Neighborhood Block Momentum

As block data is the atom of DCT-base video processing, removing or reducing frequency coefficients by a filter introduces blocky effect to video frames. To prevent undesirable block artifacts and to reduce the differences of reduction rates among neighborhood blocks, we consider a momentum measure that reflects the reduction degree of neighborhood blocks. As shown in Figure 2 (a), pre-processed neighborhood blocks for the i-th DCT block are labeled as diagonal (d), vertical (v), and horizontal (h). The fuzziness of neighborhood blocks are averaged to calculate the fuzzy neighborhood momentum as follows:

$$\mu_3 = (\mu_{R_d} + \mu_{R_v} + \mu_{R_h})/3. \tag{4}$$

Since the fuzzy momentum presents the average reduction fuzziness of the neighborhood, if a filtering method reduces bits by considering this measure, the blocky artifacts can decrease.

3.4   Fuzzy Bit-Rate Allocation

We suggest a reduction fuzziness that denotes the plausible degree of reduction. The reduction fuzziness is computed by weighted summation of fuzzy measures, which is similar to the fuzzy mean filters in [10], and it is defined as follows:

$$\mu_{R_i} = w_1\mu_1 + w_2\mu_2 + w_3\mu_3, \tag{5}$$

where $\sum_i w_i = 1$. These weights (w_i) can be adjusted for the trade-off between regulation of bit-rate and image quality. The larger value of w_1 produces the more regulated output bit-rate. On the other hand, the large value of w_2 enhances the visual quality and large w_3 reduces the block artifacts.

Using this reduction fuzziness we can allocate the bits for a DCT block. As defined in bit-rate allocation problem, given a sequence of N DCT blocks in the t-th frame with original R_t bits and a target bit budget B_t for the frame,

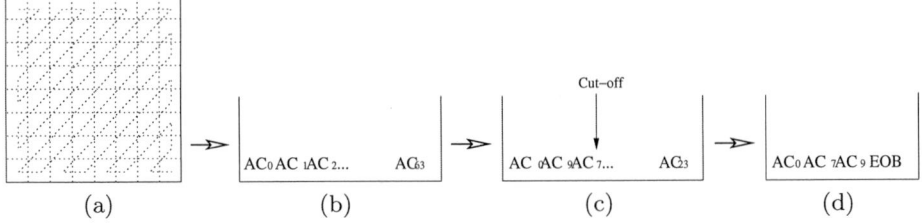

Fig. 3. Sequence of processing for the least-distortion filtering: (a) 2-dimensional DCT coefficients, (b) sequence of coefficients after zigzag scanning, (c) sequence of coefficients sorted by codeword level, and (d) coefficients filtered by LDF.

we have to determine a set of reducing bits $\Delta R_t = \{\Delta r_0, \Delta r_2, ..., \Delta r_{N-1}\}$. The amount of reducing bits for the t-th frame is calculated by $\Delta R_t = R_t - B_t = \sum_{i=0}^{N-1} \Delta r_i$, and the target bits is computed by $B_t = N_t + (B_{t-1} - O_{t-1})$, where N_t is the current bandwidth monitored by other network components, and O_t is the actual output bit-rate reduced by a transcoder. The bit-rate notation and the processing sequence are presented in Figure 2 (c).

To allocate the reducing bits (Δr_i) for the i-th DCT block, we compute the total reduction fuzziness (μ_{A_t}) in the t-th frame, and allocate Δr_i using the amount of frame reduction bits (ΔR_t) with the block fuzziness (μ_{R_i}) as follows:

$$\mu_{A_t} = \sum_{i=0}^{N-1} \mu_{R_i}, \tag{6}$$

$$\Delta r_i = \mu_{R_i} \frac{\Delta R_t}{\mu_{A_t}}. \tag{7}$$

4   DCT Block Filtering for Bit-Rate Adaptation

One of the most important functions of a transcoder is bit-rate conversion, which accepts an encoded video stream as the input and produces different bit-rate stream to meet new constraints that are not known during the encoding of original video. To reduce the bit-rate to the target bandwidth, each DCT block has to be filtered according to the allocated bits.

In this section we suggest a least-distortion filtering whose order of cut-off coefficients is arranged dynamically according to the importance of levels.

4.1   Low-Pass Filtering

In the bit domain run-length tuples are represented by variable length coded (VLC) codewords. DCT coefficients are obtained by tuple decoding and dequantization. In the low-pass filtering (LPF) high frequencies are removed for the allocated bits to match the target bits in a DCT block. Codeword selection is the similar approach to the LFP except using DCT coefficients. For a certain DCT block we start with the last codeword and remove certain number of consecutive non-zero level run-length tuples until producing a code with the target bits. If we denote the original set of L codewords as $C_i = \{w_1, ..., w_L\}$ for the i-*th* DCT block, LPF is described as follows:

$$\hat{C}_i = \{w_1, ..., w_j\}, \quad \sum_{k=j+1}^{L} Length(w_k) \geq \Delta r_i \tag{8}$$

where w_j is the j-*th* codeword and $\hat{C}_i$ is the filtered set of codewords.

We can use the LPF with the fuzzy bit allocation or without the fuzzy bit allocation technique. If we filter the high AC frequencies without using the fuzzy bit allocation technique, reducing bits for all DCT blocks are allocated equally and the distortion is independent on the local spatial contents. Otherwise, the cut-off position (j) is chosen adaptively for each DCT block according to the local contents to decrease the distortion.

4.2   Least-Distortion Filtering

Human eyes' sensitivity to spatial-temporal pattern decreases with high spatial and temporal frequency [15]. In the least-distortion filtering (LDF), we first divide each DCT coefficient level by the MPEG default quantization matrix, because this matrix weighs each coefficient according to heuristically determined perceptual or psychophysics importance. The divided coefficient levels are more related with the visual importance, which means that the relative visual influence of high frequencies decreases and that of low frequencies increases. Since the levels of low frequencies are usually larger than those of high frequencies when a dominant discontinuity exists in a DCT block and the removal effect of low frequencies is wider than that of high frequencies in the spatial domain, this division increases the quantity of a goodness metric, like PSNR.

After dividing the level of each coefficient, we sort the divided coefficients to descending order. In this sorted sequence of coefficients, the same processing with the LPF is performed, and the transcoded DCT block has little visual distortion.

If we denote the divided and sorted set of L codewords as $\tilde{C}_i = \{\tilde{w}_1, ..., \tilde{w}_L\}$ for the i-*th* DCT block, LDF is described as follows:

$$\hat{C}_i = \{w_1, ..., w_j\}, \quad \sum_{k=j+1}^{L} Length(\tilde{w}_k) \geq \Delta r_i \tag{9}$$

where $\{w_1, ..., w_j\}$ represents the original order of codewords, $\{\tilde{w}_1, ..., \tilde{w}_L\}$ denotes the sorted order of codewords, and $\hat{C}_i$ is the filtered set of codewords. The sequence of processing for LDF is depicted in Figure 3.

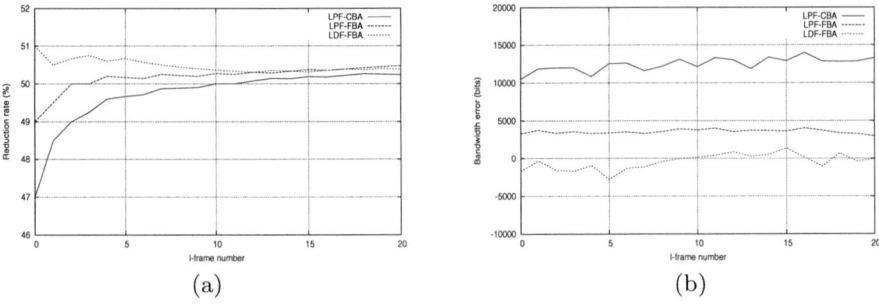

Fig. 4. Transcoded rates and bandwidth errors of Garden test video when original I-frames of average 354766 bits/frame are reduced to their 50% by three filtering methods: (a) reduction rates of each filtering method, and (b) bandwidth errors between target budgets (50% of original bit-rate) and actual outputs.

5 Experiments

In the experiments, our adaptive transcoding method applies to two test videos named "Garden" and "Tennis", which are compressed by MPEG-1 standard encoder of {I B B P B B} frame structure. Each video consists of 150 frames and 25 I-frames. To calculate the fuzzy measure of code length abundance in Eq.(1), we use $b_\alpha = 200$ and $b_{min} = 7$, which is selected empirically for MPEG-1 compression standard by experiments. To compute the reduction fuzziness in Eq.(5), we use weights as $w_1 = 80$, $w_2 = 10$, and $w_3 = 10$ for each fuzzy measure. The temporal transcoding (frame dropping for P and B frames) and the spatial transcoding (LPF or LDF) methods are combined in the experiments to achieve high data reduction and to compute the visual quality accurately.

We apply three filtering methods to test videos. LPF-CBA denotes the Low-Pass Filtering method with Constant Bit Allocation that allocates uniform bits to all DCT blocks without considering contents. LPF-FBA denotes the Low-Pass Filtering method with Fuzzy Bit Allocation that allocates bits to each DCT block dynamically with its importance. LDF-FBA means the Least-Distortion Filtering with Fuzzy Bit Allocation technique suggested in this paper. By combined with temporal dropping, each method produces the overall 80% or 78% reduction rates in "Garden" or "Tennis" video when I-frames are reduced to their 50%.

Transcoded rates by the methods are shown in Figure 4 (a). Since the LPF-CBA uses the constant bit allocation technique, the convergence to the target rate is slower than the other methods that use the fuzzy bit allocation technique. So it is desirable to use the fuzzy bit allocation technique for a network that the available bandwidth varies frequently. Without considering the amount of bandwidth errors, which can be changed according to feedback schemes, the bit-rate variation of LPF-FBA is the most stable one among three methods as shown in Figure 4 (b). Thus, by using the fuzzy bit allocation technique the output bit-rate can be stabilized and converged fast to the target bandwidth.

Table 2. SDVQ, Average PSNRs with variances and average bandwidths with variances for "Garden" (upper part of the table) and "Tennis" (lower part of the table) video samples of 50% reduction.

Filtering	Average $PSNR$	$PSNR$ σ^2	Average Bandwidth	Bandwidth σ^2	SDVQ
LPF-CBA	67.391	0.213	50.238	0.753	2.269
LPF-FBA	68.166	0.188	50.476	0.345	2.137
LDF-FBA	69.373	0.185	50.381	0.236	2.038
LPF-CBA	72.984	1.486	50.214	3.311	2.080
LPF-FBA	73.395	1.162	50.214	0.311	2.150
LDF-FBA	74.797	1.215	50.786	1.168	1.805

(a)   (b)   (c)

Fig. 5. Transcoded images of "Garden" test video: (a) original "Garden" test image, (b) transcoded image by LPF-CBA, and (c) transcoded image by LDF-FBA.

Watson has developed a digital video quality (DVQ) metric for evaluating the quality of compressed video, which incorporates models of human visual sensitivity to spatial and chromatic visual signals [15]. Since we use the frame-dropping scheme and we want to easily adapt the metric to the existing system, we simplified the original DVQ by skipping the temporal filtering techniques and the color space conversion. We use simplified digital video quality (SDVQ) metric with PSNR to evaluate the visual quality with $\beta = 4$ in the Minkowski distance for pooling the SDVQ. Reduced images with higher metrics are judged better in PSNR and lower values are judged better in SVDQ. Table 2 shows SVDQ, average PSNR, PSNR variance, average bandwidth, and bandwidth variance with different filtering methods for two test videos. After transcoding each sample to 50% of its original bit-rate, the bandwidth variance of LPF-FBA is the lowest as depicted in Figure 4 (b). SDVQ and the average PSNR of LDF-FBA is the best among three methods. Since the LDF-FBA selects the smallest level of codeword first, PSNR of this method is superior to the others about $1 \sim 2$ dB, but the variance of bandwidth increases for the sake of visual quality. As explained in Section 3.4 the weights in Eq.(5) can be adjusted to reduce the variance of bandwidth or to enhance the visual quality.

Figure 5 shows sample "Garden" images after transcoding by different methods. Since the LDF-FBA method preserves discontinuity features and neighborhood relations by fuzzy measures μ_2 and μ_3 in Eqs. (3) and (4), the image transcoded by LDF-FBA in Figure 5 (c) is finer and less blocky than the image transcoded by LPF-CBA in (b), particularly in the roof parts.

We check the processing time for each method by using a broadcasted video of 352×240 size in a LINUX system. Though the average processing time (50 millisecond) of LDF-FBA is a little expensive than that of LPF-CBA or requantization (37 or 38 millisecond), the processing time for the filtering is much cheaper than that of DCT and IDCT conversions, and since there are usually two I-frames per second in common video sequences, the processing time is short enough for real-time applications.

6   Conclusion

Transcoding methods in the DCT domain have a tradeoff between in speed and in quality. To enhance the visual quality with preserving the advantage of processing speed and bit-rate regulation, we suggest a fuzzy bit allocation technique and a least-distortion filtering method. Using code length, discontinuity bluntness, and neighborhood momentum in each DCT Block, three fuzzy measures are derived and form a reduction fuzziness for the DCT block bit allocation. Each DCT block is filtered by using the least-distortion filtering method to adjust the bit-rate for the target bandwidth, and it is suitable for real-time applications by its fast processing time. In the experiment, we show the result that the transcoded video quality by the method is better and the bandwidth is more regular than those of existing methods in both visually and quantitatively.

Acknowledgement. This work was supported by the IT Research Center (ITRC), Ministry of Information and Communication, Korea.

References

1. P. A. A. Assuncao and M. Ghanbari, "A frequency-domain video transcoder for dynamic bit-rate reduction of MPEG-2 bit stream," *IEEE Transactions on Circuits and Systems for Video Technology,* Vol. 8, No. 8, pp. 953-967, 1998.
2. S. F. Chang and D. G. Messerschmitt, "A new approach to decoding and composting motion compensated DCT-based images", *in Proc. IEEE Internat. Conf. on Acoustics, Speech, and Signal Processing, ICASSP93,* Vol. 5, pp. 421-424, 1993.
3. Francisco Garcia, David Hutchison, Andreas Mauthe, and Nicholas Yeadon, "QoS Support for Distributed Multimedia Communications," *Proceedings of IFIP/ IEEE International Conference on Distributed Platforms,* 1996.
4. M. Ghanbari, "Two-layer coding of video signals for VBR networks," *IEEE J. Select. Areas Commun.,* Vol. 7, pp. 771-781, 1989.
5. H. M. Hang and J. J. Chen, "Source model for transform video coder and its application. Part I: fundamental theory," *IEEE Transactions on Circuits and System for Video technology,* Vol. 2, No. 7, pp. 287-298, 1997.

6. Z. He, Y. K. Kim, and S. K. Mitra, "Low-delay rate control for DCT video coding via rho-domain source modeling," *IEEE Transaction on Circuits and Systems for Video Technology*, Vol. 8, No. 11, 2001.
7. ISO/IEEE 13818-2, "Information technology-generic coding of moving pictures and associated audio information-Part 2: Video," 1995.
8. TaeYong Kim and Jong Soo Choi "Content-based Video Transcoding in Compressed Domain," *Journal of Signal Processing: Image Communication,* Vol. 17, No. 6, pp. 497-507, 2002.
9. R. L. Lagendijk, E. D. Frimout, and J. Biemond, "Low-complexity rate-distortion optimal transcoding of MPEG I-frames," *Signal Processing: Image Communication*, No. 15, pp. 531-544, 2000.
10. Chang-Shing Lee, Yau-Hwang Kuo, and Pao-Ta Yu, "Weighted fuzzy mean filter for image processing," *Fuzzy Sets and systems*, Vol. 89, pp. 157-180, 1997.
11. Zhijun Lei and Nicolas D. Georganas, "An accurate bit-rate control algorithm for video transcoding," *Journal of Visual Communication & Image Representation*, No.14, pp. 321-339, 2003.
12. L. J. Lin and A. Ortega, "Bit-rate control using piecewise approximated rate-distortion characteristics," *IEEE Transactions on Circuit and Systems for Video Technology*, Vol. 4, No. 8, pp. 446-459, 1998.
13. Ligang Lu, Shu Xiao, Jack L. Kouloheris, and Cesar A. Gonzales, "Efficient and Low Cost Video Transcoding," *Visual Communications and Image Processing*, pp. 154-163, 2002.
14. I. M. Pao and M. T. Sun, "Encoding DCT coefficients based on rate-distortion measurement," *Journal of Visual Communication and Image Representation*, No. 12, pp. 29-43, 2001.
15. Andrew B. Watson, "Toward a perceptual video quality metric," *IST/SPIE conference on Human Vision and Electronic Imaging*, Vol. 3299, pp. 139-147, 1998.

A Modified Gaussian Model-Based Low Complexity Pre-processing Algorithm for H.264 Video Coding Standard

Won-Seon Song, Do-Reyng Kim, Seongsoo Lee, and Min-Cheol Hong

School of Electronic Engineering, Soongsil University, Korea
{won,krdoreyng}@vipl.ssu.ac.kr, {sslee,mhong}@e.ssu.ac.kr

Abstract. In this paper, we present a low complexity modified Gaussian model-based pre-processing filter to improve the performance of H.264 compressed video. Noisy video sequences captured by imaging system result in decline of coding efficiency and unpleasant coding artifacts due to higher frequency components. By incorporating local statistics and quantization parameter into filtering process, the spurious noise is significantly attenuated and coding efficiency is improved, leading to improvement of visual quality and to bit-rate saving for given quantization step size. In addition, in order to reduce the complexity of the pre-processing filter, the simplified local statistics and quantization parameter induced by analyzing H.264 transformation and quantization processes are introduced. The simulation results show the capability of the proposed algorithm.

1 Introduction

JVT (Joint Video Team) established by ITU-T and ISO/IEC has jointly developed H.264/AVC video coding standard to obtain higher compression gain than existing video coding standards. H.264 can be characterized by UVLC (Universal Variable Length Coding), 4×4 block-based integer transform, and variable block size motion estimation/compensation [1-3]. Due to the different coding strategies, the local statistics of coded information is different to previous standards. Therefore, any algorithms to obtain better coding efficiency or to improve visual quality should be different to other standards.

Many approaches have been explored to improve the quality of decoded images in the literature [4-7]. However, pre-processing algorithms which remove the additive noise for given bit rate or quantization step size are rarely investigated within video compression areas. When quantization step size is provided by a rate control algorithm, it is promising that the pre-processing is to modify the original video sequence so that image quality is maximized. As a general approach, a typical noise filtering approach has been employed to improve coding efficiency [8]. In Ref. [9], more sophisticated technique is introduced as a pre-processing algorithm which is operated with rate-distortion problem. Also, similar technique to incorporate knowledge of the variable length codes into the filtering decisions has been reported in Ref. [10]. The

V. Roca and F. Rousseau (Eds.): MIPS 2004, LNCS 3311, pp. 61–71, 2004.

filter attempts to maximize the resulting image quality by controlling the error residual between an original image and the motion compensated frame, given a predefined bit-rate. The above approaches only take into account of removal of blocking artifact which represents the information loss coming from quantization process, but not reducing the additive noise.

In general, video sequence captured by imaging acquisition system represents the degraded version of an original video sequence by additive noise coming from image formation system. In such case, the reconstructed video sequence using video coding standards including H.264 usually results in the loss of coding efficiency and unpleasant coding artifacts. Therefore, a filtering process is required to remove the spurious noise with preserving significant features such as edges or objects. In addition, it is desirable that the statistics of the filtering results is similar to the structure of variable length code so that the impact leads to saving of the bits required for the representation, given quantization step size determined by rate-distortion algorithm.

The statistics of variable length codes of almost video coding standards including H.264 coder has Gaussian distribution. Therefore, it is expected that the quality can be maximized when the filtering results have the similar probability to the variable length codes. In this paper, we propose a Gaussian model based pre-processing filter to maximize the quality by effectively removing the noise for given quantization parameter. Local statistics of the degraded image and quantization parameter are used to design the filter. Also, a simplified 3-tab filter without floating-point operations is addressed in order to reduce the complexity.

This paper is organized as follows. Section 2 describes the background of modified Gaussian filter. In Section 3, we present 1-D Gaussian model-based filter. Quantization noise as a parameter of the proposed filter is induced by rigorous analysis of H.264 transformation and quantization. Also, simplified pre-processing and efficient definition representing local statistics are presented. Finally, the experimental results and conclusions are described in Sections 4 and 5.

2 Background

As pointed out, typical image captured by image formation system represents the degraded version of an original image. In such case, noise filtering process should be taken place to obtain the satisfactory results and to reduce the bit-rate for given quantization step size in video compression. The final decision of the image quality is made by human viewer, and therefore HVS (Human Visual System) should be incorporated into the filtering process. Ref. [11] shows that HVS can be approximated to Gaussian model. According to it, the filtering process can be written as

$$y = h ** x, \tag{1}$$

where x and y represent the captured degraded image and the filtered image, respectively. In Eq. (1), h and ** denote two dimensional modified Gaussian impulse response and two dimensional convolution. Since Gaussian impulse response has separable property, Eq. (1) can be rewritten as two 1-D filtering form. It is

$$y = h_v * h_h * x, \tag{2}$$

where h_v and h_h are one dimensional Gaussian impulse response of vertical and horizontal directions in two dimensional imaging coordinate, and * represents one dimensional convolution. In the rest of paper, we use H instead of h_v and h_h.

When 1-D Gaussian impulse response is used to improve the performance in video coding, it is promising to take into account of local statistics and quantization noise as the parameters of Gaussian impulse response, since the statistics of the degraded image is locally different. In this paper, the following Gaussian impulse response is defined.

$$H_i = \frac{1}{Z} \exp\left(\frac{-(i^2)}{\frac{\sigma_N^2}{\sigma_B^2} k^2} \right), \tag{3}$$

where Z is the normalizing constant, and σ_B^2 and σ_N^2 represent the local variance and noise variance by quantization process, respectively. In addition, k denotes the parameter to reflect human the visible property [12].

3   1-D Gaussian Model-Based Pre-processing

In this Section, we describe various parameters of Eq. (3) in more detail. First, σ_N^2, noise variance in Eq. (3) can be induced from quantization and transformation process. The previous video coding standards such as MPEG2, MPEG4, and H.263 use 8×8 floating-point operational DCT (Discrete Cosine Transform), which leads to IDCT (Inverse Discrete Cosine Transform) mismatch problem. In order to resolve the problem, H.264 video coding standard uses the modified DCT and quantization mechanism. Typically, 4×4 block DCT transformation is defined as

$$Y = AXA^T = \begin{bmatrix} a & a & a & a \\ b & c & -c & -b \\ a & -a & -a & a \\ c & -b & b & -c \end{bmatrix} X \begin{bmatrix} a & b & a & c \\ a & c & -a & -b \\ a & -c & -a & b \\ a & -b & a & -c \end{bmatrix}, \tag{4}$$

where X is 4×4 dimensional block of the input image, and $a = 1/2$, $b = 1/\sqrt{2} \cos(\pi/8)$, and $c = 1/\sqrt{2} \cos(3\pi/8)$. Eq. (4) requires the floating-point operation, resulting in IDCT (Inverse Discrete Cosine Transform) mismatch. In order to avoid the problem, Eq. (4) can be modified as integer transform form. It is [13]

$$Y = \begin{bmatrix} 1 & 1 & 1 & 1 \\ 2 & 1 & -1 & 2 \\ 1 & -1 & -1 & 1 \\ 1 & -2 & 2 & -1 \end{bmatrix} X \begin{bmatrix} 1 & 2 & 1 & 1 \\ 1 & 1 & -1 & -2 \\ 1 & -1 & -1 & 2 \\ 1 & -2 & 1 & -1 \end{bmatrix} \otimes \begin{bmatrix} a^2 & ab/2 & a^2 & ab/2 \\ ab/2 & b^2/4 & ab/2 & b^2/4 \\ a^2 & ab/2 & a^2 & ab/2 \\ ab/2 & b^2/4 & ab/2 & b^2/4 \end{bmatrix} \quad (5)$$

$$= (CXC^T) \otimes S = W \otimes S,$$

where $\otimes$ and A^T represent the element multiplier and the transpose of matrix A, and S is utilized as a weighting matrix of quantization process, which is written as

$$Z = [Y \otimes E]/(2^{15+QP/6}) = [(W \otimes S) \otimes E]/(2^{15+QP/6}), \quad (6)$$

where E and QP denote the quantization table and the quantization index which are defined in H.264 standard [13]. Therefore, the quantization noise be determined as

$$\text{Quantization Error} = (2^{15+QP/6})/E . \quad (7)$$

In this paper, Eq. (7) is used as σ_N^2 to incorporate the quantization noise into filtering process for given QP.

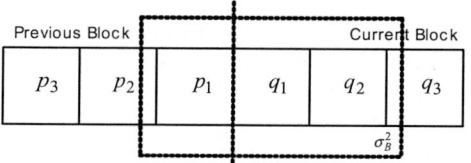

Fig. 1. Proposed 1-D filter

In our algorithm, 3-tab filter is used for low the complexity. For example, as shown in Figure 1, two neighboring pixels (p_1 and q_2) are used to represent local statistics of q_1. In Eq. (3), σ_B^2 and k are defined to describe local properties. σ_B^2 representing the local activity should take higher value in significant features including edges and objects, so that they can be preserved without over-smoothness. On the other hands, σ_B^2 should take lower value to remove the noise on flat areas. However, calculation of variance requires compute-intensive operations. For the reduction of the complexity, the following local properties are defined. They are

$$\mu_{q_1} = \frac{p_1 + 2 \times q_1 + q_2}{4}, \quad (8)$$

and

$$\sigma_B \approx \sigma_B' = \frac{\left| p_1 - \mu_{p_1} \right| + 2 \times \left| q_1 - \mu_{p_1} \right| + \left| q_2 - \mu_{p_1} \right|}{2}. \quad (9)$$

The local mean in Eq. (8) has its advantage on that it can be obtained without divider. Also, local variance can be obtained without multiplier and divider using Eq. (9). In fact, the above is one of the most important issues in practical implementation of digital filter.

The visible degree of additive noise to human viewer depends on the background as well as the local variance. For example, the noise in the bright region is more visible than that of the dark region. In order to control the visibility, the parameter, k, is defined as

$$k = \sqrt{\mu_{p_1} + 1}. \tag{10}$$

Using Eqs. (7)-(10), Q_1, the filtering result of q_1 can be written as

$$Q_1 = H_0 q_1 + H_1(p_1 + q_2). \tag{11}$$

Even though it is not expressed in detail here, the complexity of the proposed algorithm can be further reduced by making the filter coefficients integer and by pre-storing the integer coefficients into Loop-Up table.

4   Experimental Results

A number of experiments have been contacted with various sequences, resolutions, and quantization index. Among of them, QCIF "Foreman", "Container", "Hall monitor", and "Paris" sequences were used. The original sequences are degraded by Gaussian Noise, and also "Test" sequence captured by a USB camera was used. The proposed algorithm was tested with JM6 (Joint Model 6) reference code of H.264 video coding standard. For evaluating the performance of the algorithm, PSNR (Peak Signal to Noise Ratio) was utilized. For $M \times N$ dimensional 8 bits image, it is defined as

$$PSNR = 10 \log \frac{MN \times 255^2}{\left\| f - \hat{f} \right\|^2} \tag{12}$$

where $\| \cdot \|$ is the Euclidean norm, and f and $\hat{f}$ represent the original image and the reconstructed image, respectively.

In Figure 2, the noisy frames with 25 dB Gaussian noise and test frame are shown. Figures 3-6 show the corresponding reconstructed frames of Figure 2 (a)-(d) without filter and with the proposed filter. There still exists the additive noise in the reconstructed frames only by H.264 video coder. The existing noise is more visible in video sequence, since the noise is randomly scattered. On the other hands, with the proposed algorithm the noise is effectively removed. However, the edge information is a little blurred, since the Gaussian impulse response represents a kind of low-pass filter. Also, the reconstructed frames of the Test sequence are shown in Figure 7, which is more realistic case. The result shows that the proposed algorithm has the capability to remove the background noise without blurring.

PSNR and bit-rate comparisons as a function of quantization index are shown in Tables 1-4. From the tables, we observe that the proposed algorithm consistently results in PSNR gain and bit-rate saving against without filter. As the degradation is more serious, the higher PSNR gain is. Also, as quantization index is lower (lower quantization step size), the PSNR gain is higher. Therefore, the proposed algorithm can be used to obtain higher quality image at high or medium bit rate. Also, bit-rate comparison as a function of quantization index of "Test" sequence is shown in

Table 5. As expected, the proposed algorithm leads to the bit-rate saving up to 50 (%), and that the image quality is better than without filter for given quantization index.

The novelty of the proposed algorithm is that no prior knowledge about the noise and image is required to remove the additive noise, and that it has the capability to improve the coding performance.

(a)

(b)

(c)

(d)

(e)

Fig. 2. Degraded frames by adding Gaussian noises: (a) 51st frame of QCIF Foreman sequence (25 dB Gaussian noise), (b) 150th frame of QCIF Container sequence (25 dB Gaussian noise), (c) 109th frame of QCIF Hall monitor sequence (25 dB Gaussian noise), (d) 223rd frame of QCIF Paris sequence (25 dB Gaussian noise), (e) 76th frame of test sequence captured by USB camera

(a)   (b)

Fig. 3. Reconstructed frames of Figure 2-(a) (QP=24) : (a) without filter, (b) with proposed filter

(a)   (b)

Fig. 4. Reconstructed frames of Figure 2-(b) (QP=24) :(a) without filter, (b) with proposed filter

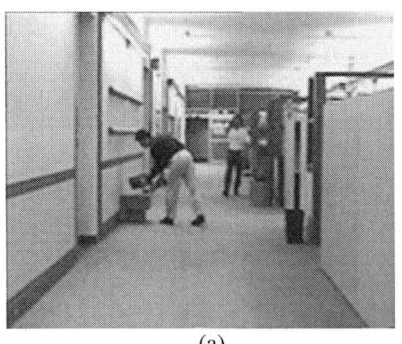 

(a)   (b)

Fig. 5. Reconstructed frames of Figure 2-(c) (QP=24) :(a) without filter, (b) with proposed filter

(a)   (b)

Fig. 6. Reconstructed frames of Figure 2-(d) (QP=24) : (a) without filter, (b) with proposed filter

(a)   (b)

Fig. 7. Reconstructed frames of Figure 2-(e) (QP=24) : without filter, (b) with proposed filter

Table 1. PSNR and bit-rate comparisons as a function of quantization index (QCIF Foreman sequence, 10 frames/sec)

	QP	without filter				with proposed filter			
		SNRY (dB)	SNRU (dB)	SNRV (dB)	Bit-rate (Kbps)	SNRY	SNRU	SNRV	Bit-rate (Kbps)
SNR 25dB	16	39.44	41.18	41.22	830.28	39.58	43.34	43.72	456.59
	20	39.08	40.87	40.80	465.96	38.85	42.50	43.11	258.40
	24	38.17	40.36	40.29	176.90	37.39	41.34	41.95	138.11
	28	35.89	40.19	39.89	83.44	35.24	39.88	40.26	75.16
	32	33.10	39.26	39.60	46.41	32.73	38.91	39.17	44.04
SNR 30dB	16	42.83	44.37	44.60	532.17	40.23	43.74	44.50	348.84
	20	41.19	43.82	44.66	251.21	39.08	42.32	43.12	205.99
	24	38.58	42.01	42.94	133.38	37.28	41.07	41.74	119.69
	28	35.86	40.26	40.85	76.33	35.10	39.68	40.08	70.79
	32	33.13	39.14	39.41	45.30	32.64	38.75	38.89	42.82

Table 2. PSNR and bit-rate comparisons as a function of quantization index (QCIF Container sequence, 10 frames/sec)

	QP	without filter				with proposed filter			
		SNRY (dB)	SNRU (dB)	SNRV (dB)	Bit-rate (Kbps)	SNRY (dB)	SNRU (dB)	SNRV (dB)	Bit-rate (Kbps)
SNR 25dB	16	40.53	42.16	42.36	696.76	40.29	44.15	44.34	303.78
	20	40.04	41.90	42.00	338.77	39.47	43.46	43.61	142.12
	24	38.72	41.16	41.36	88.93	37.98	42.28	42.13	59.32
	28	36.25	40.28	40.06	31.99	35.88	40.43	40.06	27.88
	32	33.38	39.68	39.38	15.32	33.22	38.74	38.72	14.66
SNR 30dB	16	43.66	45.39	45.46	389.47	40.68	44.56	44.88	190.22
	20	41.54	45.09	45.26	143.52	39.49	43.19	43.32	91.46
	24	38.80	42.08	43.17	61.59	37.81	41.69	41.72	48.00
	28	36.13	40.99	40.71	27.99	35.70	39.96	40.00	25.61
	32	33.28	39.43	38.84	14.63	33.15	38.64	38.22	14.17

Table 3. PSNR and bit-rate comparisons as a function of quantization index (QCIF Hall monitor sequence, 10 frames/sec)

	QP	without filter				with proposed filter			
		SNRY (dB)	SNRU (dB)	SNRV (dB)	Bit-rate (Kbps)	SNRY (dB)	SNRU (dB)	SNRV (dB)	Bit-rate (Kbps)
SNR 25dB	16	40.20	40.32	43.52	751.35	41.28	42.14	43.46	330.48
	20	39.86	40.06	42.46	393.46	40.48	41.10	42.65	156.69
	24	39.24	39.90	41.68	94.89	38.94	40.22	42.00	64.02
	28	37.10	39.35	41.25	35.02	36.74	38.88	40.96	32.14
	32	34.10	38.12	40.27	20.23	33.86	37.80	40.12	19.35
SNR 30dB	16	43.37	43.25	43.74	477.66	41.95	42.25	43.83	226.38
	20	41.77	41.81	42.84	170.43	40.69	41.10	42.79	107.22
	24	39.52	40.64	42.42	59.67	38.84	40.14	41.80	51.96
	28	36.98	39.28	41.28	31.88	36.55	38.78	40.86	30.13
	32	34.01	38.08	40.24	19.28	33.86	37.66	40.06	18.73

Table 4. PSNR and bit-rate comparisons as a function of quantization index (QCIF sequence, 10 frames/sec)

	QP	without filter				with proposed filter			
		SNRY (dB)	SNRU (dB)	SNRV (dB)	Bit-rate (Kbps)	SNRY (dB)	SNRU (dB)	SNRV (dB)	Bit-rate (Kbps)
SNR 25dB	16	39.19	39.58	39.61	902.47	39.40	40.78	40.68	602.75
	20	38.71	39.35	39.41	548.57	38.69	40.00	39.92	345.55
	24	37.41	38.99	39.15	213.74	37.01	38.40	38.39	170.57
	28	34.77	36.59	38.99	104.66	34.40	35.94	36.10	98.52
	32	31.30	34.59	34.88	59.36	31.14	34.08	34.40	57.54
SNR 30dB	16	42.67	43.18	43.25	597.49	40.59	40.42	41.19	403.44
	20	40.80	41.79	42.00	274.33	39.21	38.92	39.83	227.93
	24	37.84	39.30	39.46	153.50	36.89	36.76	37.85	143.99
	28	34.64	36.37	36.62	96.37	34.20	34.31	35.74	92.66
	32	31.26	34.38	34.60	57.18	31.10	32.44	34.11	55.75

Table 5. Bit-rate comparison as a function of quantization index (10 frames/sec, unit: Kbps)

	QP	without filter	without proposed filter
Test Sequence	16	780.26	570.04
	20	473.28	379.22
	24	278.87	250.36
	28	172.78	164.91
	32	104.62	102.19

5   Conclusion

In this paper, we propose the low complexity pre-processing algorithm to improve the performance of H.264 video coding standard. The modified Gaussian impulse response is introduced, and the local activity, quantization information, and simple visibility function are incorporated into the filtering process. The parameters are defined to control the shape of the Gaussian impulse response, so that the degree of local smoothness is adjusted by local statistics. From the experimental results, it is observed that dramatic PSNR gain and bit-rate saving are obtained with the proposed algorithm when the noise signals are added to the original video sequence. Also, it is verified that the proposed algorithm effectively removes the noise, leading to satisfactory results.

The incorporation of a robust visibility function into filtering process is under investigation. With the function, it is expected that more sophisticated formulation can be derived and better results can be obtained.

Acknowledgments. This work was supported by grant No. KRF-2003-003-D00320 from Korea Research Foundation.

References

1. ITU-T SG16/Q6, JVT-G050r1 Draft ITU-T Recommendation on Final Draft International Standard of Joint Video Specification, May 2003.
2. Iain E.G. Richardson, *H.264 and MPEG-4 Video Compression*, Wiley, 2003.
3. T. Wiegand, G. Sullivan, G. Njontegaard and A. Lutjra, "Overview of the H.264/AVC Video Coding Standard," *IEEE Trans. on Circuit and Systems for Video Technology*, vol.13, no. 9, pp. 560-576, July 2003.
4. B. Ramamurthi and A. Gersho, "Nonlinear Space-invariant Post Processing of Block Coded Images," *IEEE Trans. on ASSP*, vol. ASSP-34, pp. 1258-1268, Oct. 1986.
5. R. Rosenholtz and A. Zakhor, "Iterative Procedures for Reduction of Blocking Effects in Transform Image Coding," *IEEE Trans. on Circuits and Systems for Video Technology*, vol. 2, no.1, pp. 91-94, March 1992.
6. Y. Yang, N. P. Galatsanos, and A. K. Katsaggelos, "Regularized Reconstruction to Reduce Blocking Artifacts of Block Discrete Cosine Transform Compressed Images," *IEEE Trans. on Circuits and Systems for Video Technology*, vol. 3, no. 6, pp. 421-432, Dec. 1993.
7. A. K. Katsaggelos and N. P. Galatsanos, editors, *Signal Recovery Techniques for Image and Video Compression and Transmission*, Kluwer Academic Publishers, 1998.
8. J. C. Brailean, R. P. Kleihorst, S. N. Efstratiadis, A. K. Katsaggelos, R. L. Lagemdijk, "Noise Reduction Filters for Dynamic Image Sequences : A Review," *IEEE Proceedings*, vol.83, no.9, pp.1272-1292, Sept. 1995.
9. L. –J. Lin and A. Ortega : "Perceptually Based Video Rate Control Using Pre-filtering and Predicted Rate-Distortion Characteristics, " *Proceedings of the IEEE International Conference on Image Processing*, pp.57-60, Oct. 1997
10. C. Andrew Segall, Passant Karunaratne and Aggelos, K. Katsaggelos, "Preprocessing of Compressed Digital Video," *Proceedings of the SPIE Conference on Visual Communication and Image Processing*, vol. 4310, pp. 163-174, Jan. 2001.
11. A. B. Watson, "DCT Quantization Matrices Visually Optimized for Individual Images," Proceedings *of the SPIE Human Vision, Visual Processing and Display IV*, vol. 1913, pp. 202-216, Jan. 1993.
12. A. K. Katsaggelos, "Iterative Image Restoration Algorithms," *Optical Engineering*, vol. 28, pp. 735-748, July 1989.
13. H. S. Malvar, A. Hallapuro, M. Karczewicz, and L. Kerofsky, "Low Complexity Transform and Quantization in H.264/AVC," *IEEE Trans. on Circuit and Systems for Video Technology*, Vol. 13, no. 7, pp. 598-603, July, 2003.

Scalable Stereo Video Coding for Heterogeneous Environments[†]

Sehchan Oh, Youngho Lee, and Woontack Woo

GIST U-VR Lab.
Gwangju 500-712, South Korea
{soh, ylee, wwoo}@gist.ac.kr

Abstract. In this paper, we propose a new stereo video coding scheme for heterogeneous consumer devices by exploiting the concept of spatio-temporal scalability. We use MPEG standard for coding the main sequence and interpolative prediction scheme for predicting the P- and B-type pictures of the auxiliary sequence. The interpolative scheme predicts matching blocks by interpolating both motion predicted macro-block and disparity predicted macro-block and employs weighting factors to minimize the residual errors. To provide flexible stereo video service, we define both a temporally scalable layer and a spatially scalable layer for each eye's view. The experimental results show the efficiency of proposed scheme by comparison with already known methods and advantages of disparity estimation in the view of scalability overhead. According to the experimental results, we expect the proposed functionalities will play a key role in establishing highly flexible stereo video service for ubiquitous display environment where device and network connections are heterogeneous.

1 Introduction

Recent advancements in Internet and multimedia services have enabled immersive display, such as stereo and panoramic video display. Stereo video enables user feel more natural and immersed, but bandwidth requirement for stereoscopic transmission is twice that for conventional monocular transmission. The objective on a bandwidth-limited transmission system is to develop an efficient coding scheme that exploits the redundancies of the stereo image sequences.

A typical compression scenario of stereo video is exploiting both the effective motion compensation of individual image sequences and the reduction of disparity between the reference and target frames [1][2]. The reference (or left-view) frame is encoded by using conventional compression standard. However, the target (or right-view) frame is encoded by using disparity vector (DV) estimated from the reference frame and displacement compensated difference (DCD) instead of encoding the target frame itself. Therefore, estimating and representing DV and motion vector (MV) accurately are main objectives in a compression technique for stereo video.

[†] This work was supported by Korea Research Foundation Grant (KRF-2002-003-D00221).

V. Roca and F. Rousseau (Eds.): MIPS 2004, LNCS 3311, pp. 72–83, 2004.

Moreover, the availability of multimedia service, such as stereo video service, strongly depends on the network infrastructure and display environments of clients. For example, high quality video services require high-resolution display and broadband network. As shown in Fig. 1, there exist various types of consumer devices such as TV, LCD/CRT monitor, PDA, HMD, etc. Therefore, a new coding algorithm is needed to represent and deliver stereo video according to available network and display devices.

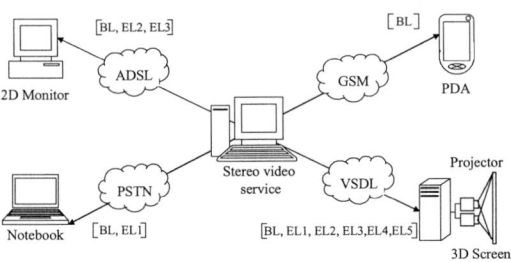

Fig. 1. Heterogeneous network and display systems with stereo video service

There are many research activities about stereo video compression using block matching algorithm (BMA) or its alternative implementations. The most general implementation method for motion/disparity estimation is BMA which is used in motion estimation of MPEG standard [3-5]. Recently, several stereoscopic video compression schemes have been developed by using multi-resolution based BMA in order to reduce a searching complexity of vectors and hierarchical disparity estimation which uses variable block size instead of using fixed block size, to raise accuracy of disparity estimation [6][7]. Most of the compression techniques of stereo video mainly focused on the developing the modified coding algorithm using the MPEG-2 multi-view profile [4][5]. These schemes modified various types of scalability to be suitable for stereo video and concentrated on improving a compression ratio using the similarity between the two views. Few research activities on coding scheme for providing flexible stereo video service have been reported [8]. However, they have a limitation of flexibility because of combining stereo coding scheme with individual scalability technique.

In this paper, we propose a highly scalable yet efficient stereo video coding method for various heterogeneous devices by exploiting the concept of spatio-temporal scalability. The proposed scheme uses MPEG-2 standard for encoding the left image and an interpolation based motion-disparity prediction scheme for predicting macro block (MB) of P- and B-type pictures of the right image sequence. Each MB of every target B-picture in auxiliary sequence is predicted by interpolating both bi-directional (forward and backward) motion predicted MB from I- and P-type pictures and disparity predicted MB from corresponding reference picture. Similarly, each MB of every target P-picture in auxiliary sequence is predicted by forward motion predicted MB and disparity predicted MB. In this scheme, we apply same weighting factor to each motion predicted MB and disparity predicted MB for estimating the best matching blocks.

To provide efficient stereo video service among heterogeneous clients, the proposed scheme uses the functionalities of spatio-temporal scalability [3][9][10]. The encoder produces one base layer (BL) bit-stream and several enhancement layer (EL) bit-streams. The BL bit-stream represents lower resolution of main sequences. The EL bit-streams provide frames of auxiliary-view as well as additional information for reproduction of the lower resolution frame with high spatial and temporal resolution. In general, spatial scalability offers flexibility of spatial resolution, but on the other hand, there is bit-rate overhead due to scalability. In stereo video coding, the spatial scalability overhead can be decreased by reducing redundancies between stereo pair.

The rest of this paper is organized as follows. In Chapter 2, we present system configuration and the proposed stereo video coder with spatio-temporal scalability. We evaluate the proposed coding scheme and analyze the experimental results in Chapter 3. Finally, some concluding remarks and possible extension of the proposed scheme are mentioned in Chapter 4.

2   Stereo Video Coding with Spatio-temporal Scalability

To deliver stereo video efficiently among heterogeneous display systems, we define one BL and several ELs as shown in Fig. 2. The BL and EL3 represent the lower resolution of left and right sequences respectively. The BL encoder performs motion compensation prediction (MCP) based encoding to remove unnecessary data, which is temporal redundancy, between current frame and previous decoded frame. The EL3 encoder performs disparity compensation prediction (DCP) based encoding as well as MCP based encoding. More specifically, it employs two types of prediction, one referencing a decoded left-view frame and the other referencing decoded right-view frames. The EL1 and EL4 generate additional data for providing full temporal resolution. The EL2 and EL5 encoder perform Intra coding to encode additional data, needed for providing full spatial resolution, without any prediction.

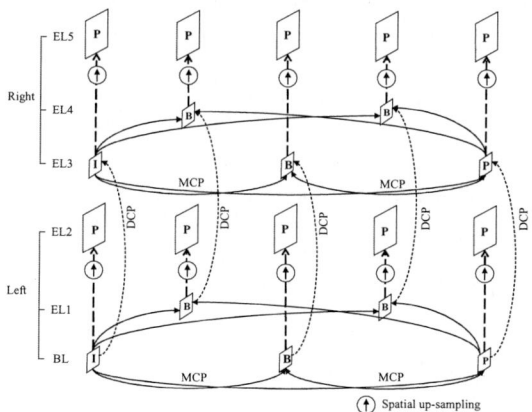

Fig. 2. Prediction for motion and disparity compensation

The proposed scalable stereo video coder is an extended version of compatible stereoscopic coding methods and can be divided into (1) stereo video coder, (2) spatial scalability coder, and (3) temporal scalability coder.

2.1 Stereo Video Coder

The structure of proposed stereo video coder is shown in Fig. 3(a). The reference frames are encoded by MPEG-2 standard. To encode the target frames, stereo encoder calculates DV with respect to temporally coincident left-view frame and estimates MV from the temporally closed right-view frames. The stereo encoder determines whether DV or MV provides higher compression efficiency, and finally, encodes both selected vector and residual information.

Disparity is the vectorial distance between the two points of a superposed stereo pair that correspond to the same point in the 3D scene. The estimation of disparity is indispensable for the prediction of the target image from the reference image. Then, the DCD is evaluated as follows:

$$DCD(x, y, DV) = I_R[x][y] - I_L[x + DV][y] \qquad (1)$$

where I_R and I_L are pixel intensity value of right and left frames. The DV is defined as:

$$DV(x, y) = \arg\min_{dv \in S} |DCD(x, y, DV)| \qquad (2)$$

where S means size of window searching area. As described in the expression (3), to find DV in real image, we calculate SAD for every macro block in the image and find a best matching-block which has minimum SAD. The DV is defined as coordinate of the point (x,y) of matching-block. However, y is around zero according to the general characteristic of DV.

Characteristics of the proposed coding scheme are as follows:

- *The reference or left video sequence are coded by non-scalable MPEG-2 video encoder.*
- *I-picture in the right-view sequence is coded by using a frame which results in disparity compensated prediction.*
- *P-picture of the right-view is coded by using the predicted frame which results in one of forward, disparity and bi-directionally (forward and backward) interpolated predictions. More specifically, the stereo encoder selects one prediction having minimum SSD to determine predicted frame.*
- *As shown in Fig. 4, B-picture is coded by using the predicted frame which results in one of forward, backward, disparity, and tri-directionally (forward, backward and disparity) interpolated predictions. Indeed, three reference frames used for prediction are the left-view frame coincidental with the right-view frame to be predicted, and the previous and next right-view frames in display order. The encoder selects one prediction according to its SSD.*

In case of P-picture, the interpolated prediction is generated by a weighted combination of a motion predicted frame and disparity predicted frame. The encoder uses interpolated prediction when a predicted macroblock with interpolated vector has minimum SSD. A macroblock predicted by the interpolated scheme is described as:

$$P_{pred}(v_f, v_d) = W_f R_{rec\,f}(v_f) + W_d R_{rec\,d}(v_d) \qquad (3)$$

where P, v, W represents predicted macroblock of P-picture, vector, and weighting factor respectively, and R_{rec} is a reconstructed macroblock of I- or P-picture. Each indexes, f and d denote forward, disparity respectively, and each weighting factors, W_f, W_d are given as 0.5 and 0.5.

In case of B-picture, the predicted macroblock is specified as:

$$B_{pred}(v_f, v_b, v_d) = W_f R_{rec\,f}(v_f) + W_b R_{rec\,b}(v_b) + W_d R_{rec\,d}(v_d) \qquad (4)$$

where B, v, W represents predicted macroblock of B-picture, vector, and weighting factor respectively. Each indexes, f, b, d denotes forward, backward, disparity respectively. In this case, the weighting factors, W_f, W_b, W_d are given as 0.25, 0.25, and 0.5.

Fig. 3(b) shows the stereo video decoder. The right-view frame is decoded with respect to the decoded left-view frame, coincidental with the right-view frame.

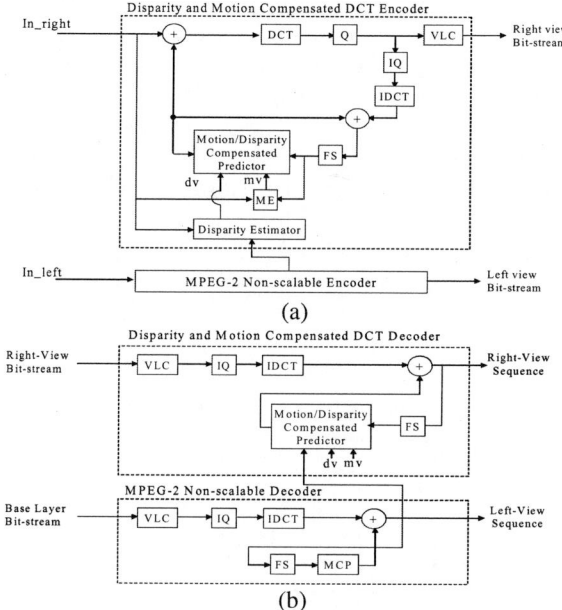

Fig. 3. Stereo video coder (a) Encoder (b) Decoder

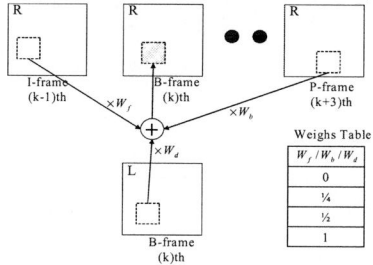

Fig. 4. Predictions in B-picture of right sequence

2.2 Spatial Scalability Coder

In addition to the bit-stream for lower spatial resolution image, spatial scalability EL encoder generates additional bit-stream for providing full resolution image as shown in Fig. 5(a). Stereo sequence is fed to the spatial scalability BL encoder after spatial downsampling. A locally decoded BL frame is spatially upsampled to the same sampling grid as the spatial scalability EL, and then subjected to EL encoder. The spatial scalability EL encoder encodes the difference between the decoded image from BL and original image at full resolution. Since the residual image loses the characteristics of natural image, it is Intra coded without MCP. In general, residual image has Laplacian distribution, which is highly peaked around zero, meaning that it can be compressed more easily than the original image. Therefore, a block having variance less than specific threshold value is skipped without encoding process in order to allot more bits to a block having large variance.

The decoding process of spatial scalability is the reverse of the encoding process as described in Fig. 5(b). To reconstruct the EL bit-stream, the bit-stream of temporally coincident frame in spatial scalability BL should be decoded first. The decoded frame in BL can be directly displayed in the client equipped with lower resolution display system. For example, when a client isn't equipped with 3D display and can provide only lower resolution display, it is efficient to provide directly lower resolution video decoded from spatial scalability BL decoder. However, to provide with full resolution video to a client equipped with high-resolution display system, it is resampled to full resolution and combines with the result from the spatial scalability EL.

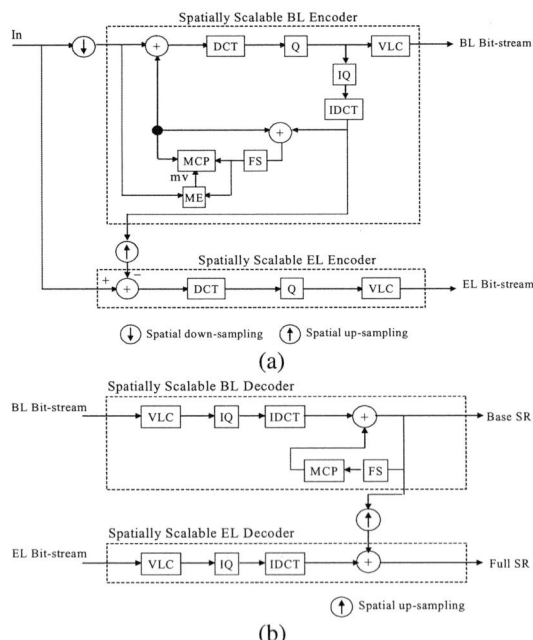

Fig. 5. Spatial scalability coder (a) Encoder (b) Decoder

2.3 Temporal Scalability Coder

Temporal scalability allows a video sequence to display as different temporal resolutions or frame rates according to the type of display and available channel capacity. In general, for temporal scalability providing a full and a half frame rate, an odd number of B-pictures is necessary [11]. In proposed temporal scalability encoder, the temporal demux (demultiplexer) splits up B-pictures, which are not used as reference frame, into temporal scalability BL and EL. For motion compensation, a BL frame is predicted only from the previous BL frames, whereas an EL frame can be predicted from both BL and EL frames. The clients can selectively receive the encoded frames of BL and ELs according to their display types and network connections. The decoded frames at the output of the temporal scalability BL can be shown by themselves at half frame rate of input video or can be temporally multiplexed in the temporal remux (remultiplexer) with the output of the EL decoder to provide full frame rate, the same as that of the input video.

2.4 Stereo Video Delivery

The proposed stereo video coding scheme aims at providing multicast-based stereo video streaming service. If a server provides stereo video service to N number of clients, unicast method requires each client to have its own video stream, separate from the others. Therefore, multiple end-to-end unicast cannot use network bandwidth efficiently because the network channel carries redundant portions of the same video stream. However, multicast-based method can remove these redundancies because the intermediate node copies the received bit-streams and sends selectively to its clients according to their capabilities.

Fig. 6 shows an example of stereo video streaming service. The video server captures stereo video from the stereo camera and then, encodes it. When an intermediate node requests one specific stereo video contents to server, the server sends relevant layered bit-streams. The intermediate node copies the received bit-streams and sends a portion or all the bit-streams to its clients. For example, client C in Figure 3.6 needs full resolution of mono video and accordingly the intermediate node provides BL, EL1 and EL2 bit-streams. The multicast clients can receive stereo video with various resolutions and can display stereo or mono video according to their types of display and bandwidth.

3 Experimental Results and Analysis

The experiments have been made with progressive scan 640×480, 24Hz refresh rate, and 4:2:0 chroma format test sequences. The picture structure is frame picture and the length of group of picture (GOP) is 15. A GOP structure includes 3 B-pictures between I- and P-pictures (M=3). Fig. 7 shows arbitrary left and right frames of the test sequence, *Laboratory* and *UbiHome*.

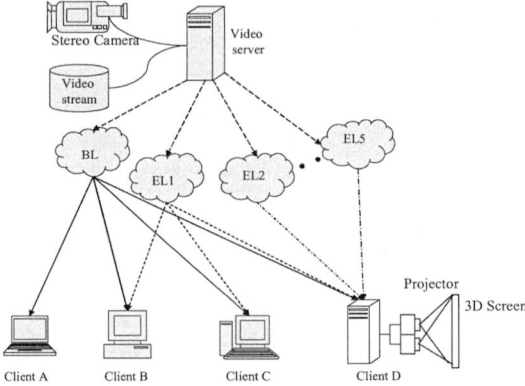

Fig. 6. An example of stereo video service using proposed stereo video coder

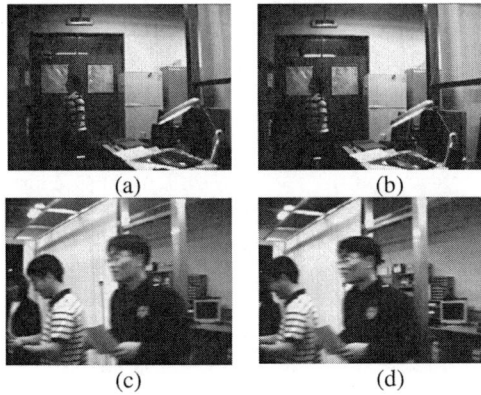

Fig. 7. Test Sequences (a) Left-view of *Laboratory* (b) Right-view of *Laboratory* (c) Left-view of *UbiHome* (d) Right-view of *UbiHome*

In Fig. 8, we compare proposed scheme with two already known methods, which are defined in MPGE-2 multi-view profile. We can easily observe that the proposed method provides a better overall performance compared with simulcast method and compatible method. In this experiment, the test sequences, *Laboratory* and *UbiHome*, are coded at 3 and 6 Mbps respectively. The bit-rates assigned to the left and right sequences are same. In this case, only spatial scalability BLs of right-view sequences are compared i.e. EL3 and EL4 layers.

In Table 1, the proposed coding scheme is compared with non-scalable MPEG-2 coder. We assigned same fixed bit-rate, 1Mpbs, to left and right sequence. For comparison of the efficiency of spatial scalability, we controlled bit-rate of the proposed scalability coder to yield the same quality. In general, the spatial scalability video coder needs more bits to provide same image quality as non-scalable coder. The additional data rate is defined as spatial scalability overhead. The proposed coding scheme combines stereo coder with spatial scalability. Therefore, it can reduce spatial scalability overhead because the quality of right sequence is improved by DCP.

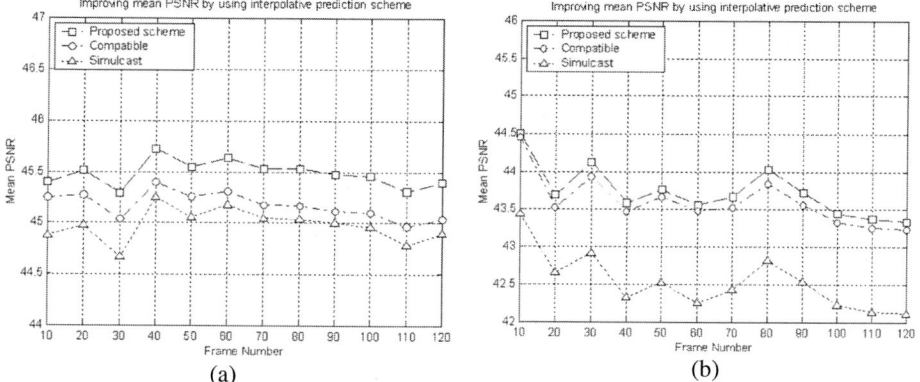

Fig. 8. Mean PSNR of right sequence with lower spatial resolution (a) *Laboratory* (b) *Ubi-Home*

The first results in the Table 1 show the spatial scalability overhead for the test sequence, *Laboratory*, whose motion is relatively slow. In this case, since most of macroblocks are coded with MV, the compression efficiency getting by the DCP is low. In case of *UbiHome* having fast motions and scene changes, however, most of macroblocks are affected by DCP instead of MCP. Therefore, the compression efficiency of sequence, *Laboratory*, is much improved when compared with that of simulcast coding. As a result, DCP in the proposed coding scheme can reduce the spatial scalability overhead.

Table 1. The spatial scalability overhead for test sequence *Laboratory*

Sequence		Laboratory		UbiHome	
		Left	Right	Left	Right
Single Layer Encoder	Bit-rate [Mbps]	1.0	1.0	1.0	1.0
	Average PSNR for luminance [dB]	41.03	40.25	36.50	36.33
	Mean PSNR [dB]	40.62		36.42	
Proposed Scalable Stereo Coder	Bit-rate [Mbps]	1.14	1.14	1.045	1.045
	Average PSNR for luminance [dB]	40.81	40.43	36.83	36.07
	Base layer bit-rate [Mb]	0.49	0.49	0.41	0.41
	Base layer bits-rate as percent of total bit-rate [%]	43	43	39.2	39.2
	Spatial scalability overhead [%d]	14		4.5	
	Mean PSNR [dB]	40.61		36.44	

Fig. 9 shows degree of disparity compensation and corresponding compression efficiency with PSNR of right sequence. Figure 9(a), shows the comparison of the results from the proposed scheme with simulcast and compatible coding, for the test sequences to correspond to a GOP. Since, the proposed stereo video scheme exploits both DCP and MCP, the compression efficiency is improved when it compared with existing schemes. The Fig. 9(b) and 9(c) show the percentage of encoded macroblock types of test sequences. In case of a slow motion sequence, Laboratory, most of macroblocks are coded with MCP instead of DCP, because of its small prediction errors, whereas most macroblocks of sequence, *UbiHome*, are coded with DCP. As a result, in case of *Laboratory*, there is a little compression efficiency when compared with simulcast coding. However, in case of *UbiHome*, the compression efficiency is improved in proportion to the number of macroblock which referred to DCP.

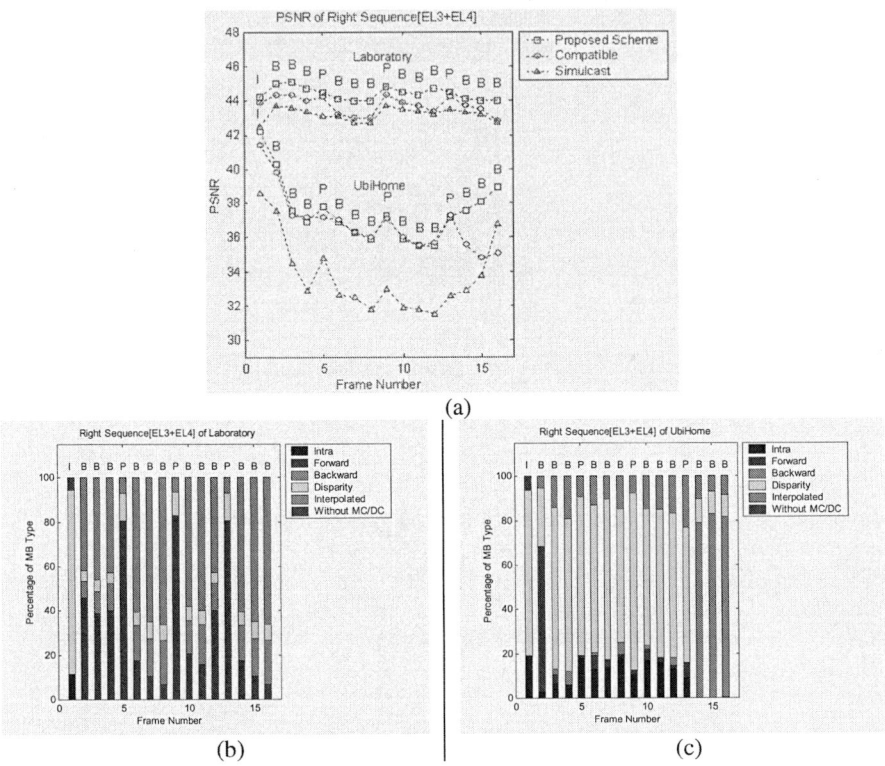

Fig. 9. The number of disparity estimated macroblock types according to the motion rate (a) PSNR comparison (b) Percentage of macroblock types for *Laboratory* (c) Percentage of macroblock types for *UbiHome*

Unlike non-scalable coding approach, the compression efficiency of scalable video coding varies according to the change of bit-rate to be allotted to each layer even if the total bit-rate is same. Therefore, it is need to find the best bit-rate partitioning point for each scalability layer. The mean PSNRs for left and right are differed according to the changes of bit allocations as presented in Table 2. In this case, the effects of the spatial scalability are not considered. Therefore, the layers for the left and right sequences are the spatial scalability BLs which are [BL+EL1] and [EL3+EL4]. A total bit-rate allotted to left and right sequences is 2Mbps. The compression efficiency is improved when we assign more bits to the layer for left sequence, instead of assigning same bits to left and right sequences. In the experiments, we can observe that the image quality is most high when the bit-rate for the left sequence is about 60% of the total bit-rate. This partition is well balanced for the overall test sequences.

Table 2. Mean PSNR according to bit partition in left and right sequences

Left[BL+EL1]/ Right[EL3+EL4]		1.4 / 0.6 [Mbps]	1.2 / 0.8 [Mbps]	1/1 [Mbps]	0.8 / 1.2 [Mbps]	0.6 / 1.4 [Mbps]
Sequence	Laboratory	43.02	43.85	43.83	43.61	42.74
	UbiHome	34.52	34.49	34.13	33.48	32.94

We can see in Table 3 and Table 4 that the mean PSNR according to the allotted bit-rate of spatial scalability BL and EL. When the total bit-rate is 1Mbps, the quality is increased as the spatial scalability BL bit-rate increases and the spatial scalability EL bit-rate decrease. However, when the total bit-rate is 6Mbps, the quality is increased as the spatial scalability BL bit-rate is reduced. If total bit-rate is low, the mean PSNR is affected by BL quality, but if total bit-rate relatively high, the mean PSNR is adversely affected by EL quality. The results suggest that to choose a partition point for the system, we need a measure which takes into account the qualities of both layers.

Table 3. Mean PSNR according to bit partition in spatial scalability base and enhancement layers (Total bit-rate = 1Mbps)

Base[(BL+EL1), (EL3+EL4)] / Enhancement[EL2+EL5]		0.7/0.3 [Mbps]	0.6/0.4 [Mbps]	0.5/0.5 [Mbps]	0.4/0.6 [Mbps]	0.3/0.7 [Mbps]
Sequence	Laboratory	35.36	35.06	34.69	33.91	33.23
	UbiHome	32.38	32.32	32.34	32.36	32.32

Table 4. Mean PSNR according to bit partition in spatial scalability base and enhancement layers (Total bit-rate = 6Mbps)

Base[(BL+EL1), (EL3+EL4)] / Enhancement[EL2+EL5]		4.2/1.8 [Mbps]	3.6/2.4 [Mbps]	3/3 [Mbps]	2.4/3.6 [Mbps]	1.8/4.2 [Mbps]
Sequence	Laboratory	45.54	45.95	46.10	46.32	46.58
	UbiHome	41.34	41.92	42.23	42.31	42.32

4 Summary and Future Work

We have proposed stereo video coding scheme for heterogeneous consumer devices by exploiting the concept of spatio-temporal scalability. The proposed scheme exploits interpolation based motion-disparity compensation and prediction to improve coding efficiency. The experimental results show the efficiency of proposed interpolative coding scheme by comparison with already known methods, compatible stereo and simulcast stereo, and the advantages of disparity estimation in terms of scalability overhead. To provide flexible stereo video service, we define both a temporally scalable layer and a spatially scalable layer for each eye-view. With this scheme, clients in the system are able to decode the stereo video based on their own display size, bandwidth availability and processing power by selectively receiving layered streams.

According to the experimental results, we expect the proposed functionalities will play a key role in establishing highly flexible stereo video service for ubiquitous display environment where devices and network connections are heterogeneous. Currently, we are pursuing a new quantizer for the residual images which are input of spatial scalability ELs.

References

[1] M.G.Perkins, "Data Compression of Stereopairs," Proc. of *IEEE Tr. on Communications*, vol. 40, no. 4, pp. 684-696, April 1992.

[2] A.Kopernik, D.Pele, "Improved disparity estimation for the coding of stereoscopic television," Proc. of *SPIE Visual Communications and Image Processing*, vol. 1818, pp. 1155-1166. 1992.

[3] M.Doma ski, S. Ma kowiak, "Modified MPEG-2 video coders with efficient multi-layer scalability," Proc. of *ICIP*, vol. 2, pp. 1033-36, Oct. 2001.

[4] A.Puri, R.V.Kollarits, and B.G.Haskell, "Basics of Stereoscopic Video, New Compression Results with MPEG-2 and a Proposal for MPEG-4," Proc. of *Image Communications*, vol. 10, pp. 201-234, 1997.

[5] Y.Song, "Improved Disparity Estimation Algorithm with MPEG-2 Scalability for Stereoscopic Sequences," Proc. of *IEEE Tr. on CE*, vol. 42, no. 3, Aug. 1996.

[6] S.Sethuraman, M.W.Siegel, A.G.Jordan, "A multiresolution framework for stereoscopic image sequence compression," Proc. of *ICIP*, vol. 2, pp. 361-365, Nov. 1994.

[7] S.Sethuraman, A.G.Jordan, M.W.Siegel, "Multiresolution based hierarchical disparity estimation for stereo image pair compression," Proc. of *the Symposium on Application of subbands and wavelets*, 1994.

[8] W.Woo, "http://vr.kjist.ac.kr/~3D/".

[9] M.Domanski, A.Luczak, S.Mackowiak, R.Swierczynski, "Hybrid coding of video with spatio-temporal scalability using subband decomposition," Proc. of *SPIE*, vol. 3653, pp. 1018-1025, 1999.

[10] U.Benzler, "Spatial Scalable Video Coding using a Combined Subband-DCT Approach," Proc. of *IEEE Tr. Circuits and Systems for Video Tech.*, vol. 10, pp. 1080-1087, Oct. 2000.

[11] M.Narroschke, "Functionalities and costs of scalable video coding for streaming services," Proc. of *Asilomar Conference on Signals, Systems, and Computers*, Pacific Grove, Nov. 2002.

Multi-source Multimedia Conferencing over Single Source Multicast: An Application-Aware Approach

K. Katrinis[1], G. Parissidis[1], B. Brynjúlfsson[2], Ó.R. Helgason[2], G. Hjálmtýsson[2],
and B. Plattner[1]

[1] Communication Systems Group
Swiss Federal Institute of Technology (ETHZ)
Zurich, Switzerland
{katrinis|parissid|plattner@tik.ee.ethz.ch}
[2] Network Systems and Services Laboratory
Reykjavik University
Reykjavik, Iceland
{bjorninn|olafurr|gisli@ru.is}

Abstract. A key issue for the establishment of large-scale multimedia conferencing in the global Internet is the provision of a ubiquitous IP multicast service. Despite extensive research and standardization efforts, worldwide availability of IP multicast is still limited. Single-source multicast constitutes an alternative that achieves to alleviate many of these problems. However, the tradeoffs in realizing multi-source sessions over single-source multicast remain poorly understood. The focal point of this paper is the realization of a multi-source distance learning conferencing platform (ET&L) over a novel single-source IP multicast protocol (SLIM). We have formulated the design space for managing trees rooted at multiple senders in a conferencing session, using both previously proposed schemes as well as new methods exploiting the application semantics of synchronous distance learning. We have prototyped each of these methods to prove their respective viability and conducted experimentation within our laboratories and over the Internet without changes or cooperation from ISPs or network administrators. The principal outcome of this work is a study of the various alternatives for implementing multi-source sessions over single-source multicast. In general, our work shows that it is feasible and efficient to use single-source multicast as a building block for implementing multi-source group communication applications.

1 Introduction

In spite of more than two decades of multicast research and experimentation, complexities have prevented successful standardization and deployment of multicast on a global scale ([1], [2]). Single-source multicast has been proposed as a promising approach to simplify and realize Internet multicast. However, important applications of multicast, such as collaborative multimedia distance learning, inherently have multiple senders. As single-source multicast is designed for distribution trees rooted at a single sender, these applications construct interesting new problem spaces.

V. Roca and F. Rousseau (Eds.): MIPS 2004, LNCS 3311, pp. 84–95, 2004.
© Springer-Verlag Berlin Heidelberg 2004

Distance learning (DL) overcomes geographical barriers in sparsely populated areas and makes it possible to achieve economy-of-scale without localizing all participants. A growing number of higher education institutions have been extending their curricula with distant education activities [3]. As a consequence, more people have access to more qualified instructors. Supporting interactivity among the participants of a learning session is crucial to make the distance learning experience more closely resemble the traditional classroom experience; moreover, interactive teleconferencing is essential to exploit multimedia and virtual reality technologies to expand collaborative learning beyond traditional teaching models. Successfully delivering such a service requires two components: a platform supporting the creation and management of learning sessions and multicast transport services to simultaneously deliver content to multiple receivers.

Easy Teach and Learn (ET&L) is a delivery platform for synchronous distance learning courses, that emphasizes on interactivity between students and teacher(s) [4]. The primary goal of ET&L is threefold: *scalability, heterogeneity* and *ease of deployment*. One of the difficulties encountered in achieving scalability and at the same time incorporating heterogeneous hosts into an ET&L session, is the limited availability of multicast [5] as an Internet service.

SLIM (Self-configuring Lightweight Internet Multicast) [6] is a novel single-source multicast protocol designed to avoid the need for multicast specific infrastructure. SLIM self-configures over the Internet, including traversal through firewalls and NATs, while exploiting available layer-2 multicast services [7] without requiring intervention by network administrators.

Using SLIM as a multicast service allows ET&L to be simplified by focusing exclusively on application-layer session management. Additionally, the interaction patterns in an educational conferencing session speak for the use of source-specific distribution trees (e.g. like trees constructed by SLIM), as it is highly probable that the lecturer(s) will monopolize the role of the sender. However, for the inherently multi-sender distance learning scenario of interest for ET&L, the single-source characteristics of SLIM are a potential challenge compared to Any Source Multicast (ASM) protocols like PIM-SM [8]. During designing the integrated prototype of ET&L over SLIM, it became apparent that current state-of-the-art does not provide us with clear guidelines on how to choose among the several possible alternatives for implementing multi-source sessions. This gap forced us to revisit the problem of using single-source multicast as a network-layer primitive to enable multi-source group communication.

The contributions of the present paper are:

1. *Formulation of the design space* for efficient management of trees rooted at numerous senders. Experimenting with multi-source conferencing using single-source multicast resulted in a layout of a design space, to unveil the tradeoffs between the various alternatives. During this process, we realized that the interaction manners in a synchronous distance learning session constitute a leading factor for efficient tree management. This lead us incorporate application-specific interaction control (*floor control*) into tree management (*Proactive Just-In-Time tree management, PJIT*). We implemented PJIT and the other reference methods into our testbed and verified their viability in multi-source scenarios through experimentation. To date, we are not aware of any former study that has collectively outlined and experimented with all these options for the purpose of

comparison. In general, similar approaches have not impacted the common perception about the viability of single-source multicast as a basic building block for higher layer services. Our planned future work will make this statement even stronger through strict performance evaluation.

2. *Clean separation of session management from underlying multicast network service.* This is achieved by having SLIM handle last hop problems and incremental deployment without cooperation from the session management layer. In contrast, a solution with replicators or an MBone-based approach would require more complicated management mechanisms to accommodate non-multicast-capable domains. In retrospective, our clean cut design approach allowed for a fast integration between ET&L and SLIM.

The rest of the paper is structured as follows. Section II discusses related work. In section III we give an overview of the ET&L platform and the SLIM protocol respectively. Section IV specifies the four reference methods and formulates the design space that these methods demarcate. In section V we delve into the implementation issues of our experimental work. Finally, we conclude with a discussion of the work presented herein in section VI.

2   Related Work

A considerable amount of work has been conducted on multi-source conferencing applications over the Internet. We classify related systems with regard to the mechanisms used to facilitate group communication.

The Classroom Presenter [12] is a representative for systems based on the ASM model as deployed in the Internet2 (PIM-SM/MBGP [14]). It uses the transport services offered by the Microsoft ConferenceXP framework [13] to realize a university-level synchronous distance learning system. In this sense, ConferenceXP assumes native IP multicast availability at every participating site. It is straightforward that the pervasiveness of platforms of this class is conditioned on the universal availability of the ASM model's infrastructure. Given the deployment problems of such protocols ([1], [2]), we propose instead a single-source multicast scheme that can be incrementally deployed. Instead of handling the complexity of tree management and address allocation on the network layer, we provide the primitives for source-specific multicast and move the overhead of supporting multiple sources efficiently to the session layer.

Beyond native IP multicast, similar systems use alternative group-scoped communication mechanisms, implemented either on the network layer (MBone) or on the application layer (Application-Layer Multicast, ALM). The INRIA Videoconferencing System (IVS) ([15], [16]) is one of the very first conferencing systems running over the MBone. The MBone is based on constructing data tunnels to traverse non-multicast enabled networks. The latter requires a considerable configuration effort, when users from new domains join a session. One needs to gain administrative access to a network and intervene in its operation to offer a multicast distance learning application to the local users. Instead, this is not a requirement for SLIM, which acts transparently by allowing point-to-point tunnels to the first hop router. The IRI-h (Interactive Remote Instruction-heterogeneous) [17] shares similar goals with the ET&L plat-form. It uses application-layer replicators to accommodate non multicast-capable hosts. Despite its ease of implementation, this solution does not scale well with the group size and maximizes packet replication on network links. Lately, numerous application-layer schemes ([18], [19], [20]) have been proposed to facilitate group-scoped media distribution. Although we do not neglect their flexibility in overcoming the deployment problems of network-layer multicast, their scalability, cost-efficiency and deployment model are still open issues.

Finally, we study prior efforts that touch on implementing multi-source sessions over single-source multicast. Holbrook ([21], [10]) suggested using a source-specific tree for each sender of the group or alternatively a session relay solution to reduce network state. In this paper we broaden the focus by comparing these alternatives with novel application-aware approaches, delivering thus a wider design spectrum. Sarac et al. introduces in [11] network-layer support for multiple senders in SSM [9] by constructing shared trees. While this solution minimizes state space and setup delay, it obfuscates the SSM protocol. In [22] the authors used multiple relays to provide multi-source support for SSM and evaluated the efficiency of their approach in terms of state space and bandwidth consumption. We survey this as one of the feasible tree management strategies; however we go further by qualitatively estimating the suitability of the strategy for our scenario by incorporating application semantics as well.

3   Systems Overview

ET&L [4] is a multimedia conferencing system tailored for synchronous delivery of university courses over the Internet. It uses the Session Initiation Protocol (SIP) [23] to facilitate typical conference signaling functions, like user location, session setup, capabilities exchange and user authorization. ET&L employs the centralized model with a dedicated server acting as the *conference focus*. A *Floor Control Server (FCS)* ensures concurrency control on who becomes the active sender. The FCS is modeled as a queue that accepts requests from active conferees and a server process that decides on the next active user according to a predefined floor-control policy.

Reliable media (distributed presentation, whiteboard, messaging) are transported to participants via a centralized Application-Layer Replicator (ALR). For interactive audio/video there appears a potential benefit for a multicast network service to allow our platform to scale to large group sizes without violating the application requirements and causing excessive bandwidth utilization, as already postulated in [24]. Specifically for our application, we need a multicast service that covers the spectrum from large-scale lectures, where audience interactivity is limited, to smaller conferences with multiple senders and high interactivity.

SLIM [6] is a single-source multicast protocol that self-configures over the Internet by exploiting the existing unicast infrastructure. As a single-source multicast, SLIM avoids the multicast address allocation problem by identifying a multicast channel with the pair $\langle S, C \rangle$, where S is the source of the channel and C the source-specific channel identifier. SLIM self-configures over the Internet by dynamically building tunnels as required and specifically addressing last-hop/first-hop issues [7], therefore avoiding assumptions about universal deployment.

Receivers send their control messages, join and leave, addressed to S using normal unicast thus avoiding the need for multicast routing. The SLIM daemon, running on SLIM enabled routers maintains no non-local information beyond what is conveyed by the join and leave control messages. Like other sparse-mode multicast protocols SLIM only forwards control messages if needed. All SLIM state is soft.

SLIM is designed to offer only the minimal set of functionality to construct and manage the topology of a singly rooted multicast distribution tree. In addition, SLIM is designed to separate the control plane functions from the data-path classification and forwarding, by defining a small set of primitives (API) through which the protocol daemon interacts with the data-path facilities. The clearly defined interface of SLIM facilitates additional advanced group management features, such as authentication, charging and security management.

Whereas the discussion below, regarding implementation and experimentation, focuses on ET&L and SLIM respectively, our formulation and conclusions apply more generally to any single-source multicast service, either implemented natively on the IP-layer (e.g. SLIM, SSM) or on the application-layer.

4   Formulation of Design Space

Implementing a multi-source application using single-source multicast requires an understanding of the appropriate construction and maintenance methods of the distribution trees. Before formulating the design space, we give a preliminary description of the methods that form the demarcation lines in the space, therefore called *reference methods*. Note that herein we limit ourselves to at most one active sender at a time. This is a reasonable assumption for many multimedia conferencing applications and is essential in our synchronous distance learning platform. Still, we intend to evaluate the impact of accommodating multiple concurrent senders to our model in our future work.

Reference Methods

The tree management methods that form the borderline of the design space are the following:

Conservative Approach: In this approach, a tree is constructed rooted at every participant that is a potential sender. The tree is created when the participant enters the conference and is kept alive throughout the entire lifetime of the conference.

Session Relay approach (SR): In the session relay approach (Figure 1A), all participants join a single multicast tree <S-R,C> rooted at a session relay (S-R) node. The active sender unicasts data packets to the session relay, which in turn relays the traffic to the multicast distribution tree.

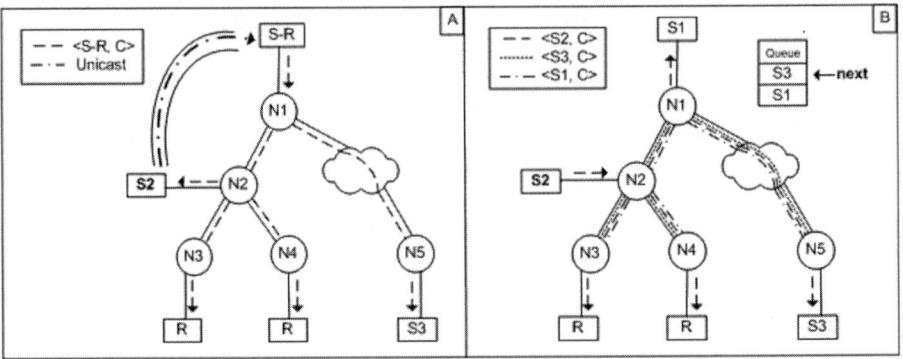

Fig. 1. Instance of an ET&L/SLIM session consisting of two/three potential senders (S1, S2, S3) and two receivers (R). Here, only five routers are shown, abstracting the rest with a cloud sign. In A) tree management conforms to the Session Relay approach with S2 being the active sender. The same scenario is shown in B) for the PJIT approach, where additionally S1 and S3 wait in the floor-control queue to become active. Therefore every participant has joined the look-ahead trees rooted at the two potential senders.

Just-In-Time Approach (JIT): This approach constructs a new tree rooted at a participant as soon as that particular participant becomes the active sender. When a new participant becomes active, the receivers are notified via a floor control message to leave the tree sourced at the relinquished participant and join the tree rooted at the new sender.

Proactive Just-In-Time Approach (PJIT): The fourth approach assumes the existence of a floor control queue that stores the identifiers of senders that have expressed their intention to become active via a floor request. It is thus possible to notify all conferees about the k next potential senders (k is a parameter of PJIT). Subsequently, every conferee joins the trees rooted at each of the k participants and maintains connectivity to each tree until the root of this tree ceases its sending activity. This method is consequently – like JIT – demand based, but with pre-allocation mechanisms (Figure 1B).

Design Metrics

Examining the entire volume of metrics that apply to tree management makes exploration of the design space rather intractable. Instead, we only focus on metrics that are critical for satisfying the requirements of the application (*application metrics*) and for economizing on network resources (*network metrics*). An exhaustive list of metrics including security considerations and resilience to failures/losses can be found in [25].

Application Metrics

Essentially, the requirements of the conferencing scenario can be collectively reduced to Quality of Service (QoS) metrics. In particular, we perceive the following two as the deciding factors:

Startup Delay: Citing the definition given in [10], this interval accounts for the delay from the time point the new sender is decided until all the group receivers start receiving data packets initiating from the sender. Startup delay affects only dynamic tree creation methods (e.g. JIT).

Data Delivery Stretch: This metric quantifies the extra delay overhead that is added to the overall data delivery delay in core-based approaches. In these, packets must first "suffer" the forwarding delay from the source to the relay and only then follow a shortest-path route to every receiver. This detour can be highly suboptimal, e.g. for receivers in the routing proximity of the sender.

Network Metrics

Our network of focus is the Internet and its best effort service model. Specifically, we perceive following metrics with regard to the network as the most significant:

Network State: Intuitively the multicast state on routers for a specific session will be a function of the number of concurrently maintained trees. Network state is an important metric for two reasons. First due to the high cost of router memory used to store multicast classifiers and second because increased memory occupancy due to classifier forwarding table size may slow down the lookups performed during forwarding.

Control Messaging Complexity: Due to soft state created by the SLIM protocol, the volume of refreshing JOIN messages for keeping trees alive is also an issue. Control messages require processing cycles on the routers they traverse and consume part of the available bandwidth.

Design Space

We construct our design space formulation using the four earlier presented reference methods, plus a well-known hybrid approach [22] that employs multiple session-relay nodes instead of a single one. Efficiency of tree management is rated with respect to the metrics presented above. A visualization of the design space is shown in Figure 2. In the diagram, each plane corresponds to each of the design metrics of interest[1] and in particular:

- x-plane: Measures the number of concurrent trees that are kept alive in a single session at every instance of time. It is straightforward, that at any time point the number of concurrent trees can span from just one single tree (e.g. JIT) to the group size N (conservative approach).
- y-plane: Captures the Relative Stretch Overhead (RSO) as given by:

$$RSO = (d_M - d_U) / d_U \qquad (1)$$

---

[1] Serving the purpose of simplifying visualization, we do not include the control message complexity in the design space discussion herein, as we expect it to exhibit behavior similar to the network state metric.

where d_M stands for the delay of data packets delivered over SLIM from the source to a single receiver and d_U stands for the unicast delay between the same source/receiver pair (delays here can be either average or maximum values). For instance, in the case of data forwarding over SLIM trees only, it follows that:

$$(d_M = d_U) \rightarrow RSO = 0 \qquad (2)$$

- z-plane: Captures the startup delay that characterizes each tree management approach, as already defined earlier in this section. The maximum value T_{max} in this axis corresponds to the largest startup delay given a particular multi-source session topology.

Foremost, the *Conservative* approach ensures for each sender the existence of a distribution tree to all receivers for any interaction pattern, i.e. ensures zero startup delay. Additionally, data delivery delay is also optimal (RSO = 0), as only shortest-path trees are used for routing data. However, this solution maximizes the state information that needs to be kept on routers, constructing maximum number (N) of static trees. The *Just-In-Time* approach is optimal in terms of network utilization, as exactly one tree exists at every instance of time. A considerable drawback is the unpredictable dispersion of startup delays. Depending on session topology and network performance, situations may occur, where the new sender starts multicasting data, although participants are still appending themselves to the distribution tree. Note also the increased engineering overhead of JIT, as its implementation depends strongly on the implementation specifics of the floor control protocol. The *Proactive Just-In-Time* approach mitigates the startup delay unpredictability of the JIT approach by constructing look-ahead trees. As participants are pre-joining only to a fraction k of future potential senders (k ∈ [0,N-1]), this approach achieves a fair tradeoff between startup delay and network state costs, and this in fact with optimal stretch. The PJIT quadrant in Figure 2 captures this tradeoff; obviously the more trees PJIT creates proactively, the higher is the probability of succeeding in creating the tree to the next sender and the larger grow the network costs. An obvious limitation is that the PJIT is bound to a queue-based implementation of the floor control service. Lastly, the *Session Relay* approach carries the advantages of the last two approaches, while entirely omitting the startup delay inconvenience of JIT. On the other hand, this approach is characterized by increased stretch due to routing data via the session relay node (which introduces also a single point of failure). In fact, Relative Stretch Overhead in Figure 2 grows unacceptably large for receivers located in the proximity of the source. Therefore, the session relay approach is rather appropriate for session topologies with sparsely populated participants. To accommodate dense groups as well, we included the hybrid approach of multiple session relays in the design space diagram. Assuming a proximity-based assignment (clustering) of receivers to SRs, the latter approach caters for lower stretch (and stress) compared to the single SR. Table 1 summarizes how each approach performs against the specified metrics. The bidirectional arrows in the PJIT row state that performance is strongly dependent on the number of pre-built/look-ahead trees (parameter k) maintained, both in terms of setup delay and state.

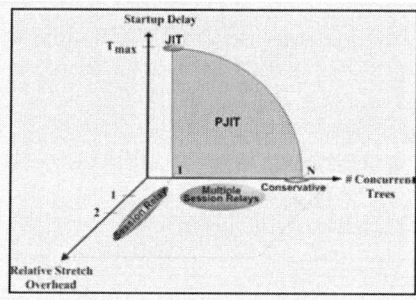

Fig. 2. Illustration of the design space for realizing multi-source sessions over single-source multicast.

Table 1. Qualitative comparison of the tree management strategies

Strategies	Stretch	Startup Delay	State	Comment
Conservative	↓	↓	↑	Ideal for session with a small number of predefined senders
Just-In-Time	↓	↑	↓	-
Proactive Just-In-Time	↓	↕	↕	Depends on floor control implementation (queue based)
Session Relay (SR)	↑	↓	↓	-

5 Implementation

In the integrated architecture, the *ET&L conference focus* manages a conference instance. SLIM is used to distribute real-time media streams (audio/video) to all active conference participants. A network setup adhering to the integrated architecture design is shown in Figure 3. In particular, ET&L exploits SLIM mechanisms to deal with the heterogeneity of the Internet. Participants acquire the necessary information to join a SLIM session from the conference focus and send out JOIN messages periodically towards the root(s) of the distribution tree(s), depending on the employed tree management strategy.

SLIM uses receiver-initiated group joins, assuming that a receiver is aware of the identifiers of the channel(s) he wishes to join. As a session progresses, the conference focus can report new session parameters to participants, like for instance a new source of a multicast stream, triggering a receiver initiated join towards the new source. Our integrated architecture enjoys a *high degree of separation* between session management and signaling of network entities. By the term "separation" we refer to the degree, to which the underlying network substrate is transparent to the session layer. Media-specific session management in our architecture does not involve signaling communication with intermediate transport/switching entities; instead only end- to-end parameters are exchanged. The rest of the session-setup process is handled by SLIM (e.g. dynamic tunneling between client and last-hop router). Application-layer replicators and MBone-based systems constitute examples that exhibit lower degree of separation compared to our scheme. In the first case, explicit signaling between the conference focus and the replicator complicates the session management protocol. In the MBone paradigm, tunnel setup adds as well to the complexity of session management.

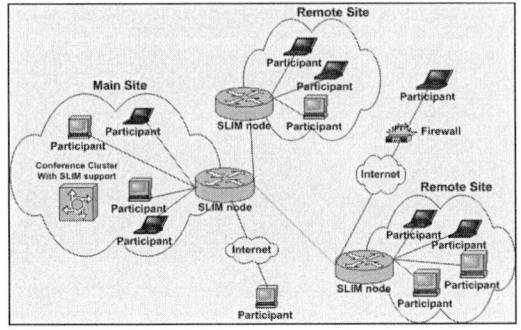

Fig. 3. Instance of the integrated ET&L-SLIM architecture. Each site connects to the Internet over a SLIM-enabled router, which provides multicast distribution. Participants not directly connected to a SLIM-enabled router can exploit SLIM mechanisms to JOIN the session. Apart from multicast JOIN messages, the rest of the conference control messaging takes place point-to-point between a participant and the conference cluster.

The above separation reduces the complexity of session management through the following features:

1. Failures in the network are handled by the SLIM protocol solely, without involvement of the session-layer or network administrators.

2. The problems that one needs to cope with due to the limited deployment of current IP multicast proposals is an obstacle for ET&L. SLIM provides incremental deployment and provides wider reach than the MBone [6].

3. The declining transparency of the maturing Internet due to firewalls and NATs is handled entirely by the SLIM protocol and avoids cluttering the session layer.

More involved details on the configuration of ET&L media transported over SLIM can be found in [25].

6   Conclusion

In this paper we have presented the integration of ET&L, a synchronous distance learning platform, with SLIM, a Self Configuring Lightweight IP Multicast protocol. As distance learning sessions inherently have multiple senders, this has proven to be an interesting exploration space for SLIM as a single-source multicast protocol. The principal result of our work so far is showing the promising role that single-source multicast can play as a building block to implement application scenarios that are inherently multi-source. In fact, we provided the design space for tree management methods in multimedia teleconferencing, considering application semantics as a major design parameter. Through the formulation of the design space we managed to demonstrate the potential value and weaknesses of each approach with regard to the metrics of interest. Additionally, we proved the viability of each method by embedding it into our prototype and testing it in various settings: in our laboratories, on production LANs and with simple scenarios over the open Internet. Our integration becomes even more valuable thanks to the strong separation between session control and management of the underlying multicast infrastructure. In fact, clean separation made the integration effort trivial, while it allowed us to implement and test different tree management methods with minimal effort. In retrospective, SLIM allowed the ET&L system to use multicast with minimum effort, without requiring any significant network infrastructure or session-layer changes.

Building on the prominent findings and achievements of this work, we are extending our research to several directions. First, it is of most importance to us to evaluate quantitatively the various tree management strategies through simulations and real life experiments using our prototype. This will allow us to characterize more accurately the efficiency of each strategy and its appropriateness for diverse application scenarios. Departing from the strict multicast context, the enhancement of ET&L with Internet multicast capabilities hands us the opportunity to further test and evaluate the overall features of our system in a wide-area scale. Our first step towards this goal is to organize university-level courses over the integrated ET&L/SLIM platform with participants located in several European cities.

Putting our findings in the context of state-of-the-art, we are unaware of any system using single-source multicast to provide multi-source services. Even though some of the reference tree management methods have already been mentioned in the literature, these have not been studied to our knowledge for the purpose of comparison. These two innovative achievements together form the principal outcome of our work: to impact the general perception of single-source multicast as a basic building block for higher layer services. From an educational point of view, our work promotes the establishment of synchronous distance learning as an Internet-wide service.

References

1. Almeroth K.C., The evolution of multicast: from the MBone to interdomain multicast to Internet2 deployment, IEEE Network, 2000, 14(1): p. 10-20.
2. Diot C., et al., Deployment issues for the IP multicast service and architecture, IEEE Network, 2000, 14(1): p. 78-88.
3. Tabs E.D., Distance education at degree-granting postsecondary institutions: 2000-2001, National Center for Education Statistics, Institute for Education Sciences, U.S. Department of Education, July 2003.
4. Katrinis K., A. Wagner, and B. Plattner, Easy Teach & Learn: an integrated synchronous distance learning platform, Technical Report, ETH Zurich, November 2003.
5. Multicast Status Online Page, http://www.multicasttech.com/status/index.html (online source), Multicast Technologies, 2003.
6. Hjálmtýsson G., B. Brynjúlfsson, and Ó.R. Helgason, Self-configuring Lightweight Internet Multicast-SLIM, to appear in the Proceedings of the International Conference on Systems, Man and Cybernetics, 2004.

7. Hjálmtýsson G., B. Brynjúlfsson, and Ó.R. Helgason, Overcoming last-hop/first-hop problems in IP multicast, Networked Group Communication (NGC), Munich, 2003.
8. Deering S., et al., The PIM architecture for wide-area multicast routing, IEEE/ACM Transactions on Networking, 1996, 4(2): p. 153-162.
9. Holbrook H. and B. Cain, Source-specific multicast for IP, Internet Draft (work in progress), Internet Engineering Task Force, 2003.
10. Holbrook H., A channel model for multicast, PhD Dissertation, Department of Computer Science, Stanford University, 2001.
11. Sarac K., P. Namburi, and K.C. Almeroth, SSM extensions: network layer support for multiple senders in SSM, in Proceedings of the 12th International Conference on Computer Communications and Networks (ICCCN), 2003.
12. Anderson R., et al., Videoconferencing and presentation support for synchronous distance learning, Frontiers in Education, 2003.
13. ConferenceXP: Conference Experience Project, http://www.conferencexp.net (online source), Microsoft Research, 2004.
14. Bates T., R. Chandra, D. Katz, and Y. Rekhter, Multiprotocol Extensions for BGP-4 (RFC 2283), Internet Engineering Task Force, 1998.
15. T.Turleti, The INRIA videoconferencing system (IVS). ConneXions - The Interoperability Report Journal, 1994, 8(10).
16. Malpani R. and L.A. Rowe, Floor control for large-scale MBone seminars, in Proceedings of the 5th ACM International Conference on Multimedia, 1997.
17. Maly R., et al., IRI-h: a Java-based distance education system: architecture and performance, Journal on Educational Resources in Computing, ACM Press, 2001, 1(1): p. 8.
18. Shi S., Design of Overlay Networks for Internet Multicast, PhD Dissertation, Computer Science and Eng. Department, Washington University in St. Louis, 2002.
19. Chu Y., S.G. Rao, and H. Zhang, A case for end system multicast, in Proceedings of the 2000 ACM SIGMETRICS International Conference on Measurement and Modeling of Computer Systems, Santa Clara, California, United States, 2000.
20. Chu Y., et al., Enabling conferencing applications on the Internet using an overlay multicast architecture, in Proceedings of the 2001 Conference on Applications, Technologies, Architectures, and Protocols for Computer Communications, 2001.
21. Holbrook H., and D.R. Cheriton, IP multicast channels: EXPRESS support for large-scale single-source applications, in Proceedings of the Conference on Applications, Technologies, Architectures, and Protocols for Computer Communication, 1999.
22. Zappala D. and A. Fabbri, Using SSM proxies to provide efficient multiple-source multicast delivery, in IEEE Global Telecommunications Conference (GLOBECOM), 2001.
23. Rosenberg J., et al., SIP: Session Initiation Protocol (RFC 3261), Internet Engineering Task Force, 2002.
24. Handley M., I. Wakeman, and J. Crowcroft, The conference control channel protocol (CCCP): a scalable base for building conference control applications, in Proceedings of the Conference on Applications, Technologies, Architectures, and Protocols for Computer Communication, ACM Press, 1995.
25. Katrinis K., et al., Multi-source multimedia conferencing over single-source multicast (extended version), Technical Report, http://www.tik.ee.ethz.ch/~etl/Docu/MSource-SSM-TechReport.pdf, ETH Zurich, June 2004.

Distributed Multi-source Video Composition
on High Capacity Networks

Yan Grunenberger[1], Phuong Hoang Nguyen[1,3], Gilles Privat[1], Jozef Hatala[2], and
Pascal Sicard[3]

[1]France Telecom R&D
28 Chemin du Vieux Chêne BP 98
38243 Meylan Cedex, France
[2]France Telecom R&D
801 Gateway Boulevard,
South San Francisco, USA
{firstname.name}@rd.francetelecom.com
[3]LaboratoireLSR-IMAG
Université Joseph Fourier
681 Rue de la Passerelle BP 72
38402 Saint Martin d'Hères Cedex, France
pascal.sicard@imag.fr

Abstract. We present the development and experimentation of a distributed system for pervasive video communication on gigabit networks. We have implemented an event-based control mechanism for the management of video sensors and their flexible dynamic configuration based on a service discovery protocol. Tests have shown that distributed acquisition of video content can be achieved with good performance by using this architecture on a high capacity network with, however, a strong dependence on the quality of the underlying networking hardware (Gigabit switches),specifically the capability of handling jumbo frames. The experiments have also validated our distributed architecture for composing complex scenes from distributed video sensors.

Introduction

Video sensors are set to become increasingly pervasive by getting the video source out of the (camera) box. These cheaper sensors can be disseminated in the environment and used either as self-contained wireless devices or attached to other equipment. As such, possibly combined with sensors in other modalities, they offer new possibilities for remote monitoring in all kinds of environments. We will concentrate here on their use for enhancing human communication interfaces towards *ambient communication* [1].

The problem addressed in this paper is the extension and adaptation of current multimedia frameworks to meet the requirements of these pervasive communication environments. The target distributed platform is predicated on network capacity increasing faster than the processing capabilities of devices. The aim of our work is to federate and manage a set of distributed video sensors on such a platform to support the distributed real-time acquisition of complex video content.

We draw for this upon techniques from the domains of virtual and augmented reality, real-time video processing, pervasive and distributed computing.

V. Roca and F. Rousseau (Eds.): MIPS 2004, LNCS 3311, pp. 96-107, 2004.

The experimental platform implemented supports an application of visual communication allowing visual "ambience sharing" between distant sites [].

1   Platform

1.1   Application

The aim of this application is to explore a visual ambience sharing approach based on a large scale, pervasive distributed video sensor system over a high capacity network. Figure 1 shows the client side: a panoramic vision, a restricted area live video, and a control pad with temporal as well as spatial control (bottom right).

Fig. 1. Client side: at the top the panoramic view refreshed every 10 seconds, at the bottom left the live video window, and at the right control commands. The client side runs on a standard web browser with a java applet and the MPEG4 player plug-in.

The system records video for up to 30 hours. In a previous version, the application was getting the video stream from each webcam connected through the USB bus. A video buffer was generated according to a static existing configuration of the camera position. Then the program generated a live video from the buffer and streamed it to a proxy, which was dedicated to distribute the content to standard web clients and to retrieve client commands for the live video window or the temporal control commands.

However, we wanted to provide the system with the capability of dynamic configuration, so we started with the analysis of its limitations. We have noticed the limit in the number of webcams that could be used: only five cameras could be connected to the computer, because the USB bus uses a bandwidth sharing mechanism. Additionally, multiple instances of the USB capture driver consumed a lot of CPU and there was a race between different threads resulting in a high CPU load that prevented other uses of computing resources.

So, we have decided to overcome these limits by distributing the video sensors over a high-speed network such as the Gigabit Ethernet and by using raw or low complexity video formats to minimize the load on the machine due to compression/decompression.

1.2   Network Platform Performance Evaluation: Gigabit Ethernet

One of the reasons for the emergence of the Gigabit Ethernet was the need to avoid the bottleneck at the network connecting high performance servers as well as the availability of high-speed media, such as optical fibers or twisted pairs of 5th category. At the beginning, only fiber cables were used, but later on, the IEEE 802.3ab [2] standard included the use of a twisted pair.

Because it is an extension of the existing Ethernet and Fast Ethernet standards, the gigabit Ethernet helped us by providing high bandwidth as well as retro-compatibility with slower networks.

As the actual MTU (Maximum transmission unit) size is only 1518 bytes, the receiving computer CPU has to be called by too many hardware interrupts to compute the IP and TCP/UDP packet checksum. It also needs to perform a memory copy of each incoming packet. To reduce the processor load, two directions are simultaneously taken by hardware vendors: increase the time interval between the frames by increasing the standard MTU size to 9000 bytes or even more (Jumbo frames) [3], or by adding more computing resources to the network adapter to perform caching and checksum validation on the adaptor [4].

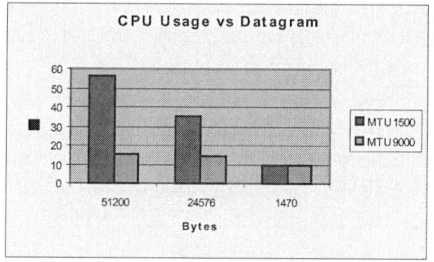 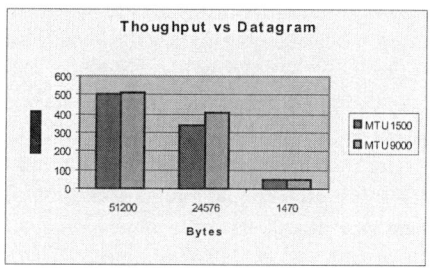

Fig. 2. Results of the Iperf tests with UDP unicast. On the left, we monitored the CPU usage in function of the size of the UDP payload and on the right, the throughput in Mbps. As we can see the CPU usage is directly linked to the size of the datagram as well as the throughput. We can notice that the use of jumbo packets (MTU is 9000 bytes long) greatly improves performance (it lowers the CPU usage and increases throughput).

We have performed extensive tests using our client cards (Intel PRO1000MT on P4 2.8Ghz under Windows XP) with our server card (3COM® 64bits on Xeon™ 2.6Ghz under Linux 2.4.25). We have first evaluated the impact of different sizes of UDP (User Datagram Protocol) datagrams on the useful throughput(Fig. 2), as well as the CPU load on the receiving system (measuring the sender system is not trustworthy, because of the high CPU usage by Iperf itself).

Next, we focus on the effect of the socket buffer size on the reception side by examining the packet loss, the jitter, and the CPU usage.

Table 1. Tests for a large UDP datagram (50Kbytes). There is no influence on the CPU usage or on the throughput, but it significantly reduces packet loss.

Reception buffer (Kbytes)	Throughput (Mbits/s)	Jitter (ms)	Datagram loss (%)	CPU usage (%)
256	502	1.8	0.91	56
1024	504	1.8	0.5	55
4096	501	1.8	0.049	55
8192	500	1.8	0	55

We think that the throughput for the small MTU size (1500 bytes) is limited because of a large number of network card interruptions. So we can say that for high bandwidth applications, it is necessary to use a large MTU, a sufficiently large socket buffer, and to fix the UDP datagram size according to the CPU usage dedicated for packet transmission.

2 Architecture

Our architecture follows a component-oriented approach in which each sensor is represented in the infrastructure by a multimedia processing unit called a *virtual device*. Assuming that sensors are used via software module wrappers, we consider that scene composition is based on streams coming from these virtual video devices. Each video stream from such a device contains both raw picture data and configuration metadata that represent the device as an element node in a network of virtual sensors. The metadata integrated in the video stream makes it possible to describe visual characteristics (direction of observation, relative co-ordinate of the zone of observation: Azimuth and Elevation) and physical attributes (fixed/mobile, optical zoom or not, rate of capture, etc.).

Our architecture is composed of system modules (cf. Figure 3) either co-located within one computer or distributed. As shown in Figure 3, the system comprises four main sub-modules: Interaction, Session, Composer Service, and Resources.

- **Interaction:** This module allows the user to configure his environment and to communicate with other applications. For example, the user can use his regular web browser to visualize composed multimedia streams or send explicit orders to configure his *Sessions* (3) (e.g. zoom using a particular camera).

- **Session:** After login (1), the user gets access rights to the distant content according to his profile and authorizations. The *Session* module is able to manage user access (3), i.e. the multi-session management. Sessions are configured and adapted automatically or interactively.

- **Resources:** This is the lowest layer in the system. Its functions are to receive requests from the service (8) with respect to the actual session (4) and access rights (5), then ask for available device (6) and extract the meta-data from the streams provided by devices (7). Next, it transfers data, metadata, and configuration information that serve for the scene composition (9). This system module supports dynamicity of multimedia streams.

- **Composer Service:** After taking into account the user profile information (10) and connecting to the appropriate video sensor manager, the corresponding view of the distant scene is formatted and sent either directly to the client (11), or to an intermediate video streaming processor that performs calibration or context adaptation functions.

The client can be a generic Web browser with appropriate plug-ins for video rendering.

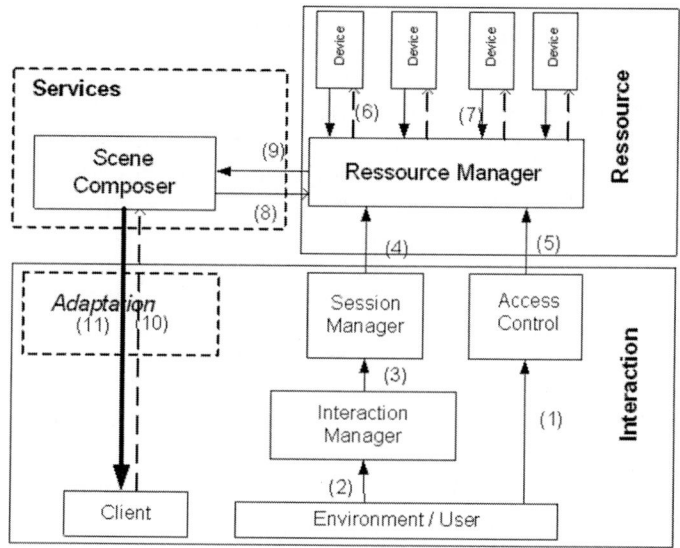

Fig. 3. Overall architecture with main modules: interaction, session, resources, and composer.

3   Implementation

We have decided to reuse some code modules of the existing application and adapt them to the proposed distributed architecture. The adaptation to the client environment uses a proxy module.

The existing code is written in C Unix and compiles both under Linux and Cygwin [5] environments.

3.1   Device Module

This module is responsible for grabbing pictures from a video sensor and making it available for the resource manager.

We have decided to use an adaptation layer to make the system independent from the capture platform (under Windows, our application uses Microsoft® Directshow and Video4Linux [6] under Linux). The adaptation layer provides access pointers to video buffers encoded in RGB 24 bits or YUV color formats. It may call some processing functions each time a new frame is available in the video buffer: this enables us to do some format conversion, such as Run Length Encoding (RLE codec) and send the compressed frame over the network.

For our first test, the network related functions are based on UDP/IP protocols, as we want to use the gigabit Ethernet. We have decided to rely on the UDP transport, since the packet loss is small according to our tests presented previously.

The UDP datagram has a size limited by the internal buffers of the operating system: for example, the size is limited to 64 Kbytes on Linux kernel-based OS. However, for a camera with 640x480 resolution, we need to fragment the frame into several small packets, so we have decided to decompose the frame line by line. In this way, any loss will result only in one or more line dropped.

Each UDP datagram contains a header with the format of the video data, the codec used, the length of the transmitted data, the offset of the current data (in this case, the line number), and a timestamp for synchronization purposes. After the header, we put the raw video data containing one or more lines of video. The number of simultaneously transmitted video lines depends on the MTU used in the system, the parameter that can be adjusted to reduce the frequency of interrupts and boost performance.

The UDP datagrams can be sent as unicast or multicast packets, the latter enables us to have multiple resource managers.

The metadata, which contains configuration data such as resolution, format, codec, is represented as a XML file and can be parsed remotely using a simple HTTP request. The XML file also contains data on video stitching so that the composer module can access the position of the camera for the video stitching.

3.2   Event-Based Dynamic Device Management

As our system relies on a set of devices distributed over a network, we have decided to use a service discovery mechanism: multicast DNS [7] with DNS-SD naming [8].

The device module can be integrated within a Java platform using the Java Native Interface – JNI. It presents the advantage of providing a simple interface for data access or modification, and enables adding an event mechanism to our modules.

We have used an open sourced implementation of multicast DNS: JmDNS [9], which was the most convenient Java implementation available for us. Multicast DNS allows easy introduction of new capture devices to the network as well as easy removing of them. This is done by an event mechanism generated on the network by multicast DNS listeners. Multicast DNS also carries the minimal configuration information, such as IP addresses and ports to enable dynamic configuration of the remote sources for the composition module. The information also includes the path of the XML file carrying the metadata, needed for high-level services such as configuration of the video-stitching.

3.3   Composition Module

The composition module has to complete different functions:
- Manage and receive video streams
- Use the composer service (Section 2) to offer a panoramic video buffer
- Extract a part of the video buffer and send it to the proxy module

To receive each video stream and to support good adaptability, the program has to rely on a multiple thread system to start or stop them at will. Each thread is charge of capturing the video stream of one virtual device, it receives a UDP packet and rebuilds a complete video frame.

We need to minimize the contention between different threads in our system, because there is little processing time per frame as we work on uncompressed or low compressed formats such as RGB 24 bits or YUV.

We decided to rely on the POSIX Thread model [10], because threads are managed at the kernel level, which allows better optimization of parallel tasks on simple or multiprocessor (SMP) systems.

The video buffer is created based on a configuration dynamically retrieved from the camera itself (we discuss this point later). The configuration basically contains the information needed to perform alpha blending rendering [11] from captured images by superposition.

Two threads use the video buffer: the embedded HTTP server able to provide a panoramic view in JPEG on demand and the streaming module that generates 352x288 MPEG4 video using the XViD codec [12]. These two threads call extraction functions responsible of refreshing the video buffer by applying alpha blending on the video stream from the capture threads.

So the extracted video is compressed in the JPEG format for the panoramic view and in the MPEG-4 format for the live video. The RTP protocol [13] is used for sending data to a RTSP [14] proxy responsible for dealing with client browsers. A SDP description file [15] is available to make the proxy aware of the format specifications.

When adding or removing a camera, the following tasks are performed: we have to launch or to stop the dedicated capture thread and adjust the size of the memory used for completing the video stitching. The information on network parameters is made available through the multicast DNS protocol, so we can open a unicast socket (or a multicast one if necessary) on a right port. The composer knows the resolution of the source as well as its format (RGB/YUV) from the XML metadata, so it can convert the color space dynamically. If a compression codec is used, it can call a decompression function, because the codec name is known.

4   Experimentation

We have led different tests in order to check the performance of our architecture and its implementation. The tests were performed on the capture and the composition platform to find out possible performance bottlenecks.

The capture platform used a range of computers with Pentium 4 processor from 1.7 GHz to 2.8 GHz running Microsoft Windows 2000/XP. The webcams were Logitech® QuickCam pro 4000, creative Pro webcam, and Philips TooCam Pro. All these computers were directly interconnected with a gigabit desktop switch (Netgear GS-108). We used Intel 100PROMT (100 Megabits/s) or 1000M/T (1 Gbits/s) network cards. Other cards such as Realtek[TM] 8169 or 3Com[TM] chipset are available, however they seem to suffer from some performance problem related to an excessive CPU usage [16] confirmed by our own Iperf tests.

The composition platform used xw8000 HP server running a 2.6 GHz Xeon with 512 Mbytes of RAM, and a 64 bits bus for the network controller, a 3com 1000 base-T server card.

4.1 Capture Evaluation

We have decided to use the Microsoft® Windows System Perfmon tools for the tests. They allow simultaneous access to the number of packets sent, the number of interrupts per seconds, and the CPU time allocated to user, privileged, or IRQ operations.

We have set up experiments to test the impact of different parameters such as resolution, format (RGB / YUV), RLE compression, or the choice of using the 9000 bytes MTU. The different parameters are shown in each graph, but the most relevant ones are indicated with arrows.

The goal of this first test was to investigate the time evolution of CPU usage. In fact, the system constantly switches between the interrupt, user, and privileged modes. We can then observe transitions between the peek of the user time activity and the privileged time activity, because socket operations are processed by the system in the privileged mode while picture processing is done in the user mode. The interrupt time is linked to the behavior of the network card: when the internal buffers are full, an interrupt is raised to signal the kernel so it can retrieve the data. The results of the tests serve to optimize the use of the different levels of buffers to guarantee a constant data stream on the receiver side.

Next, we have tried to evaluate the impact of compression on the CPU charge.

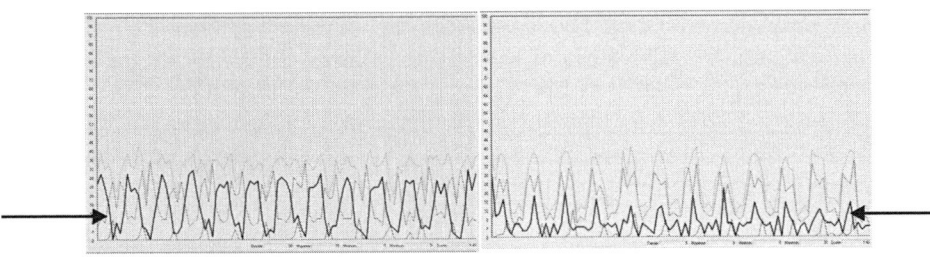

Fig. 4. Video capture at 640x480, 15 fps using RLE compression at the left and a raw stream at the right. The most relevant parameter here is the CPU user time.

As we can see in Figure 4, the CPU user time is very low when using the raw format, whereas simple compression such as RLE significantly increases the CPU usage. Such a behavior can be expected, because the RLE compression function is implemented in the module and it is executed as a user process on the host system.

As we can see in Figure 5, the CPU privileged time is smaller when using the YUV format, as the YUV packet is 2/3 the size of the RGB one. Once more this effect is expected, because packet sending is done in the kernel. So using a limited color space can be interesting, if the video sensor is itself limited in its color space.

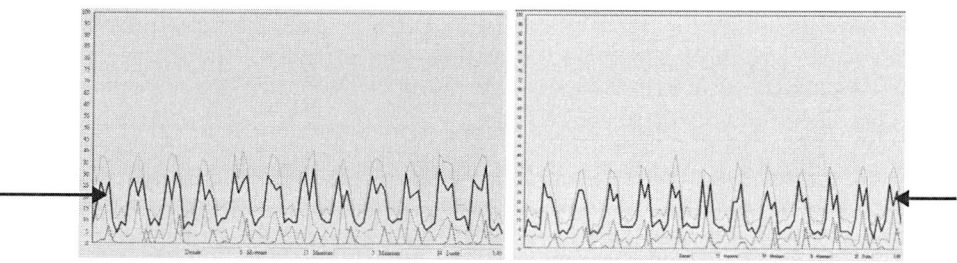

Fig. 5. Video capture at 640x480, 15 fps using RGB format at the left and YUV at the right. The most relevant parameter here is the privileged user time.

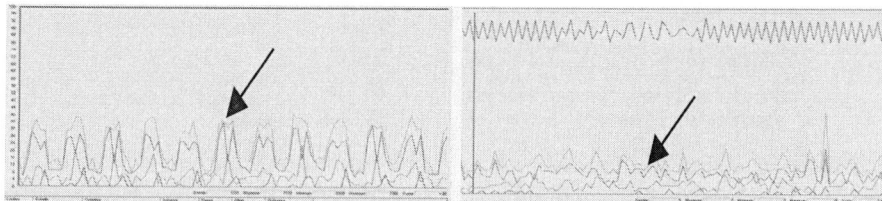

Fig. 6. Impact of using 1500 bytes MTU at the left and 9000 bytes MTU at the right. The line at the top right is the number of packets sent, which is 1/8 of the number at the left.

The use of the 9000 bytes MTU may be advantageous as it reduces the CPU consumption in every mode allowing for having more CPU resources available. This comes from the fact that there are fewer packets to send on the interface, which reduces the interrupt and privileged CPU time.

To conclude we can say that, there are a lot of parameters to tune on the capture side if we want to optimize the CPU load. We can change either the capture parameters, such as resolution or format, or we can apply compression.

4.2 Composition Evaluation

On the side of the composition module that is composed of scene composer service (Section 2) and binding service, we are interested in the frame loss, as well as in the resource consumption in terms of the CPU time or memory.

As the host machine runs a Linux-based OS, we use the Linux Top command to obtain CPU and memory usage. The composer itself monitors the frame loss statistics and the output frame rate. The host uses a 64 bits gigabit server card and we use a desktop gigabit switch to connect different sources to the composer.

As to video sources, we have used the capture module on different hardware ranging from P4 1.7 Ghz to P4 2.8 Ghz and with network controllers from 100 Mbps to 1Gbps. The different sources have output stream ranging from 50 Mbps to 75 Mbps.

Our first tests have shown that many UDP packets were lost due to the maximum size of the Linux kernel buffers (64 Kbytes for the socket buffers). We have increased them to 8Mbytes, so to have buffers setup compatible with the gigabit requirements.

We have monitored the frame rate of each capture module, the frame rate of each incoming video stream on the composer module, and the output frame rate of the live video window coded in MPEG4 (we set 9 fps as the requested target for the MPEG4 encoder). The results are presented in Table 2.

Table 2. Tests for different number of devices and different frame rates.

Capture frame rate on capture module	Frame rate of incoming video stream	Frame Loss	Output MPEG4 stream (9 fps requested)
1st cam = 9.70 fps	9.67 fps	0.3%	8.10 fps
2nd cam = 7.70 fps	7.67 fps	0.4%	
3rd cam = 10 fps	9.87 fps	1.3%	
1st cam = 9.70 fps	9.67 fps	0.3%	8.02 fps
2nd cam = 7.70 fps	7.53 fps	2.2%	
3rd cam = 10 fps	7.19 fps	28.1%	
4th cam = 6.43 fps	5.01 fps	22%	

After we start using 3 cameras simultaneously, we begin to see frame dropping, especially from the devices connected with a 100 Mbps network card. We have also monitored the system resources to locate the origin of losses. The results are presented in Table 3.

Table 3. System resources of the composer module: CPU usage in the user & system mode and memory usage.

Number of sources used by composer	CPU usage – User	CPU usage – System	Memory (512 Mbytes)
1 source = 71 Mbits/s	19.9 %	9.9%	10.8%
2 sources = 127 Mbits/s	30%	20.5%	24.6%
3 sources = 200 Mbits/s	40.7%	30.1%	44%
4 sources = 247 Mbits/s	48.6%	35.3%	69.5%

We see that the system behaves in a scalable way, as each new camera uses approximately 10% of the user CPU resources for processing, 10% of the system CPU resources (for UDP packet handling), and 15 % of memory.

As the system seems to perfectly handle the amount of data, we have started to evaluate performance of different elements of the gigabit network and especially the switch. Actually, the switch we used suffers from different problems, for instance it does not manage the mixed bandwidth (1Gbits/s and 100 Mbits/s) correctly, the effect is visible from the frame loss in Table 2. We have made some more tests using only pure gigabit controllers with the same test configurations in order to get more insight on the switch performance. We have measured the same parameters as before and this time we have obtained better results with a low frame loss as shown in Table 4.

Table 4. Tests for different number of devices and different frame rates.

Capture frame rate on capture module	Frame rate of incoming video stream	Frame Loss	Output MPEG4 stream (9 fps requested)
1st cam = 10.0 fps	10.0 fps	0%	8.05 fps
2nd cam = 7.70 fps	7.66 fps	0.5%	
3rd cam = 7.50 fps	7.30 fps	2.6%	
4th cam = 7.50 fps	7.30 fps	2,6%	

So it appears that the switch is responsible for the frame loss when using it with the mixed settings of 100Mbits/s and 1Gbits/s. Moreover, it happens that the switch were unable to handle packets with jumbo frames (MTU set to 9000 bytes) correctly, which makes us think that our implementation could have obtained better result if we used a better switch.

Conclusion

We have presented a new approach to overcome performance bottlenecks for acquisition of video scenes by using a distributed architecture on a high-speed network. It reduces the processing overhead of low-end devices and draws all potential benefits from future high capacity networks.

We have based our implementation on a video application that composes video streams sent over a gigabit Ethernet network with various packet sizes and compression formats. Our system benefits from dynamic management based on multicast DNS to provide dynamic adaptation to the number of available sources. It also makes use of flexible configuration description in XML to accommodate a large range of different resolutions and formats.

We have extensively tested the performance of our system. Our experiments show that a complex multimedia application can obtain very good performance by using a distributed architecture on a high capacity network, such as the gigabit Ethernet. However, our tests have also shown that performance strongly depends on the capabilities of the underlying networking hardware—to achieve the best performance (maximize network performance and minimize CPU usage), the system requires high quality switches capable of using large MTUs (jumbo frames). Furthermore, at the software level, kernel buffer should be set large enough to accommodate high transfer rates. We also think that service discovery protocols can be successfully used to resolve issues in dynamic configuration of communications between different system modules.

References

[1] Emile Aarts, « Ambient Intelligence : A multimedia Perspective », *IEEE Mult-imedia*, January-March 2004

[2] IEEE802.3ab, part of 802.3, IEEE 1000 Base-T Task Force http://www.ieee802.org/3/ab/

[3]   Phil Dykstra, Gigabit Ethernet Jumbo Frames, WareOnEarth Communications, Inc. Dec 1999.
[4]   Microsoft ® Inc., Network task offload, Microsoft Hardware Central, http://www.microsoft.com/whdc/device/network/taskoffload.mspx
[5]   Cygwin, a Linux-like environment for Windows, http://www.cygwin.com/
[6]   Video4Linux, a video capture API, http://linux.bytesex.org/v4l2/
[7]   Multicast DNS, Performing DNS Query over Multicast - http://www.multicastdns.org/
[8]   DNS-SD, DNS Service Discovery - http://www.dns-sd.org/
[9]   JmDNS, a Java implementation of multicast DNS - http://jmdns.sourceforge.net/
[10] POSIX Thread Programming, http://www.llnl.gov/computing/tutorials/pthreads/
[11] Alpha-blending, Duane Bong, http://www.visionengineer.com/comp/alpha_blending.shtml
[12] Xvid : http://www.xvid.org : an open-source ISO MPEG-4 compliant video codec.
[13] Schulzrinne, H., Casner, S., Frederick, R., and Jacobson, V., RTP: A Transport Protocol for Real-Time Applications; RFC 1889, 1996.
[14] Schulzrinne, H., A. Rao, Lanphier, R., RTSP: Real Time Streaming Protocol. IETF RFC 2326. April 1998.
[15] Handley, M., and Jacobson, V., RFC 2327: SDP: Session Description Protocol, April 1998
[17] Gigabit Ethernet Adapter Roundup by digit-life web site, http://www.digit-life.com/articles2/gigeth32bit

Analysis of FEC Codes for Partially Reliable Media Broadcasting Schemes

Christoph Neumann and Vincent Roca

INRIA Rhône-Alpes, Planète Research Team, France**
{firstname.lastname}@inrialpes.fr

Abstract. With many multimedia delivery schemes a client does not necessarily receive a given media content entirely (e.g. router congestions can lead some information to be lost). Mechanisms like error concealment and error resilient video coding allow the receiver to deal with partially received data and to play the content, with decreased quality though. Packet-level Forward Error Correction (FEC) can be used as a complementary technique to counter the effects of losses and reconstruct the missing packets. However if the number of packets received is too low for the FEC decoding process to finish, the received parity packets may turn out to be useless, and finally more source packets may be unavailable to the application than if FEC had not been used at all. This paper analyzes the adequacy of the LDGM Staircase, LDGM Triangle and RSE FEC codes to offer a partial reliability service for media content distribution over any kind of packet erasure channel, that is to say a service that enables a receiver to reconstruct parts of the content even if the FEC decoding process has not finished. We analyze this service in the context of a broadcasting system having no feedback channel and that offers media content distribution, like Digital Video/Audio Broadcasting.

1 Introduction

This work analyzes a FEC-based partial reliability service in the context of a media delivery system like DVB or DAB (Digital Video/Audio Broadcasting). Data broadcasting to cars (e.g. [2]), video delivery over satellites, DVB-T/H using "IP Datacast" (IPDC) [8] are possible applications for such types of transmissions. Because there is no back channel, no repeat request mechanism can be used that would enable the source to adapt its transmission according to the feedback information sent by the receiver(s). The lack of feedback channel however enables an unlimited scalability in terms of number of receivers, who behave in a completely asynchronous way. Typically the media content is transmitted in a carousel (or similar) approach, i.e. packets are transmitted cyclically and/or randomly for a significant duration that usually exceeds the transmission time of a single copy of the content. Using a reliable multicast transmission protocol like ALC [4] (which can in fact offer either a fully or partially reliable delivery

---

** This work is supported by a research contract with STMicroelectronics.

service, depending on the way it is used), along with the FLUTE [9] file delivery application, can turn out to be highly effective in this context [2].

Yet, in order to be efficient, these approaches largely rely on the use of a Forward Error Correction (FEC) scheme. After an FEC encoding of the content, redundant data is transmitted along with the original data. Thanks to this redundancy, up to a certain number of missing packets can be recovered at the receiver. More precisely k source packets (A.K.A. original or data packets) are encoded into n packets (A.K.A. encoding packets). If the FEC encoder keeps the k source packets in the set of n packets, this code is called systematic, which is the case of all three FEC codes considered here. The additional $n - k$ packets are called parity packets (A.K.A. FEC or redundant packets). A receiver can then recover the k source packets provided it receives any k (or a little bit more with LDGM/LDPC codes) packets out of the n possible. The great advantage of using FEC with multicast or broadcast transmissions is that the same parity packet can recover different lost packets at different receivers.

The FEC decoding process may not end successfully if not enough packets have been received by the receiver (i.e. less than k or a little bit more). This is the case when reception is interrupted (e.g. a carousel-based transmission may be active only for a limited duration) or is affected by a too important amount of losses (e.g. if a vehicle enters an environment with many obstacles that only enables an erratic connectivity). A partial reliability service should maximize the amount of original content reconstructed at a receiver, even if FEC decoding did not finish. The target of this work is therefore to analyze the inherent capabilities of each FEC code to maximize this partial decoding, and to find the best operational conditions to achieve this goal.

The paper is structured as follows: Section 2 gives an introduction to RSE and LDGM codes. Section 3 presents the issues with a partial reliability service for RSE, LDGM Staircase and LDGM Triangle. Sections 4 and 5 introduce experimental results in specific scenarios. Finally we conclude.

2   Introduction to RSE, LDGM Staircase, and LDGM Triangle Codes

2.1   RSE Code

Reed-Solomon erasure code (RSE) is one of the most popular FEC codes. RSE is intrinsically limited by the Galois Field it uses [10]. A typical example is $GF(2^8)$ where $n \leq 256$. With one kilobyte packets, a FEC codec producing as many parity packets as data packets (i.e. $n = 2k$) operates on blocks of size 128 kilobytes at most, and all files exceeding this threshold must be segmented into several blocks, which reduces the global packet erasure recovery efficiency (e.g. if B blocks are required, a given parity packet has a probability $1/B$ to recover a given erasure, and $B = 1$ is then the optimal solution). This phenomenon is known as the "Coupon Collector Problem" [1]. Another drawback is a huge encoding/decoding time with large (k, n) values, which is the reason why $GF(2^8)$

is preferred to $GF(2^{16})$ in spite of its limitations on the block size. Yet RSE is optimal (we say it is an MDS code) because a receiver can recover erasures as soon as it has received exactly k packets out of n.

2.2   LDGM Codes

We now consider another class of FEC codes that completely departs from RSE: Low Density Generator Matrix (LDGM) codes, that are variants of the well known LDPC codes introduced by Gallager in the 1960s [3].

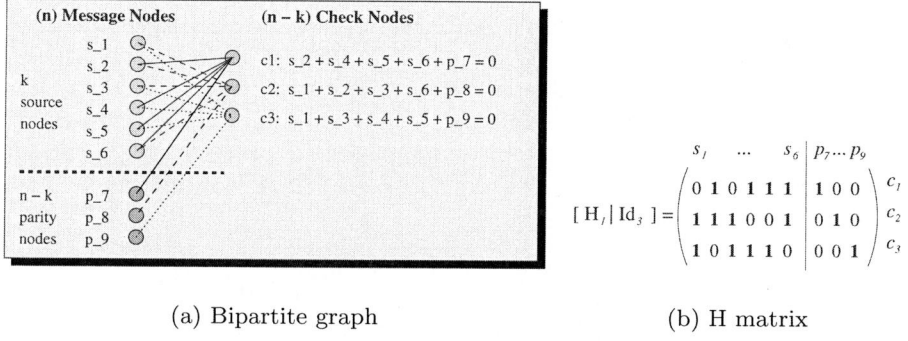

(a) Bipartite graph   (b) H matrix

Fig. 1. A regular bipartite graph and its associated parity check matrix for LDGM.

Principles: LDGM codes rely on a bipartite graph between left nodes, called message nodes, and right nodes, called check nodes (A.K.A. constraint nodes). The k source packets form the first k message nodes, while the parity packets form the remaining $n - k$ message nodes. The upper part of this graph is built following an appropriate left and right degree distribution (in our work the left degree is 3). The lower part of this graph follows other rules that depend on the variant of LDGM considered (e.g. with LDGM, figure 1 (a), there is a bijection between parity and check nodes). This graph creates a system of $n - k$ linear equations (one per check node) of n variables (source and parity packets).

A dual representation consists in building a parity check matrix, H. With LDGM, this matrix is the concatenation of matrix H_1 and an identity matrix I_{n-k}. There is a 1 in the $\{i; j\}$ entry of matrix H each time there is an edge between message node j and check node i in the associated bipartite graph.

Thanks to this structure, parity packet creation is straightforward and extremely fast: each parity packet is equal to the sum of all source packets in the associated equation. For instance, packet p_7 is equal to the sum: $s_2 \oplus s_4 \oplus s_5 \oplus s_6$. Besides LDPC/LDGM codes can operate on very large blocks: several tens of megabytes are common sizes. However LDGM is not an MDS code and it introduces a decoding inefficiency: $inef\_ratio * k$ packets, with $inef\_ratio \geq 1$, must be received for decoding to be successful. The $inef\_ratio$, experimentally evaluated, is therefore a key performance metric.

Iterative Decoding Algorithm: With LDGM, there is no way to know in advance how many packets must be received before decoding is successful (LDGM is not an MDS code). Decoding is performed step by step, after each packet arrival, and may be stopped at any time.

The algorithm is simple: we have a set of $n - k$ linear equations of n variables (source and parity packets). As such this system cannot be solved and we need to receive packets from the network. Each non duplicated incoming packet contains the value of the associated variable, so we replace this variable in all linear equations in which it appears. If one of the equations has only one remaining unknown variable, then its value is that of the constant term. We then replace this variable by its value in all remaining equations and reiterate, recursively. As we approach the end of decoding, incoming packets tend to trigger the decoding of several packets, until all of the k source packets have been recovered.

LDGM Staircase Code: This trivial variant, suggested in [5], only differs from LDGM by the fact that the I_{n-k} matrix is replaced by a "staircase matrix" of the same size. This small variation affects neither encoding, which remains a simple and highly efficient process, nor decoding, which follows the same algorithm. But this simple variation largely improves the decoding efficiency.

LDGM Triangle Code: This is a variant of LDGM Staircase where the empty triangle beneath the staircase diagonal is now filled, following an appropriate rule [12]. This rule adds a "progressive" dependency between check nodes, as shown in figure 2. This variation further increases performance in some situations, while keeping encoding highly efficient (even if a bit slower since there are more "1"s per row). Here also decoding follows the same iterative algorithm.

Interested readers are invited to refer to [11,12]. A publicly available, open source, implementation of the codes is also available at [7].

3   Partial Reliability: The Solution Space

We now explain the various possibilities when designing a partial reliability service.

3.1   Self Sufficiency Versus Interdependency

A video content is composed of I, P and B video frames, and an audio content is composed of samples also called frames. All audio frames and the I video frames

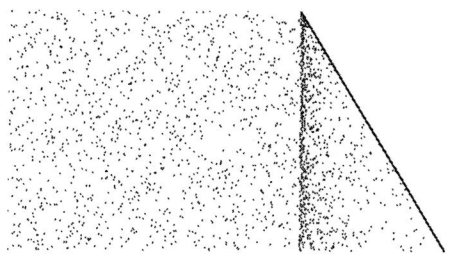

Fig. 2. Parity check matrix (H) for LDGM Triangle (k=400, n=600).

are self-sufficient entities, and the transmission scheme must absolutely preserve this self sufficiency property[1].

But P video frames depend on the previous I and P frames, and B video frames depend on the previous/next I and P frames. So in a partially reliable transmission scheme, there is an incentive to reconstruct the self sufficient frames first, before reconstructing the entire content. It leads to the idea of Unequal Error Protection (UEP) that will be considered in future works.

3.2   Object Definition

We call object an entity defined by the application and submitted to the transport layer as a whole. This definition is rather generic and therefore objects may largely differ, depending on the application: it can be an entire file, an audio track, or just a frame. If a large block LDGM-* FEC codec is used in the transport layer, the entire object, even if large, will probably be encoded as a whole (single block). But with an RSE codec, a large object has to be segmented into several blocks. As we'll see later on, this major difference will largely influence the efficiency of the partial reliability service.

3.3   Frame to Object Mapping

Several solutions exist to apply the notion of object to a media content, and control how FEC encoding is performed:

1- Map each frame to a separate object and encode them independently: the application can exploit a frame as soon as the corresponding object has been decoded (and all the other frames it may depend on with P and B video frames). Even if the frame size varies, depending on the bit rate and frame type, this size remains small, from several hundreds to a few thousands of bytes. Therefore there is no object segmentation, even with RSE. But LDGM-* codes have poor performance when operating on small block sizes, unlike RSE which benefits from its MDS feature [12]. So we can expect bad results with LDGM codes here. Additionally, since there is a high number of objects, all codes will

---

[1] This is the same requirement as when defining an RTP framing scheme with some multimedia content.

suffer from the Coupon Collector Problem (section 2.1). So we can expect all codes to have bad performance.

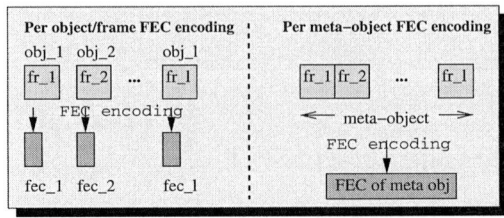

Fig. 3. Per object versus per meta-object FEC encoding.

2- Map the entire content to a single "meta object" and encode it: encoding is now performed over the whole meta-object for LDGM-*, and over blocks of maximum size for RSE. So the Coupon Collector Problem is totally eliminated with LDGM-* codes, and largely reduced with RSE (there are fewer blocks). Besides we know that LDGM-* codes perform very well with large objects [12], so their use seems rather promising.

Since a receiver must be able to reconstruct individual frames of a partially received meta-object, the underlying structure of the meta-object must be communicated. This can be achieved within the transport protocol by means of a dedicated signaling header extension (e.g. we added a private ALC header extension that contains a description of the meta-object and that is sent periodically).

Finally, operating on meta-objects means that FEC encoding is done over several seconds or minutes of the media content, or even over the entire media file. A significant latency is thus introduced in the transmission model, which is an issue in case of real-time streaming, but is not a problem with the test cases considered in this paper, or in our SVSoA proposal [6].

3.4   Benefits of an Iterative Decoding System: An Intuition

Intuitively, an iterative decoding algorithm should have a major advantage over MDS codes in partially reliable sessions. The reason is that decoding is done progressively and can be stopped at any time. A receiver can therefore exploit the subset of source packets received plus the subset of source packets already decoded. On the opposite, decoding with RSE is only possible once exactly k distinct packets have been received. If the session stops before this threshold, then no decoding is possible, and only the subset of source packets in the packets received can be exploited.

This fundamental difference is intrinsic to the FEC codec. However the intuitive conclusion that LDGM codes are more suited to a partial reliability service is not always true, as we will show later on.

4   One-to-One Frame to Object Mapping

4.1   Experimental Conditions

All the experiments use our LDPC/LDGM codec [7] and a popular RSE codec [10]. We designed a basic application on top of these two codecs, derived from the `perf_tool` of our FEC codec. Given one or more objects, the application first performs FEC encoding: to k source packets of a block, the codec produces k parity packets, all packets being 512 bytes long. Then all the source and parity packets (of all blocks) are transmitted in a fully random manner, indefinitely, by choosing one packet out of all the possibilities (to mimic a random carousel transmission). The receiving application waits until it has received enough packets to decode the block (or all blocks with RSE) and stops.

As can be noticed, there is no explicit loss model or loss rate, because random transmissions are not affected by them. Performance results could be slightly different if other transmission models were used, for instance by sending all source packets in sequence except a few of them, assumed erased, and then parity packets. But in that case we introduce additional parameters like the loss model and the loss ratios. In contrast our approach avoids these problems, is more universal, simulates lossy transmissions with a random carousel like scheme, and is in line with transmissions like data broadcasting to cars [2], DVB IP Datacast [8], or a video streaming approach like [6].

4.2   Performance with a Single Small Object

This first experiment illustrates the decoding behavior of each code. We consider a small object that fits into one RSE block (i.e. 128 source packets and 128 parity packets), and test all three codes using exactly the same packet sequence. Results are shown in figure 4 (a).

Since RSE is an MDS code, it decodes the object first, after receiving 166 packets (this value is ≥ 128 because of duplicated packet reception made possible by the random carousel transmission scheme). But RSE does not offer any partial reliability service in this scenario, and the source packets available earlier are those that have been received. The LDGM-* codes need more packets to decode the entire object. Yet source packets start to be decoded sooner, after 63 receptions for LDGM Staircase and 116 for LDGM Triangle. We also see that the Triangle version finishes the entire decoding sooner, but starts partial decoding after the Staircase version (because of extra relationships in the H matrix between all packets).

We did the same with a smaller object of size 10 packets, closer to an actual audio or video frame size. Figure 4 (b) shows the same general behavior as previously, at a smaller scale.

4.3   Performance with a Large Number of Small Objects

Here we transmit 600 objects of size 10 packets each, for a total content of about 3 MBytes, or 6000 packets of size 512 bytes. Results are shown in figure 5.

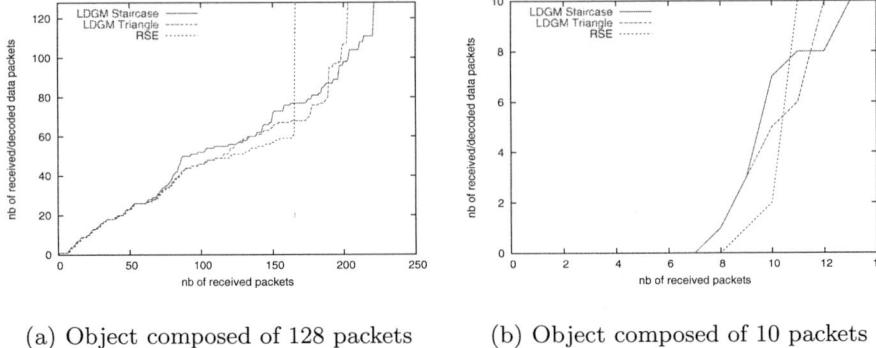

(a) Object composed of 128 packets   (b) Object composed of 10 packets

Fig. 4. Number of decoded packets with a single small object.

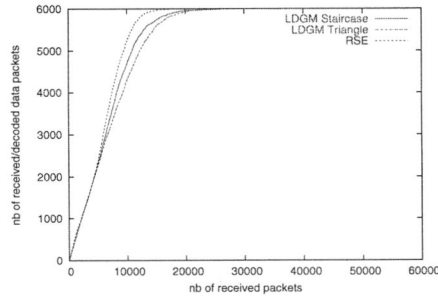

Fig. 5. Number of decoded packets with a per object encoding and a large number of small objects.

As expected, the general performance behavior is rather bad with all codes, especially with LDGM codes: RSE needs 17651 packets, LDGM Staircase 41677, and LDGM Triangle 50971. These poor results, caused by the coupon collector problem and a sub optimal use of LDGM codes, are not acceptable.

5   Frame Aggregation in a Single Meta Object

5.1   Performance Without Inter-frame Dependencies

Single Packet Frames: We first analyze each code in case of a meta object of size 6000 packets, assuming each frame forms a single packet. Results are reported in figure 6. We see LDGM Triangle finishes decoding the first (after 9403 packets received), followed by LDGM Staircase (after 9792 packets received) and then RSE (after 10282 packets received).

But partial reliability requires that decoding starts as soon as possible. LDGM Staircase starts decoding some data packets the first, after 798 received

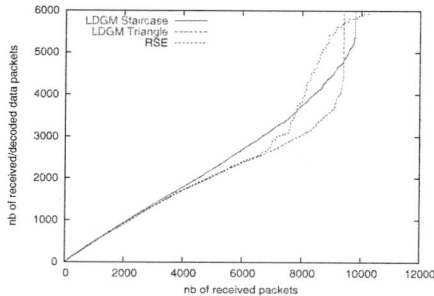

Fig. 6. Number of decoded packets of a meta-object.

packets, but it is clearly visible only after ≈ 3000 packets. At this point LDGM Staircase offers the best partial reliability service. RSE begins to decode the first block after 6634 received packets, and then quickly crosses LDGM Staircase after ≈ 7800 packets. At this point RSE offers the best partial reliability service. LDGM Triangle performs quite poorly since decoding is essentially done at the end. No partial reliability service is offered by this code, even if it exhibits the best decoding performance of all three codes!

Frames That Span Several Packets: We now consider a more realistic situation where frames span several packets. We analyzed the number of decoded frames as a function of the number of received packets. To simplify, we assumed that all the frames have the same size. If the object size is 1 packet, then the previous results of figure 6 can be used. In figure 7 we reported the results for object sizes of 2,3,4 and 10 packets respectively. We clearly see that RSE offers in all cases the best partial reliability service. The RSE performance even becomes better compared to the LDGM codes as the frame size increases.

To better understand it, we also analyzed the time at which a given packet is decoded. Results are shown in figures 8. We clearly see that with RSE, packets are decoded per block, making a group of adjacent packets available to the application at the time of the decoding. On the opposite, both LDGM codes make the packets available in a completely flat and random manner. This is an issue if a frame spans several packets, since one of them may not be available. Therefore RSE offers a better partial reliability service than LDGM codes when frames span several packets.

5.2   Performance with Inter-frame Dependencies

Finally we did similar tests but with an IPPPPI sequence of video frames, where I and P frames are respectively 10 and 5 packets long. Remind that a P frame depends on the availability of the previous I (and P if any) frame(s) of the current group of frames. The whole sequence is composed of 1000 frames, among which 200 are I frames. This test is therefore extremely close to an actual video media.

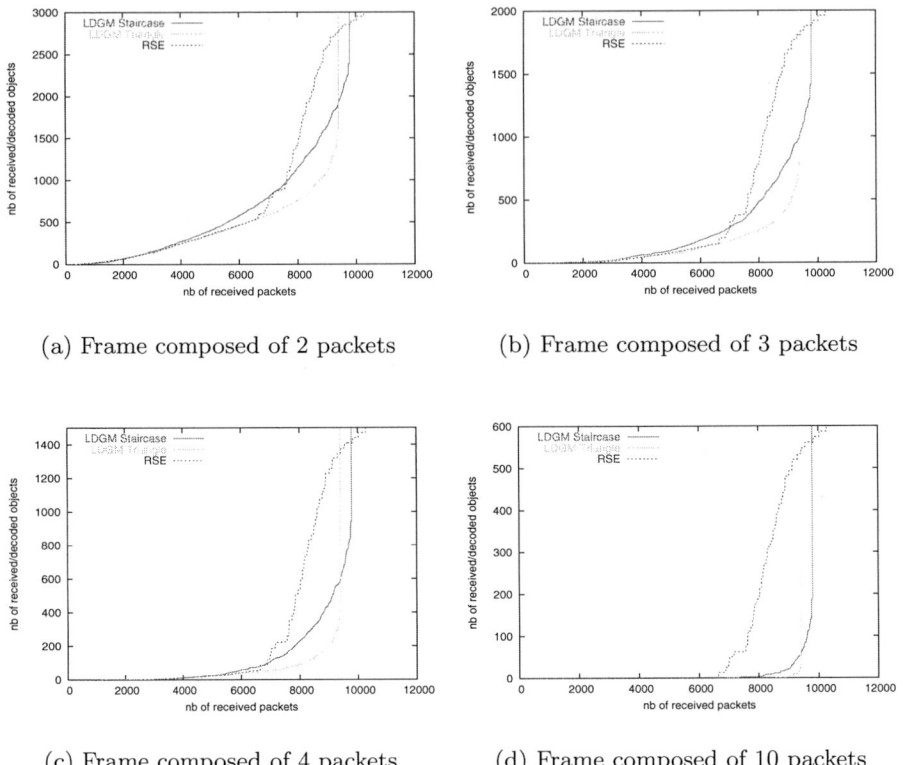

(a) Frame composed of 2 packets

(b) Frame composed of 3 packets

(c) Frame composed of 4 packets

(d) Frame composed of 10 packets

Fig. 7. Number of decoded frames of a meta-object.

Results, shown in figure 9, are rather close to that of figure 7. Here also RSE is clearly the best solution to offer a partial reliability service.

6   Conclusions and Future Works

We have analyzed the ability of three FEC codes – LDGM Triangle, LDGM Staircase and RSE – to offer a partial reliability service in case of media broadcasting schemes, that largely differ from more classical streaming approaches. We also introduced a meta-object encoding that largely improves the global efficiency of the broadcasting system. This analysis is quite innovative and we are not aware of any similar work. Traditionally only the global inefficiency ratio, which indicates how many packets must be received for decoding to complete, was considered. Our most important result is that LDGM-* codes, in spite of their iterative decoding approach and high efficiency, are not appropriate when partial reliability is desired. In contrast, RSE codes can more efficiently cope

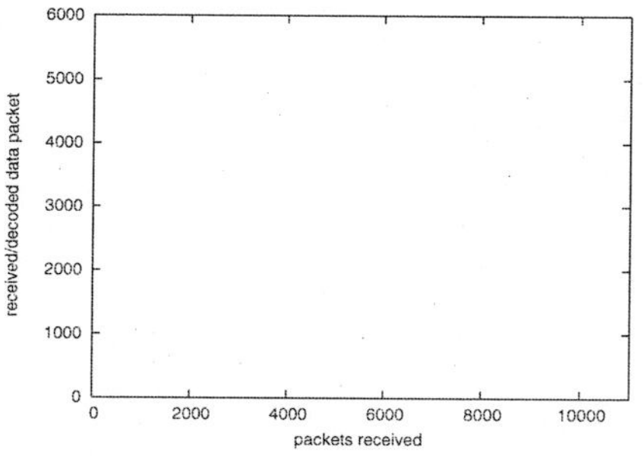

(a) RSE

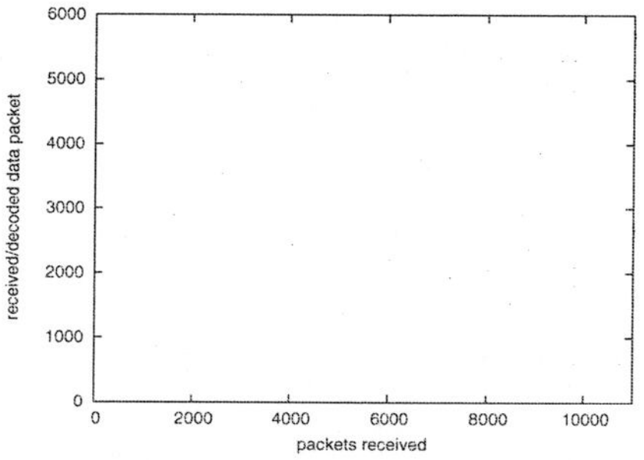

(b) LDGM Staircase

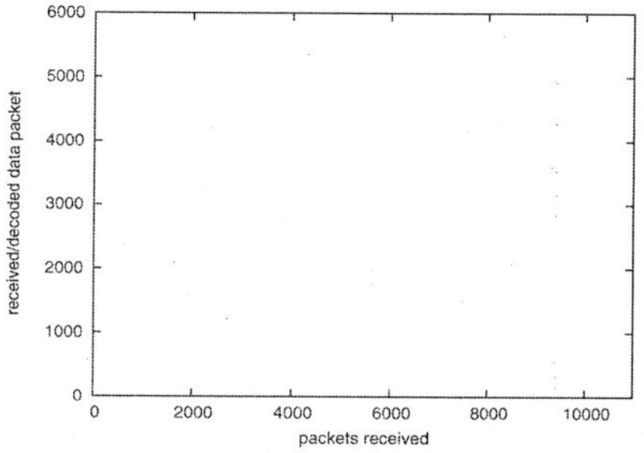

(c) LDGM Triangle

Fig. 8. Repartition of the received/decoded source packets of a meta-object as a function of time.

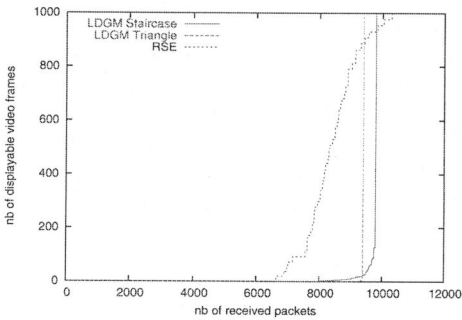

Fig. 9. Number of displayable video frames of a meta-object with an IPPPPI sequence.

with this requirement. This is rather counter-intuitive and only detailed experimentations and analyzes led us to come to this conclusion.

Finally, these results will be directly used in our Scalable Video Streaming over ALC (SVSoA) proposal [6]. Future works will also consider Unequal Error Protection FEC schemes, that should prove to be interesting to cope with interframe dependencies.

References

1. J. Byers, M. Luby, M. Mitzenmacher, and A. Rege. A digital fountain approach to reliable distribution of bulk data. In *ACM SIGCOMM'98*, Aug. 1998.
2. H. Ernst, L. Sartorello, and S. Scalise. Transport layer coding for the land mobile satellite channel. In *59th IEEE Vehicular Technology Conference (VTC'04), Milan, Italy*, May 2004.
3. R. G. Gallager. Low density parity check codes. *IEEE Transactions on Information Theory*, 8(1), Jan. 1962.
4. M. Luby, J. Gemmell, L. Vicisano, L. Rizzo, and J. Crowcroft. *Asynchronous Layered Coding (ALC) protocol instantiation*, Dec. 2002. IETF Request for Comments, RFC3450.
5. D. MacKay. *Information Theory, Inference and Learning Algorithms*. Cambridge University Press, ISBN: 0521642981, 2003.
6. C. Neuman and V. Roca. Scalable video streaming over alc (svsoa): a solution for the large scale multicast distribution of videos. In *Streaming media distribution over the Internet (SMDI04), Athens, Greece*, May 2004.
7. C. Neumann, V. Roca, J. Labouré, and Z. Khallouf. *An Open-Source Implementation of a LDPC/LDGM Large Block FEC Codec*. URL: http://www.inrialpes.fr/planete/people/roca/mcl/.
8. T. Paila. Mobile internet over ip data broadcast. In *10th IEEE Int. Conference on Telecommunications (ICT'03), Papeete, French Polynesia*, Jan. 2003.
9. T. Paila, M. Luby, R. Lehtonen, V. Roca, and R. Walsh. *FLUTE - File Delivery over Unidirectional Transport*, Dec. 2003. Work in Progress: <draft-ietf-rmt-flute-07.txt>.
10. L. Rizzo. Effective erasure codes for reliable computer communication protocols. *ACM Computer Communication Review*, 27(2), Apr. 1997.
11. V. Roca, Z. Khallouf, and J. Laboure. Design and evaluation of a low density generator matrix (ldgm) large block fec codec. In *Fifth International Workshop on Networked Group Communication (NGC'03), Munich, Germany*, Sept. 2003.
12. V. Roca and C. Neumann. Design, evaluation and comparision of four large block fec codes, ldpc, ldgm, ldgm staircase and ldgm triangle, plus a reed-solomon small block fec codec. Research Report 5225, INRIA, June 2004.

Supporting Mobility in Multicast: A Compromise Between Large and Small Group Multicast

Xavier Brouckaert * and Olivier Bonaventure

Université Catholique de Louvain, Belgium
http://www.info.ucl.ac.be {xbr,Bonaventure}@info.ucl.ac.be

Abstract. With the recent wireless boom and the IP convergence, there is a need to support efficiently a lot of concurrent small group multicast flows. In this paper, we describe how to integrate mobility support to a Small Group Multicast routing protocol named Sender Initiated Multicast (SIM). We then compare it with PIM-SSM through simulation and show that SIM requires fewer router states than PIM-SSM for small groups with static and mobile receivers.

1 Introduction

In the near future, users will expect to have rich multimedia applications on their mobile devices. This includes games, videoconferencing, instant messaging, filesharing and so on. The demand in bandwidth grows continuously and with the convergence of mobile and fixed networks to an All-IP infrastructure using IPv6, multicast technology is a way to go.

Since 15 years, protocol designers have found two ways of sending packets efficiently to a group of receivers in IP networks. The first method is to send packets to a group address that has no topological meaning [5]. Routers know how to forward the packets to the receivers that have joined the group by using a multicast routing protocol like PIM [9], CBT, DVMRP or MOSPF. The second method is to send packets to the list of the unicast receivers addresses that are placed in an additional header [2]. This method is generally considered not scalable in terms of the maximum number of receivers in a group because one can not put a lot of addresses in the header of a packet. This is why protocols using the second method belong to the "Small Group Multicast (SGM)" family (SIM [14], Xcast [2], MSC [7], [11]). In contrast, protocols using the first method belong obviously to the "Large Group Multicast (LGM)" family. Another solution, outside the scope of this paper, is to use multicast overlays [8].

Both SGM and LGM protocols have advantages and drawbacks [6]. For example, SGM protocols require far less state information in routers than LGM protocols. This is extremely important for scalability. LGM protocols can not

---

* Supported by a grant from FRIA (Fonds pour la formation à la recherche dans l'Industrie et l'Agriculture, rue d'Egmont 5 - 1000 Bruxelles, Belgium).

V. Roca and F. Rousseau (Eds.): MIPS 2004, LNCS 3311, pp. 120–129, 2004.

support millions of flows on a wide scale because each router on the tree from the source(s) to the receivers needs to maintain state for each flow passing through it. On the opposite, as in Xcast [2] for example, the receiver list is present in each packet, so that a router can forward a Xcast packet without having to maintain any state. On the other hand, Xcast suffers badly from the overhead in each packet, due to the Xcast header and the receiver list. This is even worse with IPv6, because IPv6 addresses are four times larger than IPv4 addresses. Sender Initiated Multicast [14] overcomes this limitation by letting sources attach the receiver list once in a while, but routers have to store state to be able to forward packets without receiver list. Other approaches to solve the mobile multicast problem are explained in [3] and [12].

The remainder of this paper is structured as follows. Section 2 describes the Sender Initiated Multicast protocol and our enhancements to support mobility. Section 3 presents the simulation scenario and the results for both static and mobile receivers.

2   Sender Initiated Multicast

The first version of the SIM protocol [14] was created by Vasaka Visoottiviseth and Hiroyuki Kido. It provides an unreliable connection-less source-specific multicast service. Routers that implement SIM are called SIM routers. They discover neighboring SIM routers by exchanging simple SIM Hello messages.

A SIM flow is identified by a tuple (Source, Group, Generation Count). The (Source, Group) has the same semantic as a PIM-SSM channel [1]. The 16-bit Generation Count is used to differentiate instances of the same flow : it is changed whenever the membership changes and is useful to ensure that routing is done correctly.

SIM packets are IP packets incorporating a new header, located between the IP header and the transport header. This header contains an optional receiver list, the bitmap of the active receivers in the receiver list and the flow identifier. For example, in a packet with a receiver list containing [D1,D2,D3,D4], a bitmap equal to 1001 means that the packet must be sent to D1 and D4 (bits set in the bitmap), but not to D2 and D3. D1 and D4 are called the active receivers.

SIM works in two modes : preset mode and list mode. In list mode, the source includes the receiver list in each packet while in preset mode, it only periodically includes the receiver list. List mode is similar to Xcast [2] and should only be used for short-lived flows or when the membership changes too often. Preset mode is what differentiates SIM from other Small Group Multicast protocols. In preset mode, SIM routers store forwarding information so that they can forward SIM packets that do not contain a receiver list. Obviously, the first packet of a flow must contain a receiver list. After each membership change, a receiver list is also added and the Generation Count is changed. We only analyse Preset Mode in this paper.

Packet duplication in SIM is performed by routers and not by end hosts like MSC [7] because the last hop, which is often of low bandwidth, should not be crossed twice (once for receiving the packet from a host and once for sending the packet to another host in the group). When a receiver list is present, a SIM router performs a unicast route lookup for each active receiver and finds the corresponding output interface. For each interface, a bitmap is created containing a 1 at position X when this output interface is used to send packets to receiver X, and 0 otherwise. The logical OR of all bitmaps is equal to the bitmap found in the SIM header of the packet currently being forwarded. The router sends one copy of the packet per output interface where the calculated bitmap is not all 0's. Each outgoing packet thus has a different bitmap. Also, the IP destination address field is replaced by the address of the next hop SIM router reachable on the output interface. If no SIM router is reachable via the output interface, the SIM router sends one packet per receiver (no multicast then). If only one receiver remains, the packet is sent directly in unicast. If two or more receivers are directly connected to the same interface, the IP destination address is set to the Group address found in the SIM header so that layer-2 multicast is used instead of using multiple unicast packets at the last hop.

Routers that forward a SIM packet to more than one interface are called branching routers. They set the Previous Hop field of the SIM header to their IP address and set the Skip Count field to 0, while non-branching routers (with only one output interface) increment the Skip Count field of the SIM header. With these two fields, a router that receives a SIM packet knows the last branching router and the number of IP hops between itself and this last branching router. A branching router that receives a packet with a Skip Count greater than 0 will send a SIM Redirect message to the previous branching router in order to optimize routing. The SIM Redirect message contains the ID of the SIM flow and the bitmap of the receivers that need to be redirected. When a SIM router receives a SIM Redirect, it modifies its state so that all active receivers in the bitmap of the SIM Redirect are forwarded to the initiator of the SIM Redirect. The subsequent packets will thus be sent directly to the next branching router : they will still pass through each non-branching router, but those routers will not need to maintain SIM state for this flow.

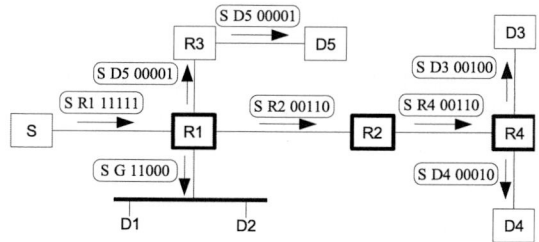

Fig. 1. SIM forwarding example

Let us consider an example to illustrate the forwarding of SIM packets. On figure 1, we only show important fields of the packets in a rounded box, namely the IP source address, the IP destination address and the SIM bitmap field. Source S wants to send a packet to [D1,...,D5]. For this, it sends a SIM packet to R1 with a bitmap equal to 11111 and a receiver list [D1..D5]. R1 performs route lookups for all destinations. D1 and D2 are directly reachable via the South interface, so a copy of the packet is sent to the group address using layer-2 multicast. D5 is the sole destination that is reachable via the North interface, and R1 sends the packet directly in unicast to D5. The remaining destinations of the packet are reachable via the East interface. A copy is sent to the SIM router on the East link, namely R2. For each copy sent by R1, the source address does not change and the bitmap is set to the active destinations for each interface (visible on figure 1). We now have 3 packets flowing respectively to D5, R2 and G. R2 will forward the packet to R4 and increment the Skip Count because it is not a branching router. R4 will finally forward in unicast to D3 and D4. Routers that actually performed SIM routing are drawn in bold on figure 1.

After this first packet, forwarding state has been created in SIM routers and R4 has discovered that there are non-branching upstream routers. It sends a SIM Redirect message to R1 containing the ID of the communication and the bitmap of destinations that should be redirected to itself. R1 processes the SIM Redirect and stores in its forwarding table the new information. Subsequent packets will be directly sent to R4. The state created in R2 will timeout and no SIM routing will be used by R2 for further packets, only standard unicast routing.

Note that state in non-branching routers is kept during 10 seconds while state in branching routers is kept for 60 seconds because non-branching routers expect to be bypassed soon when the upstream router will have received the SIM Redirect from the downstream router.

2.1   Mobility Support

Since we do not want to create a new mobility framework, we need to look at how we can integrate SIM and Mobile IPv6 [10, section 11.3.4]. Mobile IPv6 provides two basic methods to maintain (classical) multicast flows alive when a node moves across subnets. The first method is to use a bidirectionnal tunnel with the Home Agent so as to hide the mobility to other group members. The mobile node tunnels its MLD [13] messages to its Home Agent and the Home Agent acts as a group member and tunnels multicast packets to the mobile node. Clearly, bidirectional tunnels are not efficient in terms of delay and overhead in the packets. Bandwidth can also be wasted because, if two mobile nodes from the same Home network and members of the same group go to the same foreign network, two tunnels will be used. Last but not least, the Home Agent becomes a central point of failure. The second method is to join the multicast group using a multicast router located in the foreign network (assuming there is one, which is not guaranteed if we take into account the current multicast deployment). It has the disadvantage to put more stress on the multicast routing protocol but the routing is optimal, the delay better and the overhead in packets minimal.

Given the advantages of the second method, we propose to adapt it to SIM. For this, the SIM stack will rely on the state maintained by Mobile IPv6 inside each node[10]. There is no need to change anything in the Mobile IPv6 protocol.

Source mobility. Let us consider a source sending packets to a set of destinations using SIM. What should happen when the sender moves from its Home Network to a Foreign Network ? First, the mobile source detects its movement by receiving a new Router Advertisement. If the SIM router sends SIM Hello messages, the mobile will also learn the address of the SIM router. Next, it warns its Home Agent about its new Care-of Address by sending a Binding Update to it. It also sends a Binding Update to its current Correspondent Nodes. The mobile node can send SIM packets to the SIM router using its Care-of Address. Nothing more is required because the original source address (the Home Address) remains the same in the SIM Header.

The SIM stack in receivers can then replace the Care-of Address with the Home Address before sending packets to the upper layer. Thus, destinations do not need to be aware of the mobility of the source. The Home Address placed in the SIM header plays the same role as the destination option home address in unicast Mobile IPv6 packets. The SIM source address does not change and the receiver list remains the same. To avoid routing loops and other routing inconsistencies, the source should use a new Generation Count only if the SIM router advertised on the current network is different from the SIM router advertised on the previous network. The mobile node must then also attach the receiver list to the first packet sent on the new Foreign Network. If a source moves too often, it should switch to List Mode or activate the Temporary Flag in the SIM header to avoid useless state creation in SIM routers. It could also continue to use the same SIM router, even if a different one is advertised.

Since SIM state is created as soon as packets are forwarded, there is no delay for the establishment of the tree before we can send data on it. This is especially useful for source mobility support in SIM. Other protocols that need to establish the tree completely before sending on it need special mechanisms to have a slow handoff delay.

Receiver mobility. The receiver case is more complicated than the previous one. Let us consider a receiver taking part in a SIM communication that moves from its Home Network to a Foreign Network. It first discovers its movement by the received Router Advertisements. Next, it sends Mobile IPv6 Binding Updates to its Home Agent and its active Correspondent Nodes, including the source of the SIM packets. While the source is not yet aware of the new Care-of Address of the mobile node, it continues to send SIM packets to the receiver's Home Network. If the SIM packets have been converted to unicast upstream, the Home Agent will forward them to the mobile node using standard Mobile IPv6. If the Home Agent is also a SIM router and if it is aware of the Care-of Address of the mobile node, it can forward packets in unicast to the CoA. The idea is the same as with Mobile IPv6 in unicast, but here, no tunneling is used.

Once the Binding Update has reached the SIM source, it knows how to reach the mobile node directly. Thus, each mobile destination in the receiver list can be replaced by its Care-of address to ensure shortest-path routing. As the receiver list changes upon receiving a Binding Update, the source must use a new Generation Count to create new forwarding states in the network.

Additional details about the changes required to support mobility with SIM are available in [3].

3 Simulation Results

We have implemented SIM and PIM-SSM in J-Sim[15][4]. We have chosen PIM-SSM as the basis for our comparison because PIM is the most widely used multicast routing protocol [6] and because both protocols are source-specific [1]. As no SGM protocol is currently deployed, people willing to do SGM-oriented communications would use PIM-SSM. Hence, comparing SIM and PIM-SSM is interesting. The topology we used is a multi-grid network (4 grids of 10*10 routers connected by a central grid of 4*4 routers) because it can represent different mobile networks connected by a backbone. The propagation delay of all links is 100 ms and the bandwidth is 1Mbps. Around each stub grid are wireless networks attached. The multicast group always consists of one fixed source (in the upper-left grid) and 10 mobile receivers attached randomly to wireless networks. There can be a maximum of 36 IP hops between the source and a mobile receiver in our topology. The source starts sending packets at 100.0s at a constant rate of 10 packets per second. There are no transmission errors on the links. All routers implement the multicast routing protocol. A single unidirectionnal communication is performed between the source and receivers. This simplicity is necessary in order to have a minimal amount of parameters and to be able to interpret the results in a straightforward way.

Our simulations are at the IP level only. We do not take into account layer-2 wireless issues. Our mobility is an IP one. No micromobility comes into play. When a mobile node moves into another network, it attaches to an access point colocated with the access router. The mobile node creates an address valid on the foreign network immediately. We consider a fixed handoff delay of 100 ms.

Different mobility scenarios can be simulated. In a random mobility scheme, mobile nodes are moved randomly among all wireless networks. It may seem exaggerated at first because nodes can not make big jumps from one wireless network to another. But, if we consider that the four external grids are different providers in the same country, then the central grid is a backbone and the grids geographically cover the same area. Thus, a mobile node can attach at will to one or another provider or stay with the same during a communication. In contrast, a nearby mobility scheme authorizes a node to go only to the nearest and not previously visited wireless network. This represents a more classical point of view where a user walks continuously across a network. In our topology, the nearby mobility scheme will make a receiver move inside a single grid. In this short paper, we only show results using the nearby mobility scheme. Simulations using the random mobility scheme are available in the extended version [3].

We only simulated the mobility of receivers. The mobility of the source in SIM is less interesting because all the source has to do is to send packets with a new care-of address in the IP header and continue to use its Home Address in the SIM header. There is no Binding Update exchange with the receivers. As there is no flow establishment delay, the handoff time is kept to a minimum.

We first compare the scalability of PIM-SSM and SIM when no mobility occurs. Figure 2 shows the average amount of state per router for multicast flows (the total amount of state stored in the routers divided by the number of routers in the topology) as a function of the simulation time. We performed 10 simulations with receivers placed at different random places in the beginning of the simulation and we report the percentile 50 of these 10 simulations for each protocol.

We see that PIM-SSM has only one phase. Before the source starts to send packets at 100.0 s, receivers join the tree at 95.0. State exists in the network even if there is no packet sent by the source. The tree is stable until receivers leave at the end. In contrast, SIM has two phases: in the first one, all routers on the path have forwarding state until the packet is converted to unicast. During this phase, SIM Redirects are sent upstream to skip non-branching routers. After 10.0s, the states in non-branching routers time out because they are not used anymore.

Clearly, in terms of state maintenance, SIM outperforms PIM-SSM for two reasons : when timeout occurs on non-branching routers (at 110.0s), state only remains on branching routers which are far less numerous than non-branching routers. The second reason is that conversion to unicast when only one desti-nation remains is useful since the group is small (10 receivers for 416 routers): from 100.0 to 110.0 s, SIM has 2.5 times less state than PIM-SSM on average.

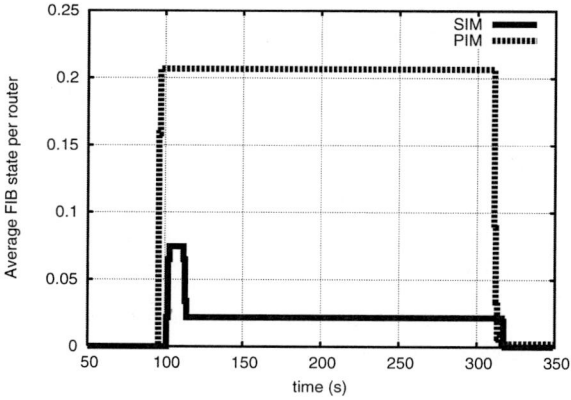

Fig. 2. Average amount of state in function of time with static receivers in PIM

Now, we evaluate the impact of a nearby mobility scenario on the amount of state to be maintained by each router. Figure 3 summaries the benefit of SIM over PIM in terms of average state stored per router. It represents the cumulative distribution of the average FIB state per router, which has the advantage of being independent of the time. Each curve is computed from ten different experiments. Each experiment lasted 200s during which 2000 packets were sent. The different curves represent the outcome for different values of the delay between two handoffs. It means that every 2 seconds (or 5 s, 10 s, ...), a randomly chosen node is moved to a nearby wireless network in the multi-grid topology. The more the curves are on the left handside of the graph, the better it is, because it means that during most of the time not much state is stored in the whole network.

SIM behaves quite well for all experiments compared to PIM-SSM, but we see that it requires more state when mobility increases. This is due to the fact that each time a membership change occurs, the source has to recreate a new flow identifier and therefore recreate new states in the network. As state is not immediately deleted for smooth transition reasons, there can be two states for a single flow during handoffs and this increases the average state.

We also see the impact of the timeout of 10 s in non-branching routers by the jump around 0.02 for inter-handoff delays greater than 10 s. It means that this value is crucial to the overal performance of the protocol and that during most of the time, the average FIB state per router is equal to this low value. When mobility increases, we have an overhead due to constant tree reconstruction and stabilization and the time needed by the Binding Update to arrive to the source becomes problematic. It may seem quite a burden to manage mobility, but remember that layer-3 mobility happens less often than layer-2 mobility and that we chose not to hide the mobility with tunnels.

PIM on the other hand is quite stable for different values of delays between two handoffs (only the 2s and the 70s curve are shown for this reason). Not much operations are needed to adapt the tree to the mobile receivers and to prune deprecated branches of the tree. With a nearby mobility scheme, the tree is never far to rejoin. The average is overal greater than in SIM because PIM creates state on each router from the source to the receivers.

The conversion to unicast when there is only one active receiver in a packet also helps SIM a lot. Imagine users sparsely disseminated on the Internet : the multicast packet will probably soon be split. It is therefore useless to use multicast routing mechanisms when only one destination remains downstream.

4   Conclusion

In this paper, we have shown how to support mobility in SIM without using tunnels to keep the shortest path from the source to the receivers. Scalability in multicast routing means that overhead in packets and the amount of states stored in routers must be kept minimal. SIM adds more overhead in packets than PIM but far less than Xcast because the receiver list is rarely present in the packets. As regards state maintenance, we have shown through simulations that

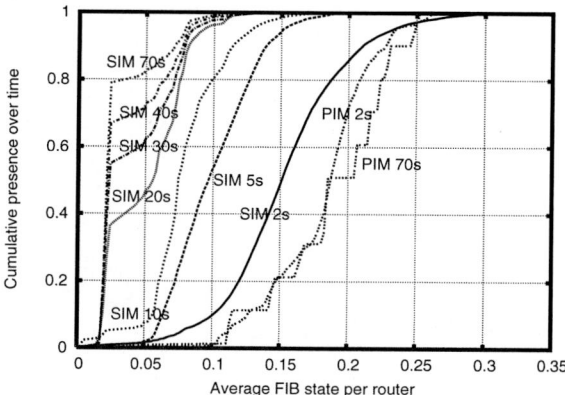

Fig. 3. Cumulative distribution of the average FIB state count for several delays between two handoffs in a nearby mobility scenario

SIM requires fewer state than PIM. Mobility obviously deteriorates performance but it is still acceptable if mobility is not too high. After 10 s of stability, state in non-branching routers times out and only branching routers have to perform SIM routing when the receiver list is not present in the packets.

We think that SIM fills a gap in the multicast landscape and provides an interesting trade-off between large group and small group multicast. With further developments, simulations and standardization, it could become a viable protocol.

There is still place for a lot of development on SIM. Currently, SIM is limited by its Hello protocol which forces SIM routers to be directly connected if we want to avoid premature unicast conversion. We are working on extensions to BGP and OSPF to support wide-scale SIM routing.

References

1. S. Bhattacharyya. An Overview of Source-Specific Multicast(SSM) Deployment. Internet RFC, RFC 3569, July 2003.
2. R. Boivie, N. Feldman, Y. Imai, W. Livens, D. Ooms, and O. Paridaens. Explicit Multicast (Xcast) Basic Specification. Internet Draft, draft-ooms-xcast-basic-spec-05.txt (work in progress), January 2003.
3. X. Brouckaert. Efficient mobility support in large and small group multicast routing protocols. Research Report RR 2004-07 INFO, June 2004.
4. X. Brouckaert. Mobile multicast simulator. http://www.info.ucl.ac.be/~xbr, June 2004.
5. S. Deering. Host Extensions for IP Multicasting. Internet RFC, RFC 1112, August 1989.
6. Diot, Levine, Lyles, Kassem, and Balensiefen. Deployment Issues for the IP Multicast Service and Architecture. IEEE Networks Magazine's Special Issue on Multicast, January 2000.

7. S. Egger and T. Braun. Multicast for small conferences : a scalable multicast mechanism based on IPv6. IEEE Communications Magazine Vol. 42 no. 1, January 2004.

8. A. El-Sayed, V. Roca, and L. Mathy. A Survey of Proposals for an Alternative Group Communication Service. IEEE Network Magazine special Issue on Multicasting: An Enabling Technology, January/February 2003.

9. D. Estrin, D. Farinacci, A. Helmy, D. Thaler, S. Deering, M. Handley, V. Jacobson, C. Liu, P. Sharma, and L. Wei. Protocol Independent Multicast-Sparse Mode (PIM-SM). Internet RFC, RFC 2362, June 1998.

10. D. Johnson, C. Perkins, and J. Arkko. Mobility Support in IPv6. Internet RFC, RFC 3775, June 2004.

11. D. Ooms. Taxonomy of Xcast/SGM proposals. Internet Draft, draft-ooms-xcast-taxonomy-00.txt (work in progress), July 2000.

12. Romdhani, Kellil, and Lach. IP Mobile Multicast: Challenges and Solutions. IEEE Communications Society Surveys and Tutorials Vol. 6 no. 1, 2004.

13. R. Vida and L. Costa. Multicast Listener Discovery Version 2 (MLDv2) for IPv6. Internet RFC, RFC 3810, June 2004.

14. V. Visoottiviseth, H. Kido, Y. Takahashi, and N. Demizu. Sender Initiated Multicast (SIM). Internet Draft, draft-vasaka-xcast-sim-01.txt (work in progress), March 2003.

15. Hung ying Tyan. J-sim. http://www.j-sim.org/.

A Comparison of Opportunistic Scheduling Algorithms for Streaming Media in High-Speed Downlink Packet Access (HSDPA)

Arsalan Farrokh[1], Florian Blömer[2], and Vikram Krishnamurthy[3]

[1] University of British Columbia, Vancouver, BC, Canada.
arsalanf@ece.ubc.ca
[2] Munich University of Technology, Germany
[3] University of British Columbia, Vancouver, BC, Canada.
vikramk@ece.ubc.ca

Abstract. High-Speed Downlink Packet Access (HSDPA) is the release 5 extension of WCDMA standard which provides high data rates (up to 10.8 Mbps) by using Adaptive Modulation/Coding (AMC) and fast packet scheduling. This paper presents a comparison of *opportunistic* scheduling algorithms for streaming media in HSDPA. We first present a discrete event model for HSDPA multimedia system and express stochastic QoS constraints that reflect the requirement for uninterrupted play-out. Next, we present a general structure of the opportunistic scheduling policies which exploit channel and/or buffer content variations for achieving the required QoS. By using computer simulations, we compare the performance of several special cases of the general policy in terms of the maximum number of the users that can be supported with the desired QoS constraint.

1 Introduction

The IMT-2000 standards for third generation (3G) wireless networks, released in 1999, promise high bandwidth efficiency for supporting a mix of real-time multimedia and high data rate traffic. Three of the five standards proposed in IMT-2000, including UMTS (the European contribution), are based on Wideband-CDMA. Though, WCDMA/UMTS specifications fully meets the IMT-2000 requirements, there is still an increasing demand for much higher downlink data rates along with a better QoS (Quality of Service). In order to meet these demands, a new packet concept, called High-Speed Downlink Packet Access (HS-DPA) was recently published in Release 5 of 3GPP UTRA-FDD specifications. HSDPA is an extension of the UMTS standard that provides an improved peak data rate and an increased throughput for world-wide cellular systems. The main features that collectively describe HSDPA are: Adaptive Modulation and Coding (AMC) schemes, Turbo codes, higher order modulation (16-QAM), CDMA multi-code operation, fast physical layer hybrid ARQ and short transmission time intervals. These features enable HSDPA to support transmission rates of

V. Roca and F. Rousseau (Eds.): MIPS 2004, LNCS 3311, pp. 130–142, 2004.

up to 10.8 Mbit/s [1] and allow high-rate downlink streaming for large number of users in a cellular system. HSDPA is a time-slotted CDMA system and requires a suitable scheduling policy to achieve the desired performance. Scheduling methodologies are not specifically defined as part of the HSDPA standard. This motivates the need to develop efficient scheduling algorithms that exploit wireless channel and/or buffer variations to achieve the required QoS.

In this paper we consider opportunistic scheduling algorithms to achieve smooth playout in HSDPA multimedia system. The term opportunistic refers to the fact that user channel and/or buffer information are exploited to schedule the service to the relatively better users (in terms of certain performance measure such as throughput) at any time slot.

In this section we first provide a brief literature review on the subject and then summarize the results of this paper.

1.1   Literature Review

Several opportunistic scheduling policies are studied in the literature. In this paper we present a single framework for these policies and customize them for the application of streaming media in HSDPA. In [6],[7], opportunistic scheduling methods are used for a general time-slotted wireless system to optimize the cell throughput while maintaining certain fairness among the users. Two categories of fairness in terms of minimum-throughput guarantee and specific shares of scheduling times are considered. However, in these papers there is no assumption on the status of the user buffers. In [3], the problem of scheduling CDMA users (one user at a time) in downlink with variable channel conditions is considered. The authors have proposed an algorithm denoted by Modified-Largest-Weighted-Work-First (M-LWWF) that considers both channel state information and user buffer state information for scheduling decisions. It is proven analytically that M-LWWF algorithm is capable of keeping the user queues stable (if stability is feasible). M-LWWF algorithm in general guarantees the QoS but may not achieve the system capacity. A general approach to support QoS of multiple real-time time data users is proposed in [4]. The scheduling algorithm is similar to [3] and can be used to maximize the number of users that can be supported with the desired QoS.

1.2   Main Results

The main results of this paper are organized as follows:
(i) In Section 2, we introduce a discrete event stochastic model for streaming users in HSDPA. The key features of HSDPA system as described in [1] and 3GPP standards are included in the proposed model. By using the discrete event model, we formulate the requirement for uninterrupted playout as stochastic QoS constraints.
(ii) In Section 3, we present a general structure for opportunistic scheduling policies that exploits channel and queue state variations. A variety of opportunistic scheduling algorithms will be considered by appropriately selecting the

parameters of the general structure. The properties of the scheduling policies will be discussed and compared in terms of optimality and/or feasibility aspects. **(iii)** In Section 4, the HSDPA transmission system for streaming users across a finite state Markovian fading channel is simulated. Two different scenarios in terms of relative locations of the users in a single wireless cell will be considered. In our simulations the relation of SNR to FER is extracted from the data provided by Nokia research [1]. Using the simulation results, the performance of the scheduling policies of Section 3 will be compared in terms of the maximum number of the users that can receive the service with the required QoS.

2 HSDPA Streaming System Model

In this section a formal stochastic model for streaming users in HSDPA is presented. We consider L users and define $\mathcal{I}$ to be the set of all users as: $\mathcal{I} = \{1, 2, \dots, L\}$.

Remark: Throughout this paper, The term Node-B refers to the Base-Station and the term UE refers to the User-Equipment (i.e., mobile station). Also note that the terms "buffer" and "queue" may be used interchangeably in this paper, however the former refers to a physical entity while the latter describes a mathematical model.

It is convenient to outline our HSDPA streaming model in terms of the following elements [1]:

Transmission Time Interval (TTI): Time is the resource that is shared among the HSDPA users. The time axis is divided into slots of equal duration referred to as Transmission Time Intervals (TTI). Let ΔT be the duration of one TTI. By convention, time slot (or discrete time) k, $k \in \mathcal{Z}_+$ is the time interval $[k\Delta T, (k+1)\Delta T)$. We assume only one user is scheduled at each time slot and scheduling decisions are made at times $k\Delta T$, $k \in \mathcal{Z}_+$. Here, $\mathcal{Z}_+ = \{0, 1, 2, \dots\}$ is the set of non-negative integers.

Power: Fast power control feature is disabled in HSDPA standards (see 3GPP UTRA-FDD in [5]). Therefore, in our model we assume that transmission power is fixed for all time slots.

Channel: Channel quality for user i at time k is characterized by the average received SNR of user i at the k'th time slot. We consider a flat fading Rayleigh channel modeled by a Finite State Markov Process (FMSP). The FMSP models have been widely used to represent continuous channel processes [2]. Define $c_i(k) \in S$ as:

$$c_i(k) \triangleq \text{Average SNR of user } i \text{ at time } k, \tag{1}$$

where, the channel space state S is obtained by partitioning the received instantaneous SNR into M intervals as:

$$S = \{s_1, s_2, \dots, s_M\} \tag{2}$$

Assume $c_i(k)$ is an irreducible discrete time finite state Markov chain with a unique stationary distribution. The Channel Quality Indicator (CQI) at time k is defined as vector $\boldsymbol{c}(k)$:

$$\boldsymbol{c}(k) = (c_1(k), c_2(k), \ldots, c_L(k)) \tag{3}$$

where L is the number of the users and $c_i(k) \in S$ represents the channel state (SNR level) of user i at time k. Note that $c_i(k)$, $i \in \mathcal{I}$ can be generally correlated across the users. Furthermore, we assume that CQI vector is measured by the users and reported to the the Node-B at each time slot. Therefore, the Node-B has a perfect knowledge of vector $\boldsymbol{c}(k)$ for $k \in \mathcal{Z}_+$.

Adaptive Modulation and Coding (AMC): In HSPDA, different modulation and error-correcting coding techniques can be dynamically selected (i.e., in real time) from a set of Modulation and Coding Schemes (MCS) [1]. We assume each MCS (Modulation and Coding Scheme) is represented by a MCS number such that MCS number belongs to the set $\mathcal{M} = \{1, 2, \ldots, |\mathcal{M}|\}$, where $|\mathcal{M}|$ denotes the cardinality (number of elements) of set $\mathcal{M}$. In case of scheduling user i at time k, define $m_i(k) \in \mathcal{M}$ as the MCS number of user i at time k. The higher the MCS number, the higher is the data rate (but the higher SNR is required as well).

Multi-Code Operation: In HSDPA, multi-code CDMA is used at each time-slot for transmission to the scheduled user. By choosing a larger number of spreading codes, higher data rates can be achieved. However, higher SNR is required for transmission with larger number of codes. Depending on the UE capabilities, different number of simultaneous spreading codes can be selected (e.g. 5, 10, 15). Let $\mathcal{N}$ be the set of possible spreading codes. In case of scheduling user i at time k, define $n_i(k) \in \mathcal{N}$ as the number of the codes assigned to user i at time k.

Discrete Transmission Rate Set: We refer to the rate chosen by the Node-B to transmit the data to the UE as transmission rate. The transmission rate at each time slot is fixed and chosen from a finite set of possible transmission rates by the Node-B. Define:

$$r_i(k) \stackrel{\triangle}{=} \text{The transmission rate to user } i \text{ at time } k. \tag{4}$$

Let R_T be the finite set of all possible transmission rates from the Node-B to the UE so that:

$$r_i(k) \in R_T, \quad i \in \mathcal{I}. \tag{5}$$

We assume that a unique transmission rate is associated with any combination of MCS (Modulation and Coding Scheme) and Multi-Code (for example see Table 1 in Section 4).

Physical-Layer Hybrid ARQ (H-ARQ): Automatic Repeat Request (ARQ) is one of the error-control procedures used in data communication systems. If the user receiver detects transmission errors in a frame, it will automatically request a retransmission from the Node-B. The Node-B retransmits the frame until it receives an acknowledgement(ACK) from the receiver or the error persists beyond a predetermined number of retransmissions. By a frame of data we mean the data that is delivered at any time slot (TTI). Note that the length of a frame in terms of bits is varying due to the different data rates at different time slots. H-ARQ (Hybrid-ARQ) uses a code with more redundancy that will correct some frame errors and hence reduces the average number of retransmissions. H-ARQ operation in HSPDA is relatively fast since it is performed between the base station and the user equipment in the physical-layer. Define $f(\cdot)$ to be a function that gives the Frame Error Rate (FER) for any combination of MCS, multi-code and channel state. Let $f_i(k)$ be the FER for user i at time k. We have

$$f_i(k) = f\left[m_i(k), n_i(k), c_i(k)\right] \triangleq \text{FER at time } k \text{ for the scheduled user } i. \quad (6)$$

Per-TTI (instantaneous) capacities: In order to quantify the achievable throughput to user i at time k, we consider a notion of per-TTI (instantaneous) capacity that incorporates the event that a frame error occurs. Define a random variable $\nu_i(k)$ to be the user i per-TTI capacity at time k. This means, in the event of scheduling user i at time k, user i receives $\nu_i(k)\Delta T$ bits at time slot k (or the entire queue i content whichever is smaller). In terms of transmission rate, we have:

$$\nu_i(k) = r_i(k)\mathbf{1}_{\{\text{No Frame Error at time } k\}}, \qquad i \in \mathcal{I}, \quad (7)$$

where $\mathbf{1}_{\{A\}}$ is the indicator function of the event A (it is equal to 1 if A occurs, otherwise is zero). Let $\tilde{\nu}_i(k, m, n)$ be the unbiased minimum variance estimate of $\nu_i(k)$, where m and n are the selected values of MCS and multi-code, respectively. $\tilde{\nu}_i(k, m, n)$ is then given by subtracting the frame error rate from the transmission rate:

$$\tilde{\nu}_i(k, m, n) \triangleq \mathbf{E}\left(\nu_i(k)|m(k) = m, n(k) = n, c_i(k)\right) \quad (8)$$
$$= r_i(k)[1 - f(m, n, c_i(k))], \quad (9)$$

where $r_i(k)$ and $f(\cdot)$ are defined in (4) and (6), respectively. $\tilde{\nu}_i(k, m, n)$ can be considered as the measure for per-TTI capacity or the achievable bit rate for user i at time k. By selecting the optimal MCS and multi-code, the maximum achievable rate at time k, denoted by $\mu_i(k)$, is obtained:

$$\mu_i(k) = \max_{(m,n)\in\mathcal{M}\times\mathcal{N}} \tilde{\nu}_i(k, m, n) \quad (10)$$

In practice the above maximization may not be possible since the exact form of function f is not known. However, a popular method for selecting MCS and multi-code based on the channel SNR and the relative thresholds is discussed in [1]. The objective is to choose an optimal MCS and multi-code in terms of certain system performance measure (for example (10) gives the optimal throughput). In our simulations, based on the available data and measurements from Nokia research, nearly optimal MCS and multi-code are always chosen [1].

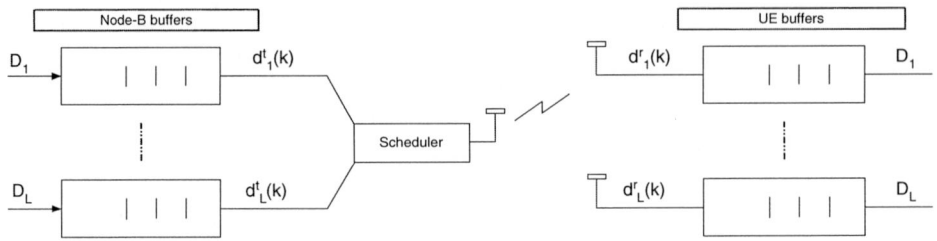

Fig. 1. Wireless Streaming Media System Model

Buffers and Streaming Rates: Figure 1 shows a queuing model for streaming user i. The streaming system is modeled with symmetric buffers, i.e., each UE buffer has an equivalent buffer in the Node-B with the same size. The data in the buffer of UE i is discharged with the constant rate D_i which is referred to as the playout rate of user i. The arriving data rate to the Node-B buffer (incoming rate from the server) of user i is also equal to D_i. The number of existing bits (unfinished work in the queue) in buffer i of the Node-B at time k is denoted by $V_i(k)$. Similarly, the number of bits in the buffer of UE i at time k is denoted by $U_i(k)$. Let B_i be the size of the buffer i of the Node-B. The relation between $V_i(k)$ and $U_i(k)$ is then given by:

$$V_i(k) + U_i(k) = B_i, \quad i \in \mathcal{I}, \tag{11}$$

where we have assumed $V_i(0) = 0$ and $U_i(0) = B_i$. Let $d_i^r(k)$ be the arriving data rate to the buffer of UE i at time k. Also let $d_i^t(k)$ be the rate that the data is discharged from buffer i of the Node-B at time k. By convention, we assume if a frame error occurs in time slot $k = s$ then $d_i^t(s) = 0$. This is reasonable since the Node-B does not transmit a new frame to the UE till it receives an ACK confirming a successful previous transmission. The effective transmission rate at time s is then zero because the same packet will be retransmitted at time $s + 1$. We conclude that $d_i^t(k) = d_i^r(k)$ and define $d_i(k) \triangleq d_i^t(k) = d_i^r(k)$. The evolution of the Node-B and the UE buffer content for user i is then given by the two following equations:

$$V_i(k + 1) = V_i(k) - d_i(k)\Delta T + D_i\Delta T \tag{12}$$
$$U_i(k + 1) = U_i(k) + d_i(k)\Delta T - D_i\Delta T \tag{13}$$

States and Scheduling Policy: The buffer content and the channel state of the users jointly constitute a Markov chain (i.e., they stochastically characterize the buffer content of the users at time $k + 1$). Therefore, the state of the system can be defined as $\boldsymbol{x}(k) \triangleq (x_1(k), x_2(k), \dots, x_L(k))$ where:

$$x_i(k) = (V_i(k), c_i(k)) \tag{14}$$

A scheduling policy Q is defined as a function (mapping) of the state space to user numbers. The mapping

$$Q(\boldsymbol{x}(k)) = i \tag{15}$$

means that user $i \in \mathcal{I}$ will be scheduled at time k if the state is reported to be $\boldsymbol{x}(k)$. Note that only one user is scheduled at each time slot.

2.1   Formulation of QoS Requirement

To achieve uninterrupted smooth real-time streaming, the situation of an empty or under-run buffer must be avoided. More precisely, the probability of having an under-run buffer at each time slot needs to be smaller than a certain threshold. This constraint can be formulated as:

$$\Pr\{U_i(k) \leq \theta_i\} \leq \delta_i, \quad i \in \mathcal{I}, \tag{16}$$

where θ_i is a threshold level for the buffer of UE i and δ_i is the threshold probability for UE i. Using (11) an equivalent constraint can be written for the Node-B buffers:

$$\Pr\{V_i(k) > \eta_i\} \leq \delta_i \quad i \in \mathcal{I}, \tag{17}$$

where $\eta_i = B_i - \theta_i$ is a threshold level for buffer i of the Node-B and δ_i is the threshold probability. Recall that B_i is the size of the Node-B (or the UE) buffer of user i.

3   Structure of the Opportunistic Scheduling Policy

We present a general opportunistic scheduling policy for HSDPA media streaming as follows:

$$Q(\boldsymbol{x}(k)) = \arg\max_{i \in \mathcal{I}} \gamma_i \frac{V_i(k)^\alpha \mu_i(k)^\beta}{\rho_i(k)^\epsilon \bar{\mu}_i{}^\kappa} \tag{18}$$

where $\mu_i(k)$ is given by (10) and $V_i(k)$ is the length (number of existing bits) of the buffer of user i in the Node-B at time k. γ_i, $i \in \mathcal{I}$ are positive real constants denoted by distribution factors (weights). If the the distribution factor of a user is increased, the scheduling priority for that user is increased as well. Therefore, distribution factors can be adjusted in a way to provide QoS fairness among the users [7]. $\bar{\mu}_i$ is the average capacity of user i in terms of bit/sec. $\rho_i(k) > 0$ is the reward factor assigned to user i packet at time k. $\rho_i(k)$ can be adjusted for each packet based on the assigned priority of the packet. α and β are non-negative real constants. We distinguish several cases based on different parameters of (18).

3.1   Channel Dependent Scheduling

The following algorithms use only channel information $c(k)$ in order to schedule a user at each time slot. These algorithms assume user data is always backlogged, i.e., for every user, there is always data available in the Node-B to send and thus the throughput is equal to the assigned bit rates. Usually, the channel-dependent scheduling is more appropriate for best-effort non-real time traffic that does not account for buffer contents and delay of the packets. Three of such policies are introduced in the following.

Maximum Signal to Interference (Max C/I): By setting $\beta = 1$, $\alpha = \epsilon = \kappa = 0$ and $\gamma_i = 1$, $i \in \mathcal{I}$, the policy in (18) reduces to

$$Q(\boldsymbol{x}(k)) = \arg\max_{i \in \mathcal{I}} \mu_i(k) \tag{19}$$

The above is a G reedy algorithm which simply serves a user at each time slot with the best available bit rate (capacity). As a result the throughput will be maximized but in the expense of starving the users with bad channels. The Max C-I algorithm can be modified by assigning different weights to different users so that certain fairness constraints are satisfied. The modified version is stated in the following.

Weighted Max C/I (Opportunistic Scheduling with Minimum Throughput Guarantee): By setting $\beta = 1$, $\alpha = \epsilon = \kappa = 0$, (18) gives:

$$Q(\boldsymbol{x}(k)) = \arg\max_{i \in \mathcal{I}} \gamma_i \mu_i(k) \tag{20}$$

It is shown in [7] that the above policy maximizes the average throughput while maintaining a minimum average throughput for each user (if possible at all). γ_i, $i \in \mathcal{I}$ are updated at each time slot in a way that the policy with the assigned weights satisfies the minimum throughput requirement for each user. However, the policy in (20) does not consider the short term throughput performance of the users (such as buffer content variations) which is an important aspect in the real-time streaming.

Proportional Fairness: By setting $\beta = 1$, $\alpha = \epsilon = 0$, $\kappa = 1$ and $\gamma_i = 1$, $i \in \mathcal{I}$, (18) gives:

$$Q(\boldsymbol{x}(k)) = \arg\max_{i \in \mathcal{I}} \frac{\mu_i(k)}{\bar{\mu}_i} \tag{21}$$

In proportional fairness scheduling the instantaneous channel capacity for each user is normalized to the average capacity of the user. Therefore, a user with the best channel relative to his own history will receive a higher priority in scheduling. This algorithm is discussed in [9] and also proposed in CDMA High Data Rate (HDR) standard [10]. However, since it does not account the buffer variations, it is shown that it may result in buffer-under-run situations [11].

3.2   Queue Dependent Scheduling

Largest-Weighted-Work-First (LWWF): Let $\beta = \epsilon = \kappa = 0$. (18) then gives the following policy:

$$Q(\boldsymbol{x}(k)) = \arg\max_{i \in \mathcal{I}} \gamma_i V_i(k)^\alpha \tag{22}$$

The only information used in the above algorithm is the queue content information. Consequently, it generally gives a poor throughput performance and may render the system unstable [11]. However, for the proposed real-time streaming problem, one might expect a better performance from (22) as compared with channel-dependent scheduling, specially for the cases where users have very different channel qualities.

3.3   Channel and Queue Dependent Scheduling

The following algorithm will take to account both the channel state and the queue content to control the buffer variations as well as achieving a possibly high throughput.

Modified Largest-Weighted-Work-First (M-LWWF): Let $\beta = 1$ and $\epsilon = \kappa = 0$. (18) then results in the following policy:

$$Q(\boldsymbol{x}(k)) = \arg\max_{i \in \mathcal{I}} \gamma_i V_i(k)^\alpha \mu_i(k) \tag{23}$$

Under the assumption that the channel is Markovian, it is shown in [3] that the above algorithm is throughput optimal which roughly means in the long run, it keeps all the user buffers non-empty (if possible with any other policy). Additionally, adjusting γ_i, $i \in \mathcal{I}$ will provide some degree of freedom to satisfy the probabilistic QoS constraint in (16). The algorithm in (23) can be viewed as a feasible solution (if any such solution exists) rather than an optimal solution. In general, one might expect this algorithm gives the best performance relative to channel-only or queue-only dependent scheduling (though it may not achieve the capacity). This is in fact the case as we observe the simulation results in the next section.

4   Numerical Results

The purpose of this section is to evaluate by numerical experiments the performance of several opportunistic policies in terms of the maximum number of the HSDPA multimedia users that can be supported with the desired QoS.

Fading channel: The channel is simulated by a Rayleigh flat fading model with a Doppler spectrum (Jake's model). Let $g_i(k)$ be the available SNR for user i at time k. Based on the Jake's model, $g_i(k)$ can be written as the transmitted SNR scaled by a fading component and a path-loss component. Expressing in dB, we have $g_i(k)_{(dB)} = (P_t/P_n)_{(dB)} - L(R_i)_{(dB)} + \{\alpha_i(k)^2\}_{(dB)}$, where P_t is the transmission power, P_n is the noise power (includes interference as well), $L(R_i)$ is the path-loss component in terms of the user i distance R_i and $\alpha_i(k)$ is the Rayleigh fading gain process for the user i (see Table 2). Furthermore, in order to apply our proposed algorithm, we model the simulated fading gain process $\alpha_i(k)$ (or equivalently $g_i(k)$) by a 10-state discrete-time Markov process.

M C S and P aram eters: In our simulations, the MCS and multi-code selection is based on the data provided from Nokia research [1]. Table 1 shows the numerical values for MCS number and spreading codes of the users and the corresponding transmission rates. In Table 2 we summarize the parameters and assumptions used for our simulations. We have tried to choose realistic parameters from common simulation assumptions for HSDPA provided by Ericsson, Motorola and Nokia in [8].

Table 1. Modulation and Coding Schemes

MCS	Modulation	Code Rate	Data Rate (1 Code)	Data Rate (15 Codes)
1	QPSK	$\frac{1}{4}$	0.6 Mbps	1.8 Mbps
2	QPSK	$\frac{1}{2}$	1.2 Mbps	3.6 Mbps
3	QPSK	$\frac{3}{4}$	1.8 Mbps	5.3 Mbps
4	16QAM	$\frac{1}{2}$	477 kbps	7.2 Mbps
5	16QAM	$\frac{3}{4}$	712 kbps	10.8 Mbps

Table 2. Simulation Parameters

Cell radius R_{cell}	500m
Max. number of users L	15
Duration of time slot ΔT	2 ms
Spreading factor	16
Channel estimation	Perfect
Carrier frequency	2 GHz
Chip rate	3.84 Mcps
FER_{max}	0.1
Transmission SNR: $\frac{P_t}{P_n}$	122dB
Fast fading model	Rayleigh-fading
Propagation model: $L(R)$	$128.1 + 37.6 \log(R)$
Channel model	10 State Markov chain
UE buffer time constant t_B	1 second (for all users)
Simulation duration	150000 ΔT (5 min)

The simulation always starts with full UE buffers. Two scenarios are considered in terms of distribution of the users in a single wireless channel. For both scenarios we determine the maximum number of users that can be supported with smooth-playout probability of $p_{success} \geq 90\%$ at a given streaming rate.

Therefore, in the QoS formulation in (16), we roughly have: $\theta_i = 0$ and $\delta_i = 0.1$ for $i \in \mathcal{I}$.

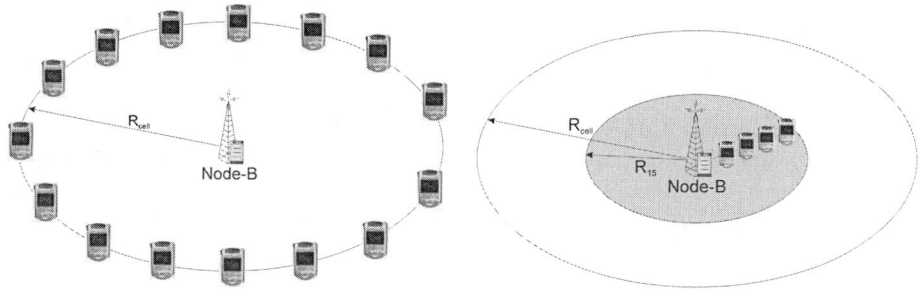

Fig. 2. Scenario 1   **Fig. 3.** Scenario 2

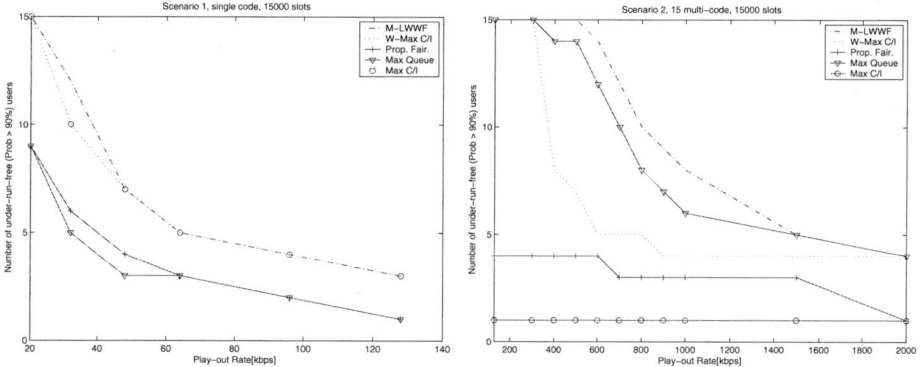

Fig. 4. Results for Scenario 1   **Fig. 5.** Results for Scenario 2

Scenario 1: The first scenario is the worst case scenario. All users are located on the cell edge ($R = R_{cell} = 500$m, see Figure 2) and consequently have relatively poor channels which only support low data rates. In this scenario single-code transmission is used.

As shown in Figure 4, in scenario 1, M-LWWF gives the best performance, however, Max C/I and W-Max C/I perform exceptionally well because of the given scenario: all users have the same distance to the Node-B with the same streaming rate. Therefore it is a relatively good solution to only pick the instantaneously best channel. Prop-Fair and Max-Queue provide lower performance compared to the other algorithms.

Scenario 2: In this scenario we use 15 multi-codes for transmission and this way increase our transmission rates significantly. Now we concentrate our simulation on the cell center where high data rates can be achieved. The distances of the users are distributed between the Node-B and the edge of the cell center $(10m \leq R \leq 250m)$ and $R_i = \frac{i}{L-1} \cdot (R_{cell} - l_{min}) + l_{min}$ where l_{min} is the minimum distance a UE has from the Node-B. It is possible to support high streaming rates to a reasonable number of users in the cell center. Again we see in Figure 5 the superiority of the M-LWWF algorithm. The Max-Queue algorithm also provides a high number of simultaneous users with large streaming rates. Prop.-Fair and Max C/I show very poor performance. The Max C/I algorithm completely fails in this scenario, because users that are further away from the Node-B are not served at all. We can see in Figure 5 that the Prop.-Fair algorithm performs slightly better than the Max C/I scheme because it serves the users with a relatively good channel compared to their channel history. Also note that by assigning appropriate distribution constants, W-Max C/I gives significantly better performance than Max C/I.

5   Conclusion

In this paper we propose a discrete event model for streaming media in HSDPA and formulate the requirement for smooth media playout as stochastic QoS constraints. We then present a single framework for opportunistic user-scheduling policies customized for HSDPA streaming system. We compare several opportunistic policies in terms of the maximum number of users that can be supported with the desired QoS. We conclude that the best result is achieved by trading-off the optimality of the algorithm to satisfy short-term QoS constraints. In this case the opportunistic policies (e.g. M-LWWF) use both channel and queue information to schedule HSDPA users.

References

1. T. E. Kolding, F. Frederiksen and P. E. Mogensen, "Performance Aspects of WCDMA Systems with HSDPA," in *IEEE 56th VTC*, vol. 1, (Vancouver, BC, Canada), pp. 477–81, 2002.
2. H. S. Wang and N. Moayeri, "Finite-State Markov Channel - A Useful Model for Radio Communication Channels," *IEEE Transactions on Vehicular Technology*, vol. 44, pp. 163–171, February 1995.
3. M. Andrews, K. Kumaran, K. Ramanan, A. Stolyar, R. Vijayakumar, and P. Whiting, "CDMA Data QoS Scheduling on the Forward Link with Variable Channel Conditions," *Bell Labs Technical Memorandum*, 2000.
4. K. R. A. S. P. W. M. Andrews, K. Kumaran, "Providing Qulaity of Service over a Shared Wireless Link," in *IEEE Communications Magazine*, pp. 150–154, February 2001.
5. "Http://www.3gpp.org,"
6. E. C. X. Liu and Shroff., "Opportunistic Transmission Scheduling with Resource-Sharing Constraints in Wireless Networks," *IEEE Journal on Selected Areas in Communications*, vol. 19, pp. 2053–2064, October 2001.

7. E. C. X. Liu and Shroff., "Transmission Scheduling for Efficient Wireless Resource Utilization with Minimum-Utility Guarantees," in *Proceedings of the IEEE VTC Fall 2001*, (Atlantic City, USA), pp. 824–828, October 2001.
8. Ericsson, Motorola, and Nokia, "Common HSDPA system simulation assumptions," in *3GPP TSG RAN WG1 Meeting #15*, (Berlin, Germany), 2000. TSGR1#15(00)1094.
9. R. J. F. Berggren, S. L. Kim and J. Zander., "22," in *IEEE Journal on Selected Areas in Communications*, vol. 19, pp. 1860–1870, October 2001.
10. P. Bender, P. Black, M. Grob, R. Padovani, N. Sindhushayana, and A. Viterbi, "CDMA/HDR: A Bandwidth-Efficient High-Speed Wireless Data Service for Nomadic Users," *IEEE Communications Magazine*, vol. 38, pp. 70–77, July 2000.
11. S. Shakkottai and A. L. Stolyar, "Scheduling Algorithms for a Mixture of Real-Time and Non-Real-Time Data in HDR," in *Proceedings of the 17th International Teletraffic Congress (ITC-17)*, (Salvador, Brazil), September 2001.

Adaptive Scheduling for Improved Quality Differentiation

Johanna Antila and Marko Luoma

Networking Laboratory of Helsinki University of Technology
Otakaari 5A, Espoo, FIN 02015, Finland
Tel: +358 9 451 6097, Fax: +358 9 451 2474
http://www.netlab.hut.fi/~jmantti3
{johanna.antila,marko.luoma}@hut.fi

Abstract. In this paper we compare the performance of static and adaptive provisioning methods with ns2-simulations. For the static provisioning case we use capacity as a provisioning parameter and for the adaptive case we use packet delay. The scheduling algorithms that we investigate are Deficit Round Robin (DRR) and delay-bounded HPD (Hybrid Proportional Delay), which is our own version of the HPD algorithm. According to our results the delay-bounded HPD algorithm is better able to achieve the targeted provisioning goal than the static DRR algorithm regardless of the load level, application mix or queue management method used.

1 Introduction

During the last decade the Internet has developed from a research network into a commercial multi service network serving heterogeneous applications (e.g. VoIP, WWW, FTP, video-conferencing) and customers. As a result, Quality of Service (QoS) provisioning has emerged as one of the most crucial problems in the Internet research. Different service architectures have been proposed for realizing QoS. Currently, the class-based Differentiated Services (DiffServ) [3] architecture seems to be the most promising solution due to its simplicity and scalability.

QoS provisioning requires new kinds of functionality from the network routers. The router has to divide the link resources between service classes and it may support class based routing or load balancing. One of the most important components in the resource allocation is a packet scheduler that determines the service order of the packets. In current routers the scheduling algorithms are static and perform the resource allocation based on the estimated traffic loads offered to different classes and the required quality level. The estimate is usually based on traffic history, from which an average load is calculated in order to predict the future load. However, in reality the loads of different classes vary quite much at a short timescale due to traffic bursts and also at a longer timescale due to traffic trends. If the resource allocation is performed in a static manner, the scheduling algorithm will not be able to adapt to dynamic load conditions. In this

V. Roca and F. Rousseau (Eds.): MIPS 2004, LNCS 3311, pp. 143–152, 2004.

paper we investigate how this problem could be solved with adaptive scheduling algorithms. The basic idea of adaptive scheduling algorithms is to dynamically adjust the class resources either periodically or on a packet per packet basis, so that the policy chosen by the operator will be fulfilled regardless of the traffic conditions. In practice, adaptivity is achieved by utilizing measurements.

In [1] we have already studied the performance of different scheduling algorithms and traffic mapping principles with relatively high abstraction level simulations. We concluded that adaptive scheduling algorithms lead to more consistent quality differentiation than static algorithms. We also identified that proper classification of traffic is at least as important as the resource allocation. The best results could be achieved when the traffic was divided into four classes: first class for real-time applications sending short packets (VoIP etc.), the second class for real-time applications sending larger packets and more bursty traffic (Video etc.), the third class for applications sending short TCP flows (Web etc.), and the fourth class for applications sending long TCP flows (FTP etc.).

In this paper we continue our previous work by comparing the performance of static and adaptive provisioning methods with ns2-simulations in a more representative network topology and with more realistic traffic models. The scheduling algorithms that we investigate are Deficit Round Robin (DRR) [7] and delay-bounded HPD (Hybrid Proportional Delay) [1].

2   Provisioning Methods and Schedulers

Provisioning at the packet level can be based on different quantities such as capacity, packet delay or packet loss. We have chosen to use capacity in the static case and packet delay in the adaptive case. Provisioning capacity is trivial: each service class is simply allocated a predefined amount of the total link capacity. For delay provisioning, we have defined a combined model of absolute and proportional delay differentiation in [1]. In the combined model the most delay sensitive class is assigned with an absolute delay bound. If this bound is about to be violated, a packet is directly dispatched from this class, otherwise the operation is based on the delay ratios between classes. The idea is that the delay bound and the delay ratios together determine the resource allocation: the more time critical the traffic, the smaller the delays.

Also other approaches for adaptive provisioning have been suggested in the literature. Christin et al.([9]) have proposed a Joint Buffer Management and Scheduling (JoBS) mechanism that provides both absolute and proportional differentiation of loss, service rates and packet delay. In JoBS the buffer management and scheduling decisions are interdependent and are based on a heuristically solved optimization problem. Liao et al.([8]) have defined a dynamic core provisioning method that aims to provide a delay guarantee and differentiated loss assurance for the traffic classes. The algorithm utilizes measurement based closed-loop control and analytic models of an $M/M/1/K$ queue for the resource allocation decisions. Our approach is different from [9] and [8] in the sense that our algorithm is more simple and completely measurement based, requiring no optimization problems or analytic models to be solved. Furthermore, in our approach the scheduling and buffer management decisions are independent, allowing different queue management mechanisms to be combined with our algorithm.

2.1   DRR

DRR is a static, frame based scheduling algorithm that aims to emulate the ideal Generalized Processor Sharing (GPS) [6] algorithm. Also other scheduling algorithms for emulating GPS have been proposed, such as Worst Case Weighted Fair Queueing (WF^{2}Q) [2] and Self Clocked Fair Queueing (SCFQ) [5]. We chose to use DRR mainly due to its simple implementation and ability to take into account variable packet sizes. In DRR each class i is assigned with a weight ϕ_i. In each service round the scheduler divides a frame of N bits among the classes in proportion to these weights. The resulting number of bits reserved for a certain class is called a quantum. In DRR, each class is also associated with a deficit counter that keeps track of the unused quantum for the class from previous rounds. Thus, packets can be transmitted from a certain class as long as there are enough bits left either in the quantum or in the deficit counter.

2.2   Delay Bounded HPD

The delay bounded HPD algorithm first checks if the packet in the highest class is about to violate its deadline. Denote by d_{max} the delay bound in the highest class, by t_{safe} a safety margin for the delay bound, by t_{in} the arrival time of the packet in the highest class queue and by t_{curr} the current time. The packet in the highest class queue is considered to be violating its deadline if

$$t_{in} + d_{max} < t_{curr} + t_{safe}. \tag{1}$$

If delay violation is not occurring, the algorithm takes into account the delay ratios between the other classes. Denote by $\bar{d}_i$ the average queueing delay of class i packets and by δ_i the Delay Differentiation Parameter (DDP) of class i. The ratio of average delays in two classes i and j should equal the ratio of DDPs in these classes

$$\frac{\bar{d}_i}{\bar{d}_j} = \frac{\delta_i}{\delta_j}, \quad 1 \leq i, j \leq N, \tag{2}$$

assuming that $\delta_1 < \delta_2 < \ldots < \delta_N$. In [4] this is interpreted so that the normalized average delays of traffic classes must be equal, i.e.,

$$\tilde{d}_i = \frac{\bar{d}_i}{\delta_i} = \frac{\bar{d}_j}{\delta_j} = \tilde{d}_j, \quad 1 \leq i, j \leq N. \tag{3}$$

In order to meet the ratios, the algorithm selects for transmission at time t, when the server becomes free, a packet from a backlogged class j with the maximum normalized hybrid delay [4]:

$$j = \arg\max \left(g\tilde{d}_i(t) + (1 - g)\tilde{w}_i(t) \right), \tag{4}$$

where $\tilde{d}_i$ denotes the normalized average queueing delay of class i, $\tilde{w}_i$ denotes the normalized head waiting time of class i and $0 \leq g \leq 1$. Thus, the algorithm uses both short and long term queuing delays for the delay measurements.

In [4] the normalized average delay of class i at time t, $\tilde{d}_i(t)$, is calculated as follows: Denote by $D_i(t)$ the sequence of class i packets that have been served before time t and by d_i^m the delay of the m'th packet in $D_i(t)$. Then, assuming that at least one packet has departed from class i before t

$$\tilde{d}_i(t) = \frac{1}{\delta_i} \frac{\sum_{m=1}^{|D_i(t)|} d_i^m}{|D_i(t)|}. \tag{5}$$

It should be noted that this kind of calculation to infinity is not feasible in practice, since the counter for the sum of delay values easily overflows. Thus, we are developing new methods to estimate $\tilde{d}_i(t)$.

3 Simulations

3.1 Simulation Scenarios

We have implemented both the DRR and the delay bounded HPD algorithm in ns2 and conducted several simulations to evaluate their performance. Figure 1 shows the topology used in our simulations. It should be noticed that the topology parameters (link bandwidths and delays) do not necessarily represent

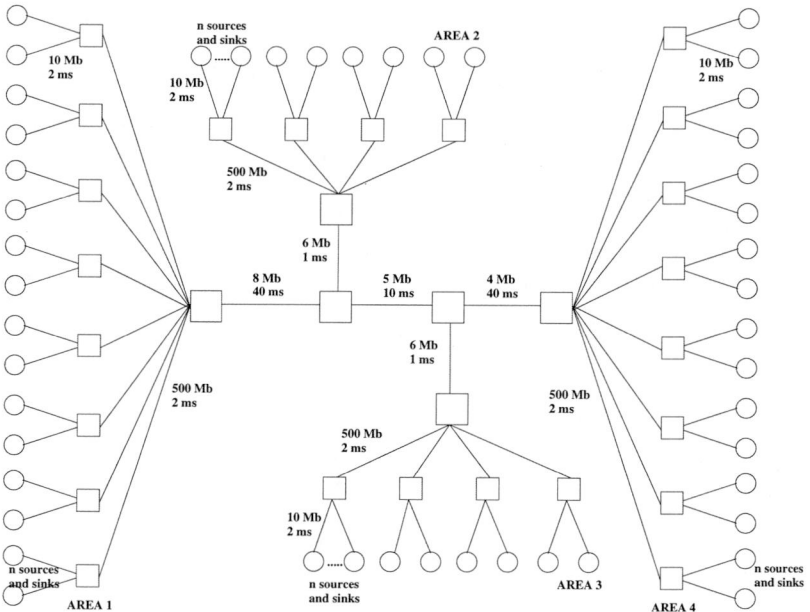

Fig. 1. Topology.

Table 1. Number of traffic sources in different areas

Application	Area1	Area2	Area3	Area4
FTP clients	4	2	2	4
HTTP clients	8	4	4	8
Video clients	4	2	2	4
VoIP clients	5	2	2	5
Control sources	2	2	2	2

any specific network technology. However, the topology and the parameters have been selected so that the most fundamental characteristics of real networks are captured: multiple bottleneck links, paths with low and high delay [1], cross traffic, bidirectional traffic etc. [10]. We have used five different traffic types in the simulations: FTP, HTTP, Video, VoIP and control traffic (small and large control messages). The generation of HTTP-traffic is based on the webcache-model implemented in ns2. In this model it is assumed that a HTTP session consists of a number of page requests, each possibly containing several objects. We have also used the webcache-model for FTP-traffic, except that in FTP there is no reading time between page requests, and a page always contains one object. Video traffic generation is based on a real trace of mpeg4 coded BBC news, from which the traffic stream has been regenerated by using a Transform Expand Sample (TES) process. In Figure 1 the topology has been divided into four separate areas. In the simulation, traffic flows from Area1 to Area4, Area2 to Area3, and correspondingly from Area4 to Area1 and from Area3 to Area2. The path between Area1 and Area4 is a high delay path, while the path between Area2 and Area3 has a low delay. Table 1 shows the number of FTP and HTTP clients, as well as video, voip and control traffic sources belonging to these areas. Notice that, for example, when there are 8 HTTP clients in Area1, there will correspondingly be 8 HTTP servers in Area4. In the simulation scenarios we have kept the simulation topology and the number of traffic classes fixed and changed the traffic load levels and the traffic shares of different applications. The mapping of traffic into different classes is performed according to the guidelines provided by our previous results that were presented in the introduction chapter. We used two scenarios for the relative traffic shares for different applications: (FTP: 9%, WWW: 71%, Video: 9%, VoIP: 10%, Control: 1%) and (FTP: 29%, WWW: 40%, Video: 20%, VoIP: 10%, Control: 1%). The first scenario corresponds to the situation today and the second scenario reflects the situation in the future, when the amount of multimedia and peer-to-peer traffic is expected to be larger. We have used three average total load levels (measured from the most congested bottleneck link) to test the algorithms: 75%, 80% and 85%.

As in [1], we dimensioned the total buffer size in the routers to be 230 packets, of which 200 packets was allocated for elastic traffic and 30 for real-time traffic. We tested the performance of the scheduling algorithms with TailDrop and with RED queue management in order to see if the differences between the scheduling

---

[1] Propagation delay is 98 ms in the long-delay path and 20 ms in the short-delay path.

algorithms are smaller when more advanced queue management methods are used. The parametrization of the DRR and the delay bounded HPD algorithm was similar as in [1]: In DRR the real-time traffic classes were over provisioned with a factor of two, and the excess capacity was divided between classes meant for elastic traffic in proportion to their expected load shares. In delay bounded HPD algorithm we set d_{max} to 5 ms and the target ratio for delays between consecutive classes to 4. The safety margin was set to be $d_{max}/10$ and g was set to be 0.875.

4   Simulation Results

We made five statistically independent iterations of each simulation scenario. The simulation time in each iteration was 3600s, including a 400s warmup period. From the simulations we measured the average aggregate throughput, packet loss and end-to-end delay for each traffic type. However, in this paper we will show only the most relevant results. In addition, for web and FTP we measured object throughput and its variability. Object throughput results are seldom presented in research papers. However, in our opinion object throughput is an important performance metric especially from the user's point of view, since the user constantly expects to receive some feedback on the proceeding of a document retrieval (i.e. the loading of independent objects).

In all the tables presented in this section the mean with a 95% confidence interval of the results are shown. The results have been averaged among all the clients belonging to the same edge area. To better asses the variability of the object throughput within a single iteration we have also recorded the relative standard deviation (rsd) of the object throughput. The rsd value reflects the peakedness of the object throughput, which is also a relevant performance metric for the end user expecting constant quality of service.

In Tables 2, 3 and 4 the results are shown for every algorithm combination with application mix 1 and with a mean load of 80%. The results with RED are shown only in Table 2, since the differences with TailDrop were so small. From Table 2 and Table 3 we can observe that the delay-bounded HPD algorithm is able to provide considerably better service for the HTTP-traffic in terms of packet losses and object throughput both in the high and the low delay paths. Also according to the rsd values the variability of the object throughput is smaller with delay-bounded HPD resulting in more constant quality of service from the user's perspective.

Naturally we can not provide considerably better service to one service class without degrading the performance of other classes. With delay-bounded HPD the performance of the FTP-traffic is worse than with the static DRR algorithm. However, since the FTP-traffic is mapped to the best-effort class that by default should not provide any quality of service we consider it more important to provide good service for the HTTP-traffic that is interactive and more delay sensitive. It should also be noticed that if we would want FTP to receive more resources from the delay-bounded HPD algorithm this could be achieved by setting the DDP values in a different way.

Table 2. Object throughput statistics, application mix 1

	Object throughputs (kbps)							
	EDGE1		**EDGE2**		**EDGE3**		**EDGE4**	
DRR-DROP	mean	rsd	mean	rsd	mean	rsd	mean	rsd
HTTP	14.6±2.0	1.0	48.2±7.9	1.1	48.6±7.9	1.1	14.6±2.0	1.0
FTP	906.4±9.6	0.52	2192.0±39.3	0.41	2100.2±119.0	0.4	881.7±21.0	0.56
DRR-RED	mean	rsd	mean	rsd	mean	rsd	mean	rsd
HTTP	14.8±1.8	0.99	48.9±7.2	1.04	49.2±7.4	1.05	14.8±1.8	0.99
FTP	905.1±14.0	0.53	2202.8±37.5	0.4	2055.8±131.7	0.41	878.1±30.5	0.55
HPD-DROP	mean	rsd	mean	rsd	mean	rsd	mean	rsd
HTTP	17.6±1.1	0.87	55.7±5.2	0.96	56.4±5.1	0.96	17.5±1.1	0.87
FTP	523.0±52.4	0.55	1062.9±109.7	0.43	992.9±102.4	0.48	535.7±63.1	0.55
HPD-RED	mean	rsd	mean	rsd	mean	rsd	mean	rsd
HTTP	17.8±0.9	0.86	56.7±4.7	0.94	57.0±4.7	0.95	17.8±1.0	0.86
FTP	530.9±48.8	0.52	1037.3±132.1	0.44	964.4±115.3	0.49	516.9±68.7	0.56

Table 3. Packet loss statistics, application mix 1

	Packet loss (%)			
	EDGE1	**EDGE2**	**EDGE3**	**EDGE4**
DRR-DROP	mean	mean	mean	mean
HTTP	6.1±2.4	3.5±1.6	4.8±1.9	6.7±2.4
FTP	0.0±0.0	0.0±0.0	0.0±0.0	0.0±0.0
HPD-DROP	mean	mean	mean	mean
HTTP	2.2±0.8	1.7±0.7	2.2±0.8	2.8±1.0
FTP	0.1±0.0	0.1±0.1	0.2±0.2	0.1±0.1

Table 4. End-to-end delay statistics, application mix 1

	End-to-end delay (ms)			
	EDGE1	**EDGE2**	**EDGE3**	**EDGE4**
DRR-DROP	mean	mean	mean	mean
Video	107.2±0.4	27.2±0.3	27.0±0.2	106.5±0.4
Voip	101.5±0.1	22.6±0.0	22.6±0.1	101.9±0.1
HPD-DROP	mean	mean	mean	mean
Video	111.5±1.2	31.5±1.9	30.2±2.0	111.9±2.0
Voip	107.4±0.3	25.9±0.1	25.5±0.3	106.9±0.4

From the result tables it can also be observed that the end-to-end delays for Video and VoIP are somewhat smaller with DRR than with delay-bounded HPD since in DRR a high over provisioning factor was used for the real-time traffic. However, the difference of a few milliseconds in end-to-end delay is not relevant in practice. All in all, the losses of the real-time traffic for both algorithms are almost zero (this is why they are not shown in the result tables) and the end-to-end delays in each edge area are perfectly tolerable. Thus both the delay-bound and capacity over provisioning are proper means to guarantee that the real-time traffic gets through with minor delays and losses. We also experimented with dif-

Table 5. Object throughput statistics, application mix 2 with different loads

	Object throughputs (bps)							
	EDGE1		EDGE2		EDGE3		EDGE4	
DRR-80	mean	rsd	mean	rsd	mean	rsd	mean	rsd
HTTP	15.5±1.9	0.92	49.4±7.6	1.04	49.4±7.9	1.05	15.4±1.8	0.92
FTP	490.2±69.3	0.66	952.1±147.8	0.58	1032.4±155.4	0.55	503.4±60.8	0.64
DRR-85	mean	rsd	mean	rsd	mean	rsd	mean	rsd
HTTP	11.1±1.8	1.07	34.0±6.7	1.24	34.3±7.2	1.26	11.0±1.9	1.07
FTP	425.4±51.0	0.66	798.8±88.8	0.60	859.8±139.5	0.61	426.0±63.1	0.68
HPD-80	mean	rsd	mean	rsd	mean	rsd	mean	rsd
HTTP	18.8±0.5	0.8	60.1±4.0	0.88	59.8±3.7	0.89	18.8±0.5	0.8
FTP	445.0±66.5	0.67	843.0±139.4	0.61	910.9±141.3	0.59	452.6±60.6	0.66
HPD-85	mean	rsd	mean	rsd	mean	rsd	mean	rsd
HTTP	17.5±0.5	0.81	52.2±3.5	0.91	52.1±3.8	0.91	17.5±0.5	0.81
FTP	358.9±48.7	0.7	628.5±69.8	0.66	673.2±113.0	0.68	353.8±51.8	0.72

ferent initial provisioning values for the DRR algorithm. Instead of using an over provisioning factor of 2 for the real-time traffic we used only an over provisioning factor of 1.1. With this provisioning the packet losses and end-to-end delays of the real-time traffic increased as was expected but the improvement for the HTTP and the FTP-traffic was negligible. In order to improve the performance of the HTTP-traffic that had large losses with the DRR algorithm we should use under provisioning for the real-time classes. However, this is not feasible since as we saw, already with 10% over provisioning the service of the real-time traffic was about to become intolerable.

It is often argued that active queue management mechanisms, such as RED can reduce the problems of static scheduling algorithms by dropping packets already before congestion situations occur. However, from all the result tables it can be seen that RED provides only minor improvements compared to the performance with simple TailDrop. We have investigated the effect of RED with two different parameter sets but with both parameter sets the advantages of RED were negligible. On the other hand, the results prove that with a proper selection of the scheduling algorithm considerable performance advantages can be achieved.

It should be noticed that in application mix 1 most of the traffic (71%) consists of short HTTP flows. When the traffic loads of different classes are biased in this way, resource allocation is a challenging task. It is not possible to provide the required service for the enormous number of web flows without severely degrading the quality of other traffic types. In Tables 5 and 6 the results are shown for application mix 2 where the loads of different traffic types are more even. Delay results are not shown since they are very similar to the results in Table 4. Now the losses in each class are considerably smaller compared with application mix 1, even with the static DRR algorithm. However, as the mean total load increases from 80% to 85% the differences between the DRR and the delay-bounded HPD algorithm become distinctive again. The losses for the HTTP-traffic with DRR are nearly 10 % while the losses with the delay-bounded

Table 6. Packet loss statistics, application mix 2 with different loads

	Packet loss			
	EDGE1	**EDGE2**	**EDGE3**	**EDGE4**
DRR-80	mean	mean	mean	mean
HTTP	2.8±1.9	1.7±1.4	2.5±1.2	3.5±1.4
FTP	0.5±0.2	0.6±0.4	0.6±0.3	0.5±0.2
DRR-85	mean	mean	mean	mean
HTTP	6.7±2.2	4.3±1.6	7.0±2.3	9.2±2.7
FTP	0.7±0.2	0.9±0.3	0.9±0.3	0.7±0.3
HPD-80	mean	mean	mean	mean
HTTP	0.1±0.1	0.1±0.1	0.2±0.1	0.3±0.1
FTP	0.6±0.4	0.7±0.5	0.8±0.3	0.7±0.4
HPD-85	mean	mean	mean	mean
HTTP	0.3±0.1	0.3±0.1	0.4±0.2	0.6±0.2
FTP	0.9±0.3	1.2±0.4	1.3±0.3	1.2±0.4

HPD algorithm are at most 0.6 %. The performance of FTP is slightly worse with delay-bounded HPD, but the difference is much smaller than with application mix 1, since now the HTTP-traffic is not totally dominating the traffic mix.

5   Conclusions

In this paper we compared the performance of static and adaptive provisioning methods with ns2-simulations in a realistic setup. We used the DRR scheduling algorithm for static provisioning with capacity and the delay-bounded HPD algorithm for adaptive provisioning with packet delay.

The simulation results showed that our provisioning goal was better achieved with the delay-bounded HPD algorithm than with the static DRR algorithm. The most distinctive difference could be observed in the service of the HTTP-traffic. We examined the adaptability of both algorithms by testing them with different load levels, application mixes and queue management methods. We observed that the delay-bounded HPD algorithm was considerably more robust in all of the cases. We also observed that provisioning with both algorithms was more challenging in a case where the traffic mix was strongly dominated by a single traffic type. This implies that load balancing or intelligent routing methods should be used to avoid large deviations in the loads of different traffic classes.

The next goal in our research is to investigate the performance of different delay estimators for the algorithm. Besides simulations we will evaluate the performance of the delay-bounded HPD algorithm with real measurements in a FreeBSD based prototype router and in an embedded router architecture to realistically asses the implementation complexity and resource overhead of the algorithm.

References

1. Antila, J., Luoma, M.: Scheduling and quality differentiation in Differentiated Services. Proceedings of Multimedia Interactive Protocols and Systems (MIPS) 2003
2. Bennett, J., Zhang, H.: WF^{2}Q: Worst-case Fair Weighted Fair Queueing. Proceedings of IEEE Infocom, 120–127, March(1996)
3. Blake, S., Black, D., Carlson, M., Davies, E., Wang, Z., Weiss, W.: An Architecture for Differentiated Services. IETF RFC 2475, December(1998)
4. Dovrolis, C., Stiliadis, D., Ramanathan, P.: Proportional Differentiated Services: Delay Differentiation and Packet Scheduling. IEEE/ACM Transactions on Networking 10:2:12–26 (2002)
5. Golestani, S.: A Self-Clocked Fair Queueing Scheme for High Speed Applications. Proceedings of IEEE Infocom, (1994)
6. Parekh, A., Gallager, R.: A Generalized Processor Sharing Approach to Flow Control in Integrated Services Networks: The Single-Node Case. IEEE/ACM Transactions on Networking, 3:1:344–357, June(1993)
7. Shreedhar, M., Varghese, G.: Efficient Fair Queueing using Deficit Round Robin. Proceedings of ACM SIGCOMM, (1995)
8. Liao, R., Campbell, A.: Dynamic Core Provisioning for Quantitative Differentiated Service. Proceedings of IWQoS, (2001)
9. Christin, N., Liebeherr, J., Abdelzaher, T.: A Quantitative Assured Forwarding Service. Proceedings of IEEE Infocom, (2002)
10. Floyd, S., Kohler, E.: Internet Research Need Better Models. Proceedings of HotNets, (2002)

Experimental Feasibility Evaluations of Heuristic Multi-layering Schemes for QoS Adaptive MPEG-4 Video Streaming

Doo-Hyun Kim[1] and Hyun-Kyu Kang[2]

[1] School of Internet & Multimedia Engineering, Konkuk University
Seoul, Korea
doohyun@konkuk.ac.kr
[2] Department of Computer Science, Konkuk University
Chung-Ju, Korea
hkkang@kku.ac.kr

Abstract. In this paper, we have shown multi-fold layered encoding mechanism that can provide fine-grained scalable storage for improving the degree of QoS adaptability. Our scheme provides both temporal scalability and fidelity scalability simultaneously, by segmenting the stored MPEG-4 data in the DCT coefficient domain as well as the temporal domain that most existing methods implement. In order to determine the reasonable set of layers, a probabilistic optimal segmenting algorithm, called J* algorithm, is defined. But, the J* algorithm is only an idealistic approach because of the huge amount of required computation time. Instead, as realistic approaches, three heuristic methods are proposed and experimentally compared with the optimal solution in terms of the average bandwidth residuum. The experiments show that the heuristics can produce relatively optimistic quality adaptation power in a simulated dynamic QoS environment, especially, the bandwidth.

Keywords: MPEG, layered video scaling, QoS adaptation

1 Introduction

With the recent growth of multimedia services over the Internet, a large amount of communication traffic has been introduced by services such as multimedia streaming and audiovisual teleconferencing. The Internet is a best-effort network, which implies that the quality of service is not guaranteed in terms of minimum bandwidth, maximum jitter, delay or data loss. This requires that the end systems should properly adapt to the modifications in network situations, especially congestion, for obtaining the desired quality of service [1-3].

There are several approaches to control the quality of streaming for the stored video encoded with MPEG-1, 2, or 4 [4-6]. The single stream adaptive encoding approach adjusts the resolution of encoding by re-unitization based on network feedback. This scheme may impose a burden on the server due to significant encoding time. The second approach is more similar to a replicated method. This is a simple

V. Roca and F. Rousseau (Eds.): MIPS 2004, LNCS 3311, pp. 153–164, 2004.
© Springer-Verlag Berlin Heidelberg 2004

extension of the single stream adaptive encoding. The server has multiple versions of the stream with different qualities. As the available bandwidth changes, the server switches among different streams according to the level of quality. This eliminates on-time encoding from the server, but requires more storage at the server side. With a layered adaptive approach [7-12], the server maintains hierarchically encoded layers of the stream. As the available bandwidth changes, the server adjusts the quality by adding or dropping active layers. This is popular due to its ability to support hetero-geneity and efficiency in terms of variable networks. On the other hand, the receivers can subscribe to a subset of video layers depending on their processing power and available bandwidth [11]. This scheme supports a finer granularity of quality control and thereby, higher utilization of available bandwidth compared with single or repli-cated stream approaches. This paper deals with the layered scheme that provides the statistically optimal layering algorithm and heuristic alternatives, which can reduce the computation complexity of the optimal layering algorithm.

1.1   Related Works

Most of the works conducted for improving scalability with layered encoding schemes are so confined to the QoS adaptive layered selection using one base layer and only one or two enhancement layers [13-17]. Thus, it seems not to be fine-grained enough to reflect the QoS dynamicity. For instance, the base layer may still be large to transfer over current Internet in a timely manner in the case of network congestion. This may cause the degradation of the quality of service required by users.

Scalable-layered MPEG-2 Video Multicast method [17] divides the DCT coeffi-cient into several layers. The base layer contains the DC coefficient, AC coefficients with lower frequency, and header information for decoding. So the decoding of base stream is possible with some degradation of video quality. The other layers contain the other AC coefficients with pre-determined rule and the information for merging. This method is similar to our approach in the point that total bit-stream size is not increased compared with original MPEG stream and that it can flexibly adapt to the dynamic QoS variation without overloading the server. However, it cannot support sufficient adaptability since it does not provide fidelity adaptation and only provides time domain adaptation. In this paper, we propose the fine-grained layering scheme combining temporal scalability and fidelity scalability.

Unlike MPEG-1 and MPEG-2, MPEG-4 decomposes the scene into objects, called audio-visual objects (AVO's)[6]. Thereby, we can construct the scene configuration with scalability and omit the object when secure bandwidth is not ensured. Com-pared with the object scalability of MPEG-4, the proposed algorithm corresponds to the natural scene coding standard rather than object coding. However, the algorithm can be applied to facilitate object scalability. That is, it can be applied for the object-level scalable transmission by sending some parts of an object not by send or drop.

1.2 Our Contributions

This paper also deals with the layered scheme in that we use both the temporal scalability and fidelity scalability simultaneously, by segmenting the stored MPEG-4 data in the DCT coefficient domain as well as the temporal domain that most existing methods use. But, while the previous discussion [15-19] related to the layered scheme uses simple 2~3 layer slicing methods without comparison with other alternative layering methods, we focus on studying the feasibilities of other possible methods.

At first, we define the optimal set of layers and a probabilistic optimal segmenting algorithm, called J* algorithm, and explain that J* algorithm is only an idealistic approach because of the huge amount of required computation time. Consequently, we propose three heuristic methods as realistic approaches, and conduct experimental comparison with the optimal solution in terms of the average bandwidth residuum. The experiments are conducted on the simulated network environment with MPEG-4 test bit-stream and compared the feasibilities for the QoS adaptation. The experimental results show that these heuristics can produce relatively optimistic quality adaptation power in a simulated dynamic QoS environment, with high bandwidth optimization.

This paper is organized into the following sections. In section 2, the concepts of QoS Adaptation, definition of residuum problem and optimal layering are introduced. In section 3, we present an algorithm for finding the statistically optimal layer set for the generation of fidelity layer. Section 4 proposes three different heuristics for minimizing computation complexity. In section 5, we experimentally show that the heuristics can produce relatively optimistic quality adaptation power in a simulated dynamic QoS environment, especially, the bandwidth optimization.

2 Quality Adaptation and Residuum Problem

2.1 Layered Video Scaling

Temporal Scaling. The temporal scaling is used to reduce the resolution of the video stream in the time domain by decreasing the number of video frames transmitted within a time interval. The temporal scaling extracts the pictures from the MPEG-4 video stream according to the picture types, I-, P-, and B-VOP, and consequently, results in three intermediate representations of an I-substream, a P-substream, and a B-substream. I-substream can be decoded by itself at the MPEG-4 decoder. In this case, the frame rate is reduced since all of P-pictures and B-pictures are dropped. But, the P- and B-substreams are not self-contained. Thus, in order to decode the P-substream, it should be merged with the I-substream at first. Similarly, B-substream should be merged with both the I-substream and the P-substream.

Fidelity Scaling. The fidelity scaling degrades picture quality by reducing the number of DCT coefficients applied to the compression of a picture. 2-D(8x8) DCT coefficients of a block are converted to 1-D sequence by using either zig-zag scanning or alternate scanning rules. So, the sixty-four 1-D ordered DCT coefficients are decomposed into proper number of DCT groups. The DC term must be included in the first group. Fidelity scaling can be applied to inter-pictures(I-substream) as well as intra-pictures(P-substream and B-substream). There are several pictures in a GOV, and many blocks in a picture. Each block for each VOP type of every GOV is accumulated into one accumulated block.

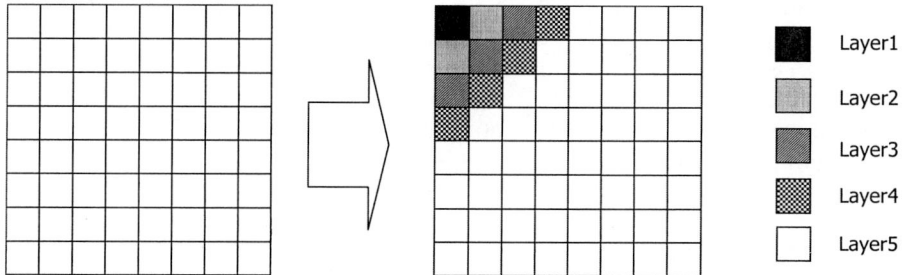

Fig. 1. Concept of fidelity scaling

```
BandwidthAdaptation(nextBandwidth){
    foreach e ∈ {I-VOP, P-VOP, B-VOP}
    {
        if(sizeof(e) =< nextBandWidth)
        then {
            send whole of e;
        } else {
            accumulatedSize=0;
            foreach f ∈ { all fidelity layers of e }
            {
                accumulatedSize = accumulatedSize + sizeof(f);
                if(accumulatedSize > nextBandwidth)
                then {
                    return;
                } else {
                    send f;
}}}}}
```

Algorithm 1. Bandwidth Adaptation Algorithm

2.2 Bandwidth Adaptation

To represent the network environment, network bandwidth, delay, jitter and loss rate have to be considered. But we consider only the bandwidth factor provided by QoS Monitor [12], because of simplicity and effectiveness in representing network QoS. The 'predicting the bandwidth variation for the short intervals' has attracted the attention of many researchers, but is not relevant of scope of this paper.

The adaptation to bandwidth in fluctuation means selecting and transmitting appropriate layers for the given bandwidth at a time as described in Algorithm 1. For the QoS adaptation, the temporal adaptation is applied first, and then the fidelity adaptation is conducted to fill-out the residuum resulted from the temporal adaptation. While the previous works classified the status of network into LOADED/UNLOADED/CONGESTED status, our algorithms use the bandwidth value for fine-grained adaptation. The Algorithm 1 can use AIMD(Additive Increase and Multiplicative Decrease) and MIMD(Multiplicative Increase and Multiplicative Decrease) for enhancing the adaptability[7-11].

2.3 Residuum Problem

A video layering scheme may divide a VOP at random points and easily generate, for example, five layers. However, these layers may not fully utilize the available network bandwidth and result in large bandwidth residuum. Thus, constructing layers that possibly minimize the residuum becomes one of the most important issues in layered QoS adaptation. If each VOP has the set of optimal layers, we will be able to optimally adapt to network variation. We define the Optimal Layer Set, J*, as follows:

Definition 1. *Optimal Layer Set*, J*: When the bandwidth distribution and the number of layers are given, the optimal set of layers is defined as set of layer which has smaller average residuum than any other set of layers.

3   J* Algorithm – A Probabilistic Optimal Layering Algorithm

3.1 Accumulated Block

There are several pictures in a GOV, and many blocks in a picture. Fidelity layering is to classify all blocks with the same type of VOP in a GOV into a dummy layer. To decide which layer is used for a certain DC, the total number of bits with the same DC, which is from all blocks in the same type of VOP should be found. As illustrated in Fig. 2, each accumulated block for each VOP type of every GOV is constructed for this purpose, and the accumulated block is used for fidelity layer decision.

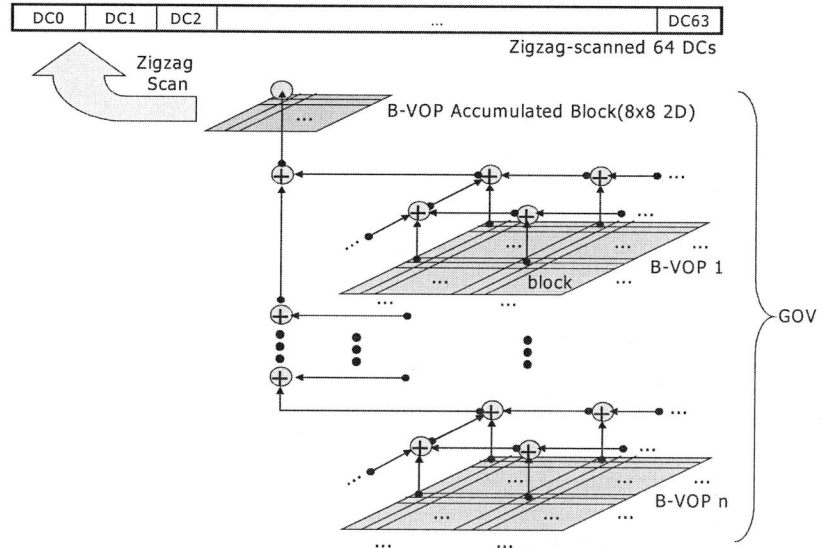

Fig. 2. Finding an accumulated block

3.2 J* Algorithm

J* algorithm produces the optimal layer set, that is, J* which can be found on the case of minimum value of the average residuum. The average residuum values are calculated by the Algorithm 2 on the given bandwidth distribution, **S,** for each possible

```
FindJstar (K){
    foreach R∈ {64Ck-1} {
        for S = 0 to S = Smax {
            for I = 1 to K {
                do {

                    Layer_I =  ∑_{i=J_I}^{J_{I+1}-1} DC_i ;

                        TotalLayerSum = TotalLayerSum + Layer_I;
                } while (TotalLayerSum < S)
            }
            residuum_s = (S - TotalLayerSum) * Pr(S);
        }
        residuum_R =  ∑_{s=0}^{Smax} residuum_s ;

    }
    return  J* = min_{R∈{64C_{K-1}}} (residuum_R) ;
}
K: the number of DC groups, R: the combination of DC groups
S: the available bandwidth: 0 <= S <= Smax,
J_I: DC index where the I^{th} layer begins.
Smax: the bandwidth required for transmitting all 64 DCs
Layer_I: the size of ith DC group, DC_i = the size of i-th DC
Residuum_s: the residuum for given bandwidth S
Residuum_R: the total residuum of combination R
Pr(S): the probability of S, J*: The optimal AC/DC groups
```

Algorithm 2. J* algorithm

layer set among $\binom{64}{K-1}$, where K is the number of layers. The J* algorithm will find the J* theoretically, but it costs much time, and has much computational complexity due to processing for the every case and the whole probability distribution. Therefore, we propose, in next chapter, three layering heuristics that generate a semi-optimal set of layers. The bandwidth distribution, **S,** can be assumed to be a normal distribution without loss of generality, since it can be replaced with an appropriate measured distribution of the real-world network environment.

4   Heuristic Approach

In this chapter, three heuristic algorithms, $J^{uniform}$, J^{slice}, and J^{DC} are proposed. While these algorithms can not guarantee that it must find J*, they have very fast response and produce a layer set close to J*, that consequently can be easily applied to the real situation. The Fig. 3 shows an illustration where these three heuristic algorithms are applied to an example macro-block.

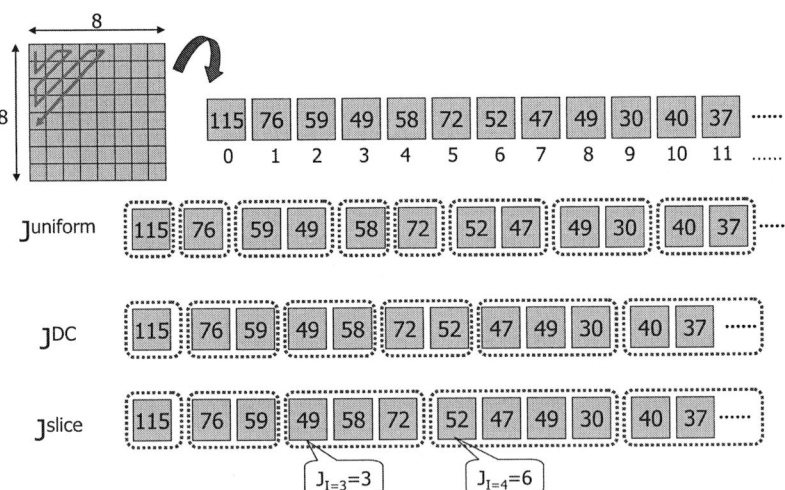

Fig. 3. Illustration of heuristic layering algorithms by applying to an example macroblock

4.1 $J^{uniform}$ Algorithm

In $J^{uniform}$ scheme, one base layer and all enhancements layers are possibly of the same fixed size, where the number of groups, K is given. That is, the *size of layer$_i$* should be the closest to the average layer size, $\dfrac{\sum\limits_{i=1}^{64} accumulated\_DCT\_block_i}{K}$.

4.2 J^{slice} Algorithm

In J^{slice} scheme, which most existing schemes adopt [17-19]. The 64 accumulated DCT blocks are grouped in a diagonal manner as illustrated in Fig. 4.

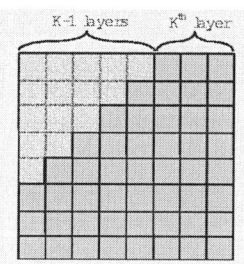

4.3 J^{DC} Algorithm

J^{DC} is a grouping mechanism wherein the first block, that is, the accumulated DC block, becomes the referential size and becomes a base layer. The enhancement layers must be grouped at the size up to the closest to the size of the base

Fig. 4. J^{slice} scheme

layer. In this scheme, the number of layers and their size are variable and depends only on the accumulated DC block size.

Table 1 shows a result produced by applying the four layering schemes to an actual I-VOP from "suzie.mpg." The Fig. 5 shows the number of DC's and the size of each layers produced by each layering scheme, including J* algorithm, to an actual I-VOP from "suzie.mpg." The number of layers was set to 7, that is, K=7, for J*, $J^{uniform}$, and J^{slice}. Meanwhile, eight layers were automatically generated in J^{DC}.

Table 1. Number of DC's and size of each layer

	Layer	1	2	3	4	5	6	7	8
J*	DC No.	0-1	2-4	5-6	7-9	10-14	15-20	21-63	N.A.
	bits	5494	5047	2722	2198	2930	2833	8353	N.A.
J^{DC}	DC No.	0-1	2-3	4-6	7-10	11-17	18-25	26-40	41-63
	bits	5494	3788	3921	3441	3321	3207	3173	1973
J^{uniform}	DC No.	0-1	2-3	4-6	7-10	11-18	19-30	31-63	N.A.
	bits	5494	3788	3921	3441	3828	3942	3904	N.A.
J^{slice}	DC No.	0	1-2	3-5	6-9	10-14	15-20	21-63	N.A.
	bits	2330	5267	4407	4005	2687	2676	6946	N.A.

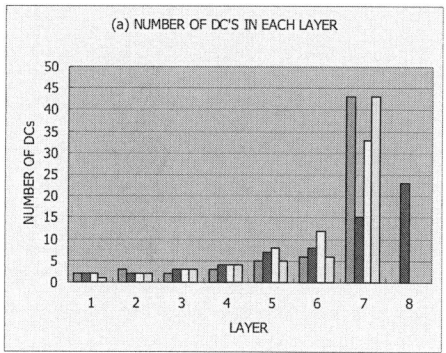
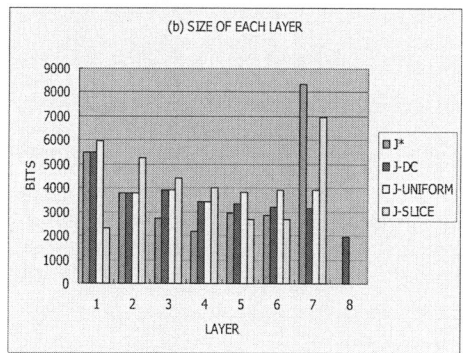

Fig. 5. Number of DC's and size of each layer

5 Experiments

5.1 Configurations

As shown in Fig. 6, the configuration of experimental simulation consists of the bandwidth generator and the residuum evaluator. The bandwidth generator simulates the varying nature of network bandwidth according to the normal distribution with $\mu = 50$, $\sigma = 15$. The rate controller takes the output values from the bandwidth generator, and decides which GOV layers are to be transmitted using the meta-data. The media sender module takes this decision information, and extracts the actual data, which is appropriate for the selected layer, from the stored decomposed MPEG-4 file with reference to the meta-data. Meanwhile, the residuum evaluator compares the output value from the bandwidth generator with the amount of the data from the media sender. The actual data is sent to the merger to produce the scaled MPEG-4 file, which is used for the comparison of the image quality.

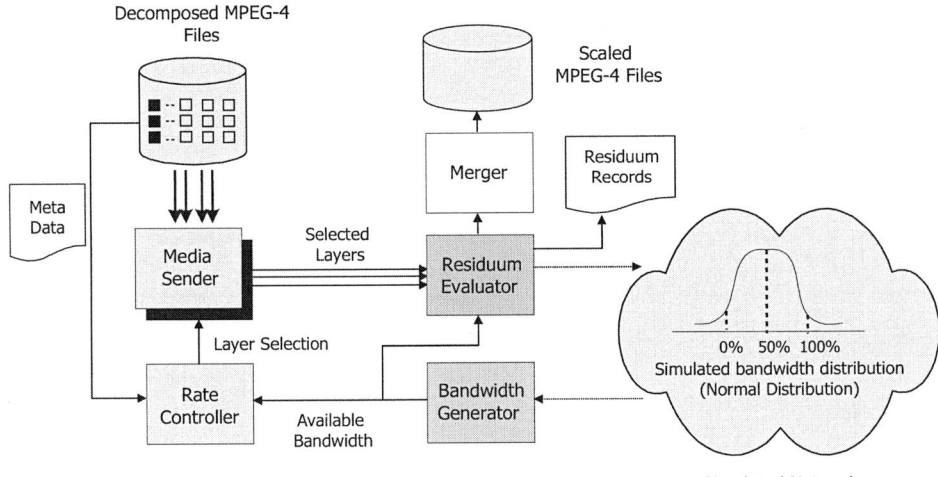

Fig. 6. Configuration of Simulation

5.2 Average Residuum Comparisons

Table 2 shows the average residuum of the four layering methods applied to the "foreman.mpg." Each column represents the average of the residuum produced by applying the bandwidth adaptation algorithm, i.e., the Algorithm 1 for each possible layer, K. These average values are experimental results after applying to each K hundredth different bandwidth generated from the normal probability distribution function. Meanwhile, the column for J^{DC} cannot be given the average for each K because the J^{DC} determines the number of possible layers; hence, the number of layers for each GOV is different. Instead, the first column of the J^{DC} represents the average residuum for each GOV with the same number of layers, K. The second column shows the average of the all-average values, and the third has the average of layers, that is, the average of K. In case of K=6, while more than one day was required by using Pentium III 500MHz PC for calculating the optimal set of layers, J* with Algorithm 2, only minimum computation time was needed for the heuristic methods.

The experimental results show that the ratio of residuum between the J* and heuristic algorithms are smaller as K increases. This is because the residuum is smaller when the size of the layer becomes smaller as K increases. The main result of this experiment is that the $J^{uniform}$ in overall has smaller residuum than the J^{slice}, inspite the differences become insignificant as the number of layers increases. The J^{DC} was not performing better than $J^{uniform}$ or J^{slice} when these were compared with the same number of layers.

Table 2. Average residuums in "foreman.mpg"

K	J*	$J^{uniform}$	J^{slice}	J^{DC} Avg. Res.	Total Res.	Avg. Layer	K	J*	$J^{uniform}$	J^{slice}	J^{DC} Avg. Res.	Total Res.	Avg. Layer
2	23672	26336	34079	-			11	NA	5620	5551	6770		
3	13015	18665	27953	-			12	NA	5155	5407	8144		
4	10666	14176	22585	-			13	NA	4835	5363	7697		
5	8049	11304	17809	-			14	NA	4559	5352	5881		
6	7400	9772	13937	6567	6724	10.98	15	NA	4279	5350	3731		
7	NA	8397	10732	6313			16	NA			2955		
8	NA	7665	8333	7933			17	NA			-		
9	NA	6747	6735	7793			18	NA			2329		
10	NA	6120	5915	7005			19	NA			2873		

5.3 Bandwidth Adaptability Comparisons

Table 3 shows the comparison results among the actual residuum values of the I-VOP at the first GOV in the layering into four layers from the test image sequence of "suzie.mpg." In case of using Algorithm 2 for the J*, there are some cases when even a single layer cannot be transmitted at 30%, or 10% of the maximum bandwidth. This is because the optimization was performed with the assumption of normal distribution. Meanwhile, the method of J^{slice} can transmit only DC values even for the smallest bandwidth because this method has the first layer that consists of only DC values, and the layer size is as small as possible. The transmission was possible even with only 10% of the bandwidth required for full quality. Therefore, we could say the J^{slice} method shows the better survivability. But, $J^{uniform}$ mostly outperforms the J^{slice} proposed in [17-19]. The K=11 for JDC has been determined by the J^{DC} algorithm and the J^{DC} does not allow four layers, hence it is hard to compare the J^{DC} with others, even though the picture quality of the J^{DC} is definitely the best.

Table 3. Picture Quality adaptation and for an I-VOP in the first GOV of "suzie.mpg."

Bandw.	J* (K = 4)	J^{uniform} (K = 4)	J^{slice} (K = 4)	J^{DC}(K=11)
35,899 (70%)				
Resid.	4,550	9,982	19,084	1,644
25,642 (50%)				
Resid.	8,647	8,827	11,045	3,857
15,385 (30%)	N.A. (Not Generated)			
Resid.		788	5,164	2,709
5,128 (10%)	N.A. (Not Generated)	N.A. (Not Generated)		N.A. (Not Generated)
Resid.			2,039	

6 Conclusions

In this paper, we have described a mutli-fold layered encoding mechanism that can provide fine-grained scalable storage and thereby improves the degree of QoS adaptability. Our scheme provides both temporal scalability and fidelity scalability simultaneously, by segmenting the stored MPEG-4 data in the DCT coefficient domain as well as the temporal domain that most existing methods are rely on. In order to determine the reasonable set of layers, a probabilistic optimal segmenting algorithm, called J* algorithm, is defined. But, since the J* algorithm is only an idealistic approach because of the required computation amount, three heuristic methods, instead, are proposed and experimentally compared with the optimal solution in terms of the average bandwidth residuum. The experiments show that the heuristics can produce relatively optimistic power in the quality adaptation to the simulated dynamic network QoS, especially, the bandwidth optimization.

The multi-layering heuristics proposed in this paper can be applied to many other fields requiring the scalability of MPEG contents, as well as to the streaming services. Currently, each layer is stored into separate files with the file structure provided by the commercial OS[19]. But, in future, this preliminary storage mechanism will be enhanced with the improved file structure more appropriate for the operation of the video servers.

Acknowledgement. This work was supported by the faculty research fund of Konkuk University in 2004.

References

1. H. Vin, "Heterogeneous Networking", IEEE Multimedia, Vol. 7, 1995, pp. 84-88.
2. B. Blair and J-B. Stefani, Requirements of distributed multimedia applications (ch 3), Open distributed processing and multimedia, Addison-Wesley, 1997.
3. J.C. Bolot, "Characterizing end-to-end packet delay and loss in the internet," Journal of High Speed Networks, Vol. 2, No. 3, 1993, pp.289-298.
4. D. LeGall, "MPEG: A video compression standard for multimedia applications," Communication ACM, Vol. 34, No. 4, pp. 46-58, Apr. 1991
5. International Standard ISO/IEC 13818-2, Information technology - Generic coding of moving pictures and associated audio information: Video, 1996.
6. Battista, Stefano; Casalino, Franco; Lande, Claudio, "MPEG-4: A Multimedia Standard for the Third Millennium," IEEE Multimedia, October-December 1999, 74-83.
7. D. Wu, Y.T. Hou, and Y.Q. Zhang, "Transporting Real-Time Video over the Internet: Challenges and Approaches", Proceedings of the IEEE, Vol. 88, No. 12, 2000.
8. N. Shacham, "Multipoint Communication by Hierarchically Encoded Data," IEEE Infocom'92, pp. 2107 – 2114, 1992.
9. J.C. Bolot and T. Turletti, "A rate control mechanism for packet video in the Internet," IEEE INFOCOM'94, 1994, pp.1216–1223.
10. S. McCanne, V. Jacobson, and M. Vetterli, "Receiver-driven layered multicast," ACM SIGCOMM'96, 1996, pp. 259-266.
11. X. Li, S. Paul, and M.H. Ammar, "Multi-Session Rate Control for Layered Video Multicast", Proceedings of Symposium on Multimedia Computing and Networking, 1999.
12. R. Rejaie and M. Handley, "Quality Adaptation for Congestion Controlled Video Playback over the Internet," ACM SIGCOMM'99, 1999, pp.189-200.
13. Puri, L. Yan, and B.G. Haskell, "Syntax, semantics and description of Temporal Scalability," ISO/IEC JTC1/SC29/WG11 MPEG Doc. 93/795, 1993.
14. R. Arvind, R. Civanlar, and R. Reibman, "Packet Loss Resilience of MPEG-2 Scalable Video Coding Algorithms", IEEE Trans. circuit and systems for video tech., Vol.6, No.5, 1996, pp. 426-435.
15. N. Cranley, and L. Murphy, "Adaptive Quality of service for streamed MPEG-4 over Internet," 14th UKTTS 2001, Dublin, Ireland, May 16-18, 2001, pp. 1206-1210.
16. Eleftheriadis, and D. Anastassiou, "Optimal data partitioning of MPEG-2 coded video," Proc. of the 1st Int'l conference on image processing(ICIP-94), Austin, Texas, Nov. 1994.
17. F. Ruijin, L. B. Sung, and A. Gupta, "Scalable Layered MPEG-2 Video Multicast Architecture," IEEE Transactions on Consumer Electronics, Vol. 47, No. 1, pp. 55-62,
18. D0H. Kim, J. H. Yang, J. Y. Kwak, S. H. Ahn, W. J. Yoo, and H. C. Kim, "QoS Adaptive Streaming Based On Enhanced MPEG-2 Scaling", Proceeding of IMSA2001, Hawaii, 2001.
19. Ahn, S. H., Kang, M. G., Kim, D. H. and Kim, H. C., "QoS Adaptive MPEG-2 Streaming based on Scalable Media Object Framework", ICOIN-15, pp. 683-688, 2001.

Residual Motion Coding Method for Error Resilient Transcoding System

June-Sok Lee[1], Goo-Rak Kwon[1], Jae-Won Kim[1],
Kyung-Hoon Lee[2], and Sung-Jea Ko[1]

[1] Department of Electronics Engineering, Korea University, Seoul, Korea
[2] National Security Research Institute, Daejeon, Korea
Tel: +82-2-3290-3228
sjko@dali.korea.ac.kr

Abstract. For video adaptation in universal multimedia access systems, video contents have to be transcoded according to the requirements of network bandwidth and the performance of clients. However, the transcoded video streams are vulnerable to channel disturbances when they are transmitted over the wireless channels. In this paper, we present an error resilient transcoding system using the *residual motion coding* method for wireless visual communications networks. The residual motion coding method interleaves into the bitstream the *residual motion vector* that is auxiliary motion information to be used at the decoder to recover the lost motion vector. With the residual motion vector, the decoder can precisely reproduce the motion vector of the corrupted macroblock and reconstruct the lost macroblock by using error concealment based on temporal prediction. Experimental results show that the proposed method can not only achieve the error robustness of the motion vector, but also improve the performance of error concealment at the decoder.

Keywords: Transcoding, Error resilience, Error Concealment, Residual motion coding.

1 Introduction

With the generalization of mobile communication services, recent interest has been focused on the transmission of real time video over wireless communications networks. To provide this service, video contents need to be transcoded for a variety of users who have different network conditions ro terminal devices with different display capabilities [1]. Fig. 1 shows a video transcoding environment for wireless communications networks. The transcoding to the reduced bitrate is performed by the transcoding system to match the bandwidth of wireless networks and user's requirements. The transcoding system can be applied at the edge of the access network and incorporated with the network routers.

The transcoded video bitstreams, however, can be corrupted by transmission errors such as random bit errors and packet losses which are frequently occurred

V. Roca and F. Rousseau (Eds.): MIPS 2004, LNCS 3311, pp. 165–174, 2004.

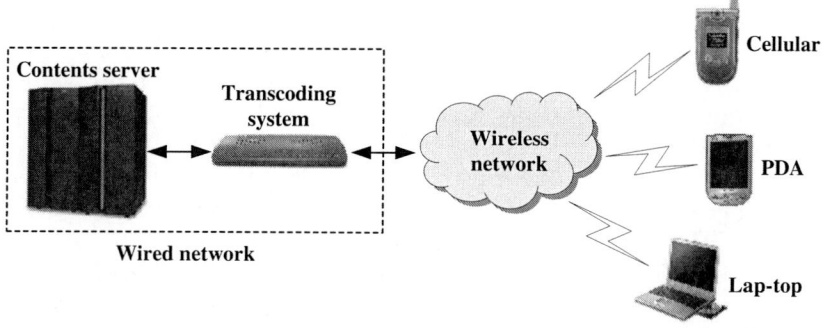

Fig. 1. Video transcoding environment for wireless communications

in the wireless networks [2]. The transmission errors can lead to the burst error, that is the loss of several slices or even a whole frame. For compressed video sequences, the burst errors not only corrupt the current decoded frame, but they are also propagated to successive frames.

Many error resilience [3]-[4] and concealment [5]-[7] techniques have been proposed to cope with the transmission errors. The error localization method is used to prevent the spatial error propagation by inserting re-synchronization markers in the bitstream at certain intervals [3]. If an error has been detected in the bitstream, the decoder discards the data between the re-synchronization markers. This method, however, is useless to the burst packet errors since the markers may be lost together with the corrupted packets. In [4], an intra refresh method is used to eliminate the error propagation by inserting periodic intra frames and intra macroblocks (MB's). However, the intra refreshing is too bit-consuming to be used for the low bitrate communications. In [5]-[7], the error concealment (EC) method is used to reconstruct the corrupted MB at the decoder. They estimate the lost motion vector (MV) by using the correctly received MV's of the neighboring MB's. However, the EC method tends to introduce visual degradation at the reconstructed frame when motion compensation is performed with inaccurately predicted MV's.

In this paper, we present an error resilient transcoding system using the residualm otion coding (RMC) method for wireless visual communications networks. The RMC method interleaves into the bitstream the residualM V (RMV) that is auxiliary motion information to be used at the decoder to recover the lost MV. With the RMV, the decoder can precisely reproduce the MV of the corrupted MB and reconstruct the lost MB by using EC based on temporal prediction.

The rest of this paper is organized as follows. In Section 2, the proposed transcoding system is introduced. Section 3 describe the proposed error resilience and concealment methods in detail. Section 4 shows the experimental results and conclusions are given in Section 5.

2   Proposed Transcoding System

Figure 2 illustrates the proposed transcoding system consisting of three main processing units: source loader, transcoding engine, and multimedia streamer. The source loader opens the compressed video streams in the local computer or concatenates the packets received from the network to reconstruct the original bitstream. The transcoding engine generates transcoded bitstream which satisfies the network conditions, system requirements, and user preferences. In addition, the error resilience information is encoded during the transcoding process. The multimedia streamer performs data partitioning to enhance the error resilience of transcoded data and fragments the bitstream into packets. Then, the packets are transmitted over wireless communications networks.

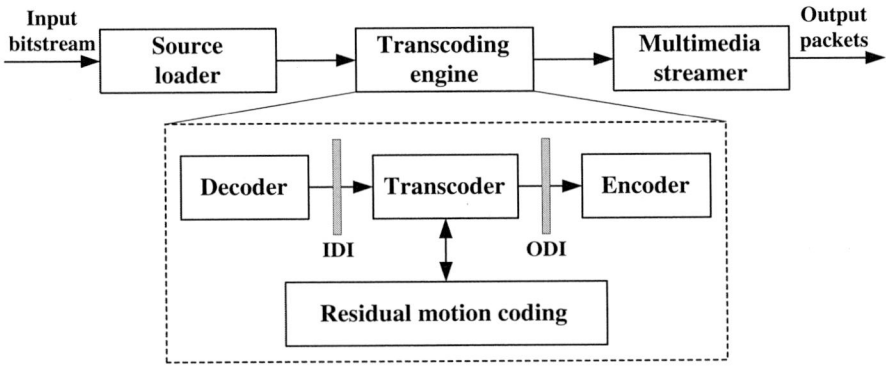

Fig. 2. Structure of the proposed transcoding system

The transcoding engine is the most important part of the proposed transcoding system. It consists of data interface (DI), decoder, transcoder, encoder, and RMC blocks. The DI unit manages all information used for transcoding process. With the DI unit, the transcoder can be separated from the encoder and decoder so that various transcoding algorithms can be adapted according to the requirements of users. The input data interface (IDI) retrieves the significant information from the compressed bitstream for the transcoder. The transcoder analyzes and modifies the extracted data and carries the transcoded video information to the output data interface (ODI). Then, the encoder generates a bitstream by using the data contained in ODI. During transcoding process, the RMC block communicates with the transcoder to insert error resilience information into the bitstream. In the following section, we describe the functionality of the RMC block in detail.

3   Proposed Error Resilience Method

In the low bitrate video coding, the significance of MV's increases in terms of error recovery as the encoding bitrate decreases, since the MV's take up a large portion of the bitstream. Table 1 shows the significance of the MV in the

Table 1. Significance of MV's in the low bitrate video coding

Bitrate [Kbps]	Bits for MV [Kbytes]	Total [Kbytes]	Ratio [%]
128	18.0	313.9	5.7
64	18.4	154.0	11.9
32	18.0	74.0	24.2

low bitrate video coding. The standard video sequence "Foreman" was used to examine the significance of MV at different bitrate in the low bitrate coding. The test sequence of 300 frames with QCIF resolution (176×144) was encoded to produce a H.263 bitstream with 15 fps. The ratio of the bits for MV to the total bits for the test sequence is compared at the encoding bitrate of 128, 64, and 32 Kbps. This results indicates that the loss of MV tends to result in more serious video quality degradation at the decoder as the bitrate decreases.

To solve this problem, we propose the RMC method. The proposed RMC method adds auxiliary motion information, i.e. the RMV, into the bitstream to improve the error resilience of the MV. The RMV is the motion information to recover the MV of the corrupted MB. To achieve the robustness of the RMV, the RMV of the current frame is inserted into the bitstream of the next frame. If two adjacent frames are not corrupted at the same time, the RMV of the current frame is preserved even though the current frame is corrupted by the burst errors. The RMV is obtained by subtracting the original MV from the predicted MV. Fig. 3 shows the concept of the proposed RMC method. Let F_n be the nth encoded frame at the encoder. The RMV of the ith MB in F_{n-1} is given by

$$R_i^{n-1} = V_P^{n-1} - V_i^{n-1}, \qquad (1)$$

where V_P^{n-1} and V_i^{n-1}, respectively, are the predicted and original MV's of the ith MB in F_{n-1}. We interleave the RMV R_i^{n-1} for F_{n-1} into the bitstream of F_n. Since the RMV is encoded with a frame delay, the error resilience information for F_{n-1} can be preserved even though a whole frame is corrupted by the transmission errors.

Next we review the inter frame motion prediction (IFMP) method to obtain the V_P^{n-1}, that was proposed in our previous literature [8]. In the most of the video coding methods such as MPEG-4 and H.263, the intra frame motion prediction method is used to obtain the predicted MV. For example, in the H.263 video encoder, the predicted MV of the current MB is the median value of three candidate MV's of the left, upper, and upper right MB's. However, the MV's of the neighboring MB's may not be available when the MB's are corrupted by the

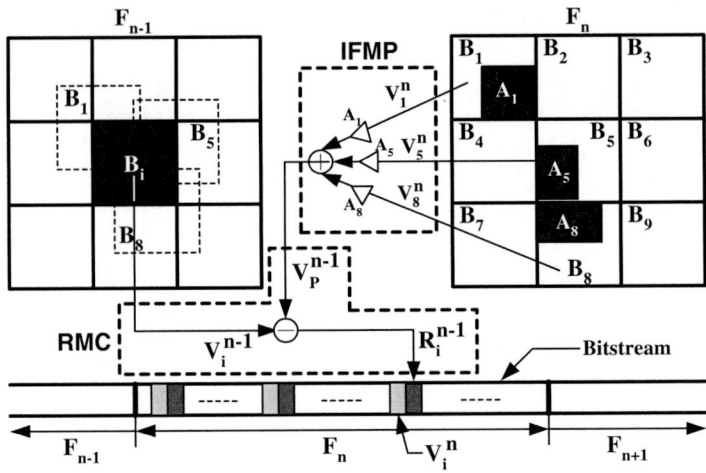

Fig. 3. Concept of the proposed RMC method

burst errors. Unlike the intra frame motion prediction, the IFMP method uses the temporal correlation between the MB's to calculate the predicted MV. In Fig. 3, the MB's B_1, B_5, and B_8 in F_n have temporal correlations with the B_i in F_{n-1} since the three MB's are motion compensated from the dashed squares in F_{n-1}. V_P^{n-1} is determined by taking a weighted average of the MV's of the temporally correlated MB's in F_n, which is given by

$$V_P^{n-1} = \frac{\sum_{i=1}^{9} A_i^n \cdot V_i^n}{\sum_{i=1}^{9} A_i^n}, \qquad (2)$$

where V_i^n is the MV of the ith MB in F_n and A_i^n is the weighting factor for the V_i^n. We set the weighting factor to be equal to the number of pixels in the area A_i of F_n, which indicates the measure of the temporal correlation.

After calculating the R_i^{n-1}, the RMV is encoded to generate the bitstream. The bitstream syntax for the RMV consists of code word and k bits of data. The code word indicates the coding mode of the RMV and the data is the encoded bits of the RMV. Fig. 4 illustrates the decision rule to select the coding mode of the R_i^{n-1} and the operation for each code word is explained as follows:

- **Code word 00**: If R_i^{n-1} is equal to the V_P^{n-1}, the code word "00" is inserted into the encoded bitstream. Thus, the code word "00" represents that the V_i^{n-1} is a zero MV and the MV of the corrupted MB can be replaced with a zero MV.
- **Code word 01**: If R_i^{n-1} is equal to the zero MV (V_0), the code word "01" is allocated to encode the R_i^{n-1}. The code word "01" denotes that V_P^{n-1} is

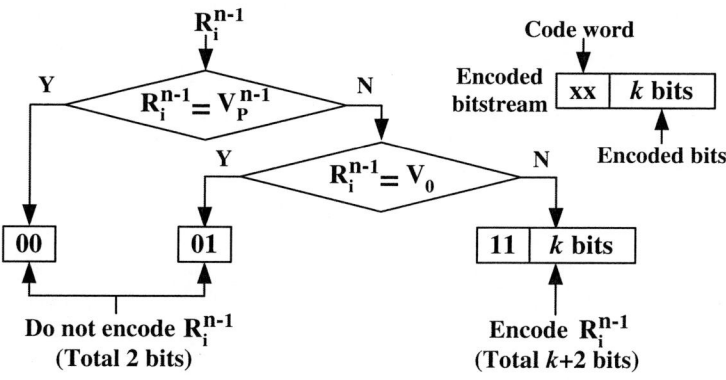

Fig. 4. Coding mode decision for the RMV

equal to the V_i^{n-1} so that the MV of the corrupted MB can be replaced with the V_P^{n-1} at the decoder.

– **Code word 11**: If the R_i^{n-1} is inconsistent with all the above conditions, the code word "11" is encoded with the k bits of encoded R_i^{n-1}. The code word "11" denotes that the MV of the corrupted MB is replaced with the V_i^{n-1} reconstructed by using the R_i^{n-1}.

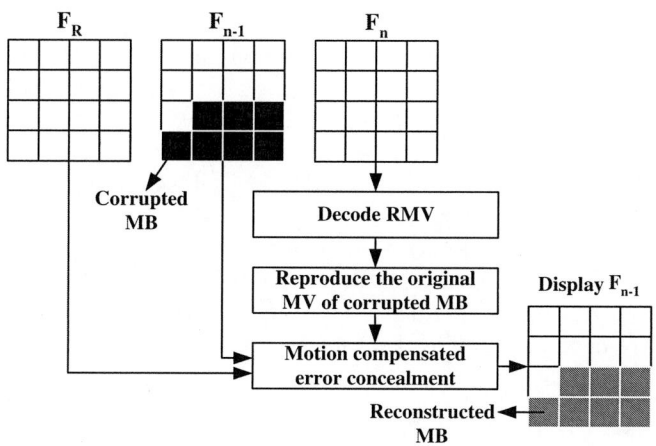

Fig. 5. Error concealment using the RMV

Adding the RMV into the bitstream can significantly improve the robustness of MV, but it may cause the picture quality degradation or distortion since encoding of the RMV increases the transmission overhead and the DCT coefficients are coarsely quantized to comply with the bitrate constraint. Thus, we adopt the Lagrangian rate-distortion (R-D) method to minimize the distortion for the given rate constraint. The R-D function is given by

$$J = D + \lambda(B_{RMV} + \sum_{i=1}^{N} B_i), \qquad (3)$$

where D is the total distortion, B_{RMV} is the bits allocated for the RMV's, N is the number of MB's in the frame, and $\sum_{i=1}^{N} B_i$ is the bits allocated for N MB's. The optimal quantization parameter for each MB is obtained by minimizing (3) subject to

$$(B_{RMV} + \sum_{i=1}^{N} B_i) \leq B_{Frame}, \qquad (4)$$

where B_{Frame} is the bitrate constraint for one frame.

In the proposed RMC method, the RMV is inserted into the bitstream to enhance the robustness of the MV. At the decoder side, the proposed EC is performed by using the RMV to reconstruct the corrupted MB as shown in Fig. 5. Since the RMV is encoded with a frame delay, the decoder requires additional frame buffer. Assume that the decoder has received frame F_n and is about to display frame F_{n-1} with corrupted MB's. Since F_{n-1} is corrupted by the transmission error, the decoder extracts the RMV from the bitstream of F_n and then the original MV of the lost MB is recovered by using

$$V_i^{n-1} = V_P^{n-1} - R_i^{n-1}. \qquad (5)$$

The V_P^{n-1} can be obtained by using the IFMP method as the same way in the encoder. With the V_i^{n-1}, the corrupted MB's are motion compensated from the reference frame F_R.

To reconstruct the corrupted MB's in the intra coded frame, we use the intra 16×16 prediction mode of the H.264/AVC video coding standard [9]. This mode is originally used to enhance the coding efficiency of the intra frame at the H.264/AVC encoder, but we adopt it to predict the corrupted MB in the intra coded frame at the decoder. Fig. 6 shows the previously decoded pixels and the pixels in the corrupted MB. Each pixel of the corrupted MB is predicted from the previously decoded samples by using the linear 'plane functions' in [9].

4   Experimental Results

The standard video sequences, "Foreman", "Silent", "Stefan", and "Table tennis", are used to evaluate the performance of the proposed RMC method. Each test sequence of 300 frames with QCIF resolution (176×144) is encoded to the H.263 bitstreams of 64 Kbps with 7.5 fps.

Table 2 shows the results of the proposed RMC method that include the bitrate overhead caused by adding the RMV, the proportion of the RMV to

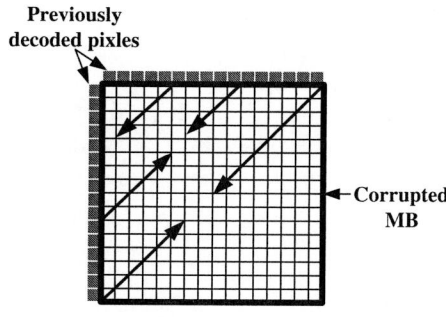

Fig. 6. Plane mode of intra 16×16 prediction

Table 2. Result of the residual motion coding

Sequences	Overhead [bits]	Proportion [%]	PSNR variance [dB]
Foreman	716	7.8	−0.004
Silent	481	5.3	+0.005
Stefan	714	7.8	−0.007
Table tennis	599	6.5	+0.010

the bitstream, and the variance of PSNR with the RMV adopted. As shown in Table 2, the RMC method causes a small amount of bit overhead since the RMV is interleaved into the encoded bitstream. The overhead depends on the motion activities of the object in the test sequence. For the "Silent" sequence with small motion activity, the overhead is 5.3% of the encode bitstream. On the other hand, in the "Foreman" and "Stefan", both of which have high motion activity, the overhead takes up 7.8% of the bitstream. Although inserting the RMV into the bitstream causes some overhead of encoding bitrate, the picture quality degradation caused by the proposed method is negligible in terms of PSNR variance as shown in Table 2. This is so because the proposed R-D optimization minimizes the distortion subject to the bitrate constraint.

The performance of the proposed EC method is compared with that of the TCON method that is the EC method of the H.263 test model TMN-10 [10]. The corrupted frames are selected by using a two state Markov model with a 10^{-3} average packet loss rate. The selected frames are partially damaged with the corruption ratio of 50%. Fig. 7 shows the PSNR result of the TCON and proposed EC methods. Notice that the proposed method exhibits higher PSNR performance than the TCON method for the entire frames.

5   Conclusions

In this paper, the RMC method was proposed to support the error resilience and concealment of the transcoded bitstream. Since the RMC encodes the RMV with a frame delay, the original MV of the corrupted MB can be reproduced at the decoder even though a whole frame is corrupted by the transmission error. Moreover, the reproduced original MV remarkably improves the performance of the EC at the decoder. Experimental results indicate that the proposed RMC method can be effectively adopted for the transcoding system to enhance the error robustness of the bitstream.

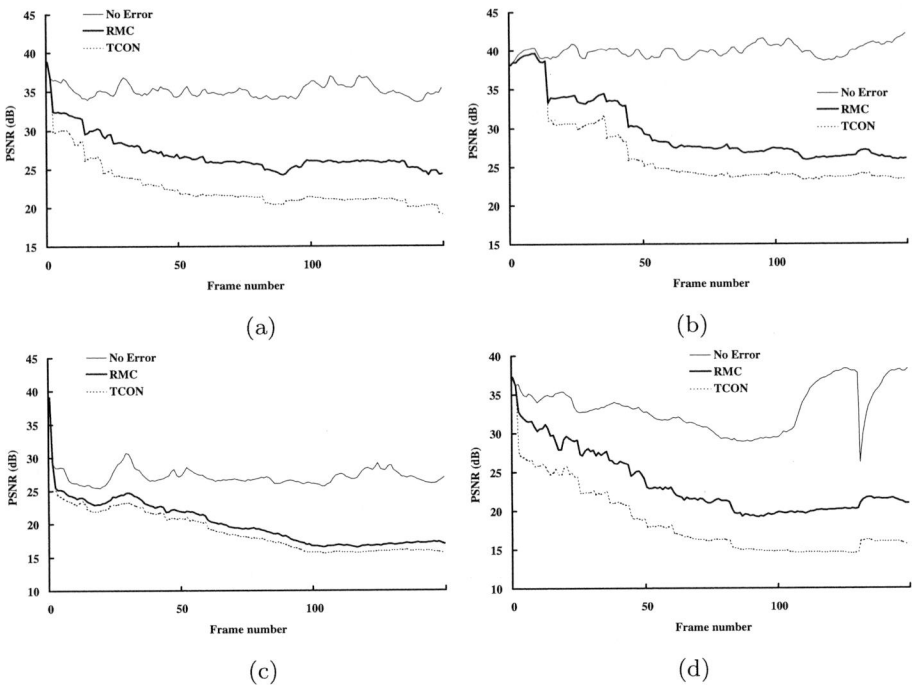

Fig. 7. Comparison of PSNR corresponding to the test sequence (a) Foreman, (b) Silent, (c) Stefan , (d) Table tennis

References

1. Vetro, A. (ed.): MPEG-21 requirements for for digital item adaptation. ISO/IEC JTC1/SC29/WG1 N4684, Korea (2002)
2. Wang, Y., Zhu, Q. F.: Error control and concealment for video communication: a review. Proceedings of IEEE, Vol. 86. (1998) 974–997
3. Wenger, S., Knorr, G., Ott, J., Kossentini, F.: Error resilience support in H.263+. IEEE Trans. Circuits Syst. Video Technol., Vol. 8. (1998) 867–877
4. ITU-T Recommendation H.263, Video coding for low bit rate communication (1998)

5. Chen, M. J., Chen, L, G., Weng, R. M.: Error concealment of lost motion vectors with overlapped motion compensation. IEEE Trans. Circuits Syst. Video Technol., Vol. 7. (1997) 560–563
6. Zhang, J., Arnold, J. F., Frater, M. R.: A cell-loss concealment technique for MPEG-2 coded video. IEEE Trans. Circuits Syst. Video Technol., Vol. 10. (2000) 659–665
7. Wang, Y. K., Hannuksela, M. M., Varsa, V., Hourunranta, A., Gabbouj, M.: The error concealment feature in the H.26L test model. IEEE International Conf. Image Processing, Vol. 2. (2002) 729–732
8. Lee, J. S., Lee, Y. A., Park, W. S., Ko, S. J.: A burst error concealment technique for visual communications in a mobile environment. Proc. IEEE Int. Conf. Consumer Electronics, Los Angeles, CA (2003) 18–19
9. Draft ITU-T Recommendation H.264 and Draft ISO/IEC 14496-10 AVC. Joint Video Team of ISO/IEC JTC1/SC29/WG11 & ITU-T SG16/Q.6 Doc. JVT-G050, T. Wieg, Ed., Pattaya, Thailand (2003)
10. Signal Processing and Multimedia Lab., Univ. British Columbia. TMN-10 (H.263+) Codec. (1998)

A Fast Motion Estimation Algorithm Based on Context-Adaptive Parallelogram Search Pattern

Chuanyan Tian, Chengdong Shen, and SiKun Li

Office 607, School of Computer Science, National University of Defense Technology
410073 ChangSha, China
tianchuanyan@163.com   skli@nudt.edu.cn

Abstract. Block Motion Estimation is vital to the coding (compression) of natural video images, and it is one of the most important components of the standards of video compression technology. In order to reduce the high computational complexity of Motion Estimation, this paper proposed a context-adaptive parallelogram search pattern, and developed a fast motion-estimation algorithm based on the search pattern. Parallelogram search pattern contains five search models, and the search model may be chosen from a selection of five search models in the process of Motion Estimation. The experimental results show that the proposed algorithm can provide a remarkable computational complexity reduction in Motion Estimation(ME), while it maintains high compression efficiency.[1]

1   Introduction

BMA (Block-Matching ME Algorithm) aims at exploiting the strong temporal redundancy information between successive frames, and is vital to video-coding standards, such as MPEG-4 [1] and H.264 [2,3]. BMAs attempt to find one block from the reference frames (forward or backward) that best matches the predefined block in the current frame. The minimizing matching criterion, used in most cases, is the mean absolute error between this pair of blocks. The best-matched block producing the minimum distortion is searched within the search window in the reference frames. The displacements of the current block with respect to the best-matched reference block in the x and y directions compose the motion vector of this current block. If the full search (FS) is used to evaluate all possible candidate blocks in search window, experiments demonstrate that block-matching motion estimation consumes up to 80% of the computational power of the encoder. Therefore, fast algorithms are highly desired to significantly speed up the search process without sacrificing the compression efficiency seriously. Besides the three-step search (TSS), new three-step search (NTSS) [4] , four-step search (4SS) [5], block-based gradient descent search (BBGDS) [6] and diamond search (DS) [7] algorithms, some computationally efficient variants were developed, among which are typically HEXBS algorithm [8] and UMHexagonS algorithm [9,10].

---

[1] Supported by the National Natural Science Foundation of China under Grant No. 90207019

V. Roca and F. Rousseau (Eds.): MIPS 2004, LNCS 3311, pp. 175–186, 2004.

In TSS, NTSS, 4SS, BBGDS and DS algorithms, square-shaped or diamond-shaped search patterns with different sizes are employed. These algorithms performed well in relatively small search range and low-resolution video sequences. But in many applications, such as SDTV or HDTV, the search range should be large enough for high coding efficiency. HEXBS algorithm adopts a hexagon-shaped search pattern, and experiments demonstrate that it has faster processing with the similar distortion compared with TSS, NTSS and 4SS. But in some sequences such as Stefan, HEXBS algorithm is likely to drop into a local minimum in the early stages of search process, and result in a bad PSNR value. The main reason of the disadvantage of HEXBS algorithm is the singleness of its search pattern and its search strategy. UMHexagonS algorithm, considering of multiple prediction modes, multiple reference frames of H.264, is developed as the integer pel motion estimation algorithm for H.264. Studies show that UMHexagonS algorithm can achieve further improvement in motion accuracy than HEXBS algorithm; but its computational complexity is larger than HEXBS algorithm, the reason for which will be specified in Section 2.

In this paper, we provided a new fast motion estimation algorithm based on context-adaptive parallelogram search pattern (CAPS), The CAPS algorithm resolves the aforementioned contradiction by using context-adaptive parallelogram search strategy. And it is used to be compatible with the multiple prediction modes and multiple reference frames characteristics of H.264.

The remainder of the paper is organized as follows. In Section 2, we briefly analyze the UMHexagonS algorithm. The proposed search pattern and the developed ME algorithm are described in Section 3. Experiment results are presented in Section 4. Section 5 is conclusion of this paper.

2   Analysis of UMHexagonS Algorithm

UMHexagonS algorithm is a hybrid Unsymmetrical-cross Multi-Hexagon-grid Search algorithm for integer pel motion estimation. And UMHexagonS algorithm adopts a kind of hierarchical motion search strategy. UMHexagonS algorithm uses three kinds of search patterns: Unsymmetrical-cross search pattern, Uneven Hexagon-grid search pattern and Extended Hexagon-based Search pattern. In Fig.1, we use step 2 to indicate an unsymmetrical-cross search pattern, where horizontal search range equals W and vertical search range equals W/2 (W equals to 16), step 3-2 shows an uneven Multi-Hexagon-grid search pattern, which is constructed by extend 16-HP (Sixteen Points Hexagon Pattern), the search process in step 3-2 starts from the inner hexagon to the outer hexagon. Extended Hexagon-based Search pattern visualizes in Step 4 in Fig.1.

UMHexagonS algorithm includes three steps with different kinds of search patterns: 1) Unsymmetrical-cross search; 2) Uneven Multi-Hexagon-grid search; 3) Extend hexagon-based search. Fig.1 shows a typical search procedure of motion vector (4, 10) in a search window, whose search range is 32×32, and (0, 0) vector is assumed the initial search point. From Fig.1, we can derive the total number of search points per block can be expressed by formula (1):

$$N_{UMHexagonS}(m_x, m_y) = N_1 + N_2 + N_3 \tag{1}$$

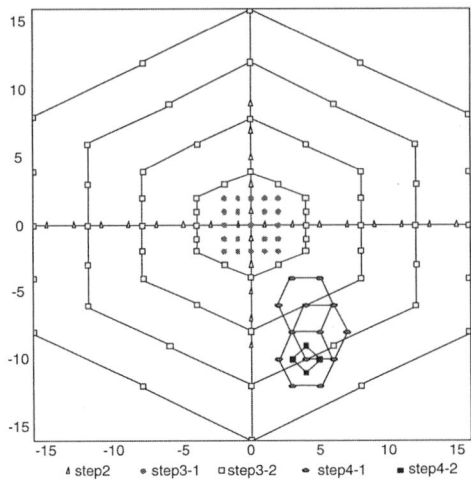

Fig. 1. Search process for the motion vector (4, 10) by UMHexagonS algorithm, search radius equals 16

Where (m_x, m_y) is the final motion vector of the predefined block in the current frame, N_1 is the execution times of the Unsymmetrical-cross search pattern used in the search process, N_2 is the execution times of the Uneven Multi-Hexagon-grid search pattern used in the search process, and N_3 is the execution times of the Extend hexagon-based search pattern used in the search process.

In order to prevent the search points from getting into a local minimum, in the earlier stages of UMHexagonS algorithm, global search is performed with many search points that can cover the overall search area, such as step 2 and step 3. Compared with HEXBS algorithm, UMHexagonS algorithm can reduce its susceptibility to get stuck in local optima. But this maybe is a shortcoming of UMHexagonS algorithm that motion searching is not flexible to the scenes with different motion degrees, especially in the beginning of lower resolution search. Experiments show that there is little difference between the number of search points of the smaller motion vector (2,3) and that of the larger motion vector (4,10).

Here, we assume reasonably that the global minimum has a monotonic distortion, that is, the nearer to the global minimum, the smaller the distortion in all directions within a small neighborhood around the global minimum. Based on this theory, we can see that many candidate points in Fig.1 are not indispensable to the searching of the motion vector (4,10). So, there is usually point-redundancy among the search points in search process, especially in the beginning of lower resolution search. Point-redundancy is the main reason for high computational complexity of motion estimation.

In our view, the rootstock of the aforementioned disadvantages in UMHexagonS algorithm is that the candidate points in search patterns are distributed irrationally, and the motion tendency, obtained in former phase of search process, is unexploited in the searching process of motion vectors. In

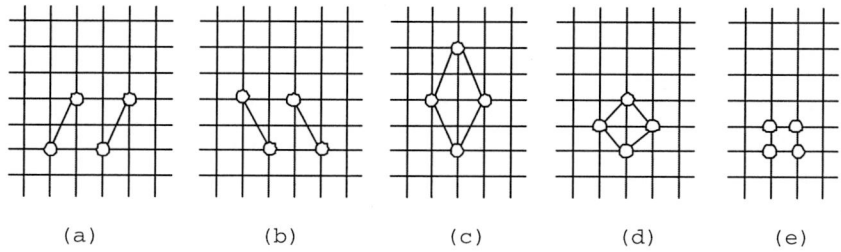

Fig. 2. CAPS Pattern (a)(b)(c) CAPS models (d)(e)EPS models

short, UMHexagonS algorithm is inefficient in terms of the speedup of motion estimation. Thus, we develop a new fast motion estimation algorithm, which is based on a novel search pattern called Context-Adaptive Parallelogram Search Pattern (CAPS).

3   Context-Adaptive Parallelogram Search Algorithm

3.1   CAPS Pattern

The CAPS pattern of our proposed algorithm has five kinds of parallelogram search models. Because the search model is chosen from a selection of the five search models depending on the recentaly-obtained motion vector in the process of Motion Estimation, we call the parallelogram search pattern CAPS (Context-Adaptive Parallelogram Search) pattern.

We divide the five search models into two groups: one is called CAPS model group, the other is called EPS (extended parallelogram search) model group. The CAPS model group contains three search models, which are shown by (a), (b) and (c) in Fig.2 A CAPS model includes four candidate points, and two neighboring points have a distance of 1 or 2 in horizontal direction and a distance of 2 or 0 in vertical direction. The EPS model group contains two extended parallelogram search models. Fig.2 (d) and (e) depict the two extended parallelogram search models. The EPS model indicated by Fig.2 (d) covers five candidate points (left, right, up, and down dots around the center with distance 1) in the motion field, and two points in five search points are the Minimum Block Distortion (MBD) point and the Secondary Minimum Distortion (SMBD) point of the latest search step. The EPS model illustrated in Fig.2 (e) covers four search points, and one point among them is the MBD point of the latest search step; the two neighboring points have a distance of 1. The EPS models are finally used in the process of motion estimation. Note that the EPS model shown by (d) in Fig.2 includes the same checking points as the end-diamond pattern.

3.2   Algorithm Development

With the designed CAPS pattern, we develop the search procedure of CAPS algorithm by four search stages described as follows . In the first stage, median

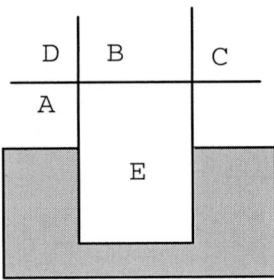

Fig. 3. Reference block location for prediction of motion vectors

prediction of the motion vector is done for the predefined block. The median value, among the motion vectors of the blocks on the left, top, and top-right (or top-left) of the predefined block, is calculated to predict the motion vector of the predefined block (as Fig.3 shows). Corresponding to the median value, the initial search point of the predefined block can be achieved. So, median prediction is also called as an initial search point prediction. The median prediction is expressed as following formula:

$$pred\_mv = median(Mv\_A, Mv\_B, Mv\_C) \qquad (2)$$

In the second stage, depending on the direction of $pred\_mv$, the search model may be chosen from a selection of search models, and is used for searching the motion vector of the predefined block. The optimum search point of this search step can be obtained. If the optimum search point of this search step is as same as the optimum search point of the latest search step, block-matching search turns to execute the fourth stage; otherwise, it goes to execute the third stage. In the third stage, based on the relative location between the optimum search point and the secondary optimum point, one of CAPS models is selected adaptively to search fine motion vector. If the optimum search point, obtained in this search step, is as same as the latest optimum search point, block-matching search goes into the fourth stage; otherwise, it keeps on searching in the third stage. In the fourth stage, based on the relative location between the optimum search point and the secondary optimum search point, one of EPS models is selected adaptively to search more fine motion vector. With the optimum search point obtained in the fourth stage, we can obtain the final motion vector of predefined block. Note that only three or two new candidate points will be evaluated in each searching step. Fig.4 shows an example of the searching process of the motion vector (4, 10). There are 21(=4+3+2+3+2+2+2+3) search points evaluated in the searching process of motion vector (4, 10).

The proposed CAPS algorithm can be described by the following steps in detail.

1. Using formula (2) to predict the initial motion vector of the current block (namely the predefined block), the predicted motion vector is $pred\_mv$. We use the value as the initial motion vector of the current block.

2. The initial motion vector make an angle with horizontal direction. The angle (α) can be worked out.

$if((\alpha >= 0)\vartheta\vartheta(\alpha < 30))$
{

The CAPS model shown in Fig.2 (c) is chosen and used for searching along the negative direction and the positive direction of the initial motion vector respectively.

}
$if((\alpha >= 30)\vartheta\vartheta(\alpha < 90))$
{

The CAPS model shown in Fig.2 (b) is chosen and used for searching along the negative direction and the positive direction of the initial motion vector respectively.

}
$if((\alpha >= 90)\vartheta\vartheta(\alpha < 150))$
{

The CAPS model shown in Fig.2 (c) is chosen and used for searching along the negative direction and the positive direction of the initial motion vector respectively.

}

Seven new candidate points are checked, then the MBD point and the SMBD point are obtained. If the MBD point is as same as the initial search point, the searching goes to Step 4; otherwise, it goes to Step 3.

3. Suppose there have finished n searching steps. An x-, y- pair specifies horizontal and vertical direction. $search\_range$ denotes search radius, $P_1(x_1, y_1)$ and $P_2(x_2, y_2)$ are the MBD point and the SMBD point of the $n-1$ searching step. $P_0(x_0, y_0)$ is the MBD point of the $n-2$ searching step. The vector from P_0 to P_1 can be expressed by formula (3), and the vector from P_2 to P_1 can be expressed by formula (4).

$$P_0P_1 = (\Delta x_0, \Delta y_0) = (x_1 - x_0, y_1 - y_0) \tag{3}$$

$$P_2P_1 = (\Delta x_1, \Delta y_1) = (x_1 - x_2, y_1 - y_2) \tag{4}$$

$for(i = 0; i < search\_range; i++)$
{
$if((\Delta y_0 = 0)\vartheta\vartheta(\Delta y_1 = 0))$
{

the CAPS model shown in Fig.2 (c) is chosen, and the searching direction can be expressed as following:
$if(\Delta x_0 < 0)$
searching direction is the positive x-direction;
$if(\Delta x_0 > 0)$
searching direction is the negative x-direction;

}
$if(((\Delta y_1 = 0)\vartheta\vartheta(\Delta y_0 \times \Delta x_0 < 0))\|((\Delta y_0 = 0)\vartheta\vartheta(\Delta y_1 \times \Delta x_1 < 0)))$
{

the CAPS model shown in Fig.2 (a) is chosen, and the searching direction can be expressed as following:

$if(\Delta x_1 > 0)$

searching direction is the negative y-direction;

$if(\Delta x_1 < 0)$

searching direction is the positive x-direction;

$if(\Delta x_0 > 0)$

searching direction is the positive y-direction;

$if(\Delta x_0 < 0)$

searching direction is the negative x-direction;

}

$if(((\Delta y_1 = 0)\vartheta\vartheta(\Delta y_0 \times \Delta x_0 > 0))\|((\Delta y_0 = 0)\vartheta\vartheta(\Delta y_1 \times \Delta x_1 > 0)))$

{

the CAPS model shown in Fig.2 (b) is chosen, and the searching direction can be expressed as following:

$if(\Delta x_1 < 0)$

searching direction is the negative y-direction;

$if(\Delta x_1 > 0)$

searching direction is the positive y-direction;

$if(\Delta x_0 < 0)$

searching direction is the negative x-direction;

$if(\Delta x_0 > 0)$

searching direction is the positive x-direction;

}

$if((\Delta y_1! = 0)\vartheta\vartheta(\Delta y_0! = 0)\vartheta\vartheta(\Delta x_1! = 0)\vartheta\vartheta(\Delta x_0! = 0))$

{

the CAPS model shown in Fig.2 (c) is chosen, and the searching direction can be expressed as following:

$if((\Delta y_1 < 0)\vartheta\vartheta(\Delta x_0 < 0))$

searching direction is left-up direction of P_1;

$if((\Delta y_1 > 0)\vartheta\vartheta(\Delta x_0 < 0))$

searching direction is left-down direction of P_1;

$if((\Delta y_1 > 0)\vartheta\vartheta(\Delta x_0 > 0))$

searching direction is right-down direction of P_1;

$if((\Delta y_1 < 0)\vartheta\vartheta(\Delta x_0 > 0))$

searching direction is right-up direction of P_1;

}

}

Three or two search points are checked, and the MBD point and the SMBD point are refreshed. If the MBD point is still the MBD point of the former searching step, the searching is switched to Step 4; otherwise, it executes Step 3 repeatedly.

4. Assume that $P_1(x_1, y_1)$ and $P_2(x_2, y_2)$ are the MBD point and the SMBD point of the latest search step respectively. The vector from P_2 and P_1 is expressed as formula (4).

$if(\Delta y_1 = 0)$

the EPS model shown in Fig.2 (d) is chosen;

$else$

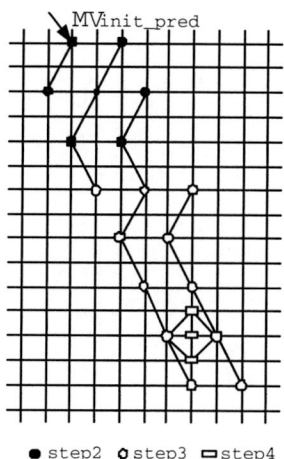

● step2 ○ step3 ▫ step4

Fig. 4. Search process for the motion vector(4, 10) by CAPS algorithm, search radius equals 16

the E P S m odel shown in F ig.2 (e) is chosen;
Three new search points are checked. The new MBD point is obtained by comparing the latest MBD point with the three new candidate points. The new MBD point is the final MBD point of the predefined block. From the final MBD point, the final motion vector of predefined block is obtained.

The search process of the proposed CAPS algorithm can be applied to each block of video frames, and we can call it as ME progress. From the ME process, the total number of search points required for per block is expressed as following formula:

$$N_{CAPS}(m_x, m_y) = 7 + 2n_1 + 3n_2 \qquad (5)$$

where, (m_x, m_y) is the final motion vector of the current block, n_1 is the execution times of the search models shown by Fig.2 (a) and Fig.2 (b), and n_2 is the execution times of the search models shown by Fig.2 (c), Fig.2 (d) and Fig.2 (e).

3.3   Analysis of the Proposed CAPS Algorithm

Compared with the UMHexagonS algorithm and other algorithms, we will examine the proposed CAPS algorithm in terms of the number of search points required in the ME process. To the motion estimation in video-coding, computational complexity can be measured by the number of search points required for the motion estimation. Fig.1 and Fig.4 are the search process charts of the motion vector (4, 10) for UMHexagonS algorithm and CAPS algorithm respectively. Seeing from Fig.4 and Fig.1, we can draw that the proposed CAPS algorithm evaluates 21(=4+3+2+3+2+2+2+3) candidate points; whereas the UMHexagonS algorithm checks beyond one hundred candidate points.

Under locating the same motion vector (m_x, m_y) , formula (5) and formula (1) express the possible minimum number of search points $N_{CAPS}(m_x, m_y)$ required by the proposed CAPS algorithm and $N_{UMHexagonS}(m_x, m_y)$ required by UMHexagonS algorithm respectively. Subtracting the corresponding number of search points required in CAPS algorithm from that in UMHexagonS algorithm, we can obtain the number of search points saved by the proposed CAPS algorithm as following formula:

$$\Delta N = N_{UMHexagonS} - N_{CAPS} \tag{6}$$

While locating the same motion vector, the speed improvement ratio (SIR) of CAPS algorithm over UMHexagonS algorithm is defined as follows:

$$SIR = \frac{N_{UMHexagonS} - N_{CAPS}}{N_{UMHexagonS}} \tag{7}$$

Formula (5) shows that: the larger and faster motions are, the more search points the proposed CAPS algorithm can save.

4   Experimental Results

For comparing the objective performance of the proposed algorithm with FS, HEXBS and UMHexagonS algorithms, many experiments were carried out. All the concerning block matching algorithms were integrated within version 7.6 of the JVT software [11], and were executed on Pentium 3 with 1G(Hz) CPU and 256M RAM.

Even though we have examined several different resolution sequences, we have selected only four sequences in this paper, and emphasized on the performance of the proposed CAPS algorithm. These are CIF sequences Foreman, Coast_guard, Mobile and SIF sequence Stefan. Main parameters, used in our experiments, are shown in Table1.

Table 1. Main parameters in experiment

video sequence (format, fps, fn)	Foreman(30, 300), Coast_guard(30, 300), Mobile(30, 300), Stefan(30, 300)
video format	CIF, SIF
video sequence style	IPBPBP···
QP	28, 32, 36, 40
search range	32×32
reference frames	5
block size	16×16 , 8×8 , 4×4
others	RDO method, Hadamard transform, CABAC entropy coder

To simplify our comparison, we have obtained some important performance parameters such as PSNR (Peak Signal-to-Noise Ratio), the ME time and the

Table 2. SIR values in the ratio of our CAPS algorithm'speed to FS, UMHexagonS and HEXBS algorithms'speeds (Search Range: 32 × 32)

SIR (%)	CAPS/FS				CAPS/UMHexagonS			
	QP=28	QP=32	QP=36	QP=40	QP=28	QP=32	QP=36	QP=40
Foreman(cif)	92.14	91.25	89.85	88.19	41.35	37.85	30.12	11.32
Stefan(sif)	92.88	92.83	92.09	90.98	45.64	48.75	46.17	39.31
Coast_guard(cif)	93.52	92.24	91.66	90.77	59.66	42.19	41.46	25.34
Mobile(cif)	93.39	91.99	91.72	91.80	75.43	50.34	51.31	56.59

speedup results (SIR). These results can be seen from Table2 and Fig.5 respectively.

Table 2 shows the average SIR improvement in percentage of CAPS algorithm over FS, and UMHexagonS algorithms. In Table 2, "CAPS/FS" indicates the value in the ratio of CAPS algorithm's ME proceeding speed to FS algorithm's, "CAPS/UMHexagonS" indicates the value in the ratio of CAPS algorithm's ME proceeding speed to UMHexagonS algorithm's. "SIR (%)" represents speed improvement ratio. When the quantizer value equals 28: For the Foreman sequence, CAPS algorithm achieves 92.14 percent in SIR to FS algorithm, achieves 41.35 percent in SIR to UMHexagonS algorithm; For the Mobile sequence, CAPS algorithm achieves 93.39 percent in SIR to FS algorithm, achieves 75.43 percent in SIR to UMHexagonS algorithm; For the Stefan sequence, CAPS algorithm achieves 92.88 percent in SIR to FS algorithm, achieves 45.64 percent in SIR to UMHexagonS algorithm. When quantizer value equals 32: For the Foreman sequence, CAPS algorithm achieves 91.25 percent in SIR to FS algorithm, achieves 37.85 percent in SIR to UMHexagonS algorithm; For the Mobile sequence, CAPS algorithm achieves 91.99 percent in SIR to FS algorithm, achieves 50.34 percent in SIR to UMHexagonS algorithm. Apparently, the experimental results substantially justify that the proposed algorithm has effectively reduced the computational complexity of the ME component in the encoder.

We all know that high value of PSNR represents good quality of the encoded bitstream, and the quality of the encoded bitstream is directly affected by the precision of the ME algorithm used. So, the PSNR is a guideline for the estimation accuracy. From the RD-plot described in Fig.5 (a) and Fig.5 (b), we observe that the RD-plot of UMHexagonS algorithm nears the RD-plot of FS algorithm, which just verifies our analisis in Section 2. We also see that the RD-plot of CAPS algorithm is very close to the RD-plot of FS algorithm, which indicates that CAPS algorithm has a relatively small reduction in PSNR (Peak Signal-to-Noise Ratio).

The above-mentioned experimental results indicate that CAPS algorithm has considerably reduced the computational complexity of ME comp in video-coding while keeping similar video quality as compared to FS algorithm.

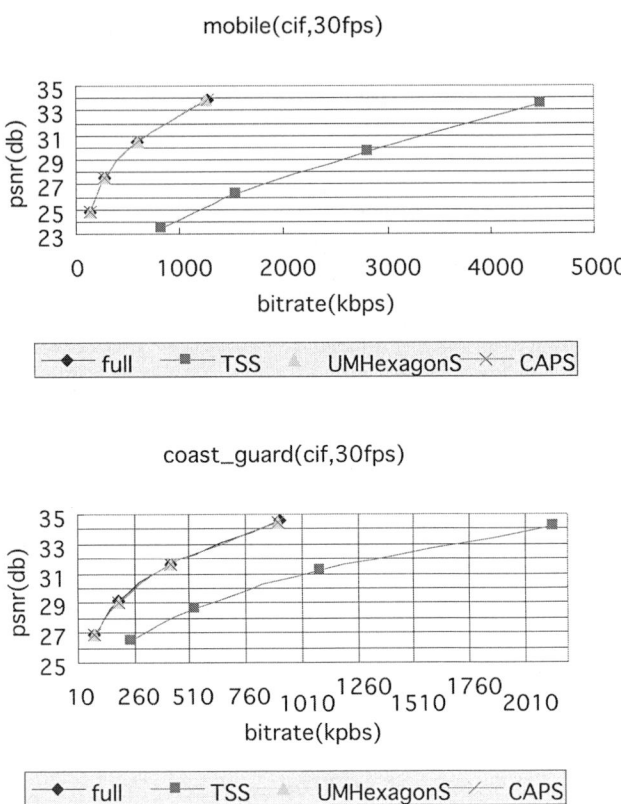

Fig. 5. RD performance plot (a) sequence Coast_guard. (b) sequence Mobile.

5   Conclusions

In this paper, we present a fast ME algorithm based on CAPS patterns. The proposed CAPS algorithm is compatible to variable block sizes and multiple reference frames of H.264. Experiments demonstrate that it can further reduce the computational complexity than other algorithms, such as UMHexagonS algorithm. Especially, the larger and faster the motion is, the more significant the speedup gained by the CAPS algorithm is.

Acknowledgements. We would like to thank the anonymous referees for their careful reading of the submitted version of the paper and for their comments.

References

1. A.M.Tourapis,O.C.Au,and M.L.LIOU,"Highly efficient predictive zonal algorithms for fast block-matching motion estimation",IEEE Transactions on Circuits and Systems for Video Technology,Vol.12,Iss.10,Pages:934-47,Oct'02.
2. Peng Yin et al.,"FAST MODE DECISION AND MOTION ESTIMATION FOR JVT/H.264",ICIP 2003.
3. ITU-T Ree.H.264/ISO/IEC 11496-10,"Advanced Video Coding",Final Committee Draft, Document JVT-G050,March 2003
4. R.Li,B.Zeng,and M.L.Liou,"A new three-step search algorithm 4Kavita Ravi and Fabio Somenzi. Minimal Assignments for Bounded Model Checking. Tools and Algorithms for the Construction and Analysis of Systems (TACAS), March-April 2004.
5. L.M.Po and W.C.Ma,"A novel four-step search algorithm for fast block motion estimation",IEEE Trans. Circuits Syst. Video Technol.,Vol.6,pp.313-317, June 1996
6. L.K.Liu and E.Feng,"A block-based gradient descent search algorithm for block motion estimation in video coding", IEEE Trans. Circuits Syst. Video Technol.,Vol.6,pp.419-423, Aug 1996
7. S.Zhu and K.-K. Ma,"A new diamond search algrithm for fast block-matching motion estimation", IEEE Trans. Image Processing , Vol.9,pp.287-290, Feb 2000
8. Ce Zhu, Xiao Lin and Lap-Pui Chau,"Hexagon-Based Search Pattern for Fast Block Motion Estimation", IEEE Transaction on Circuits and Systems for Video Technology, Vol.12, NO.5 ,may 2002
9. Zhibo Chen, Peng Zhou and Yun He, Fast Integer Pel and Fractional Pel Motion Estimation for JVT, ISO/IEC JTCI/SC29/WG11 and ITU-T SG16 Q.6, December 2002
10. Zhibo Chen, Peng Zhou and Yun He, Fast Motion Estimation for JVT, Joint Video Team(JVT) of ISO/IEC MPEG & ITU-T VCEG 7th Meeting,7-14 March,2003
11. http://bs. hhi. de/~suehring/tml/download/old_ jm/jm7.6.

A Fast Successive Elimination Algorithm for Multiple Reference Images

Hyun-Soo Kang

Graduate School of AIM, Chung-Ang University,
221, Heuksuk-dong, Dongjak-ku, Seoul, 156-070, Korea
hskang@cau.ac.kr

Abstract. This paper presents a fast full search algorithm for motion estimation. The proposed method is a successive elimination algorithm (SEA) for multiple reference frame applications. We will show that motion estimation for the reference images temporally preceding the first reference image is less intensive in computation compared with that for the first reference image. Simulation results explain that our method reduces computation complexity although it has the same quality as the full search algorithm (FSA).

1   Introduction

Motion estimation (ME) have been widely adopted in video systems, since ME is very effective to exploit temporal redundancy of video signals. There is still a lot of need for the methods that can find out motion vectors more accurately and faster. Of ME algorithms, full search algorithm (FSA) yields the optimal motion vectors but requires much computation. To relieve the computational problem, there have been many algorithms such as 2-D logarithmic search algorithm, three step search algorithm, conjugate direct search algorithm, cross search algorithm, four step search algorithm, and diamond search algorithm [1][2][3][4]. Recently, there have been some input documents for motion estimation for H.264. In [5], adaptive hexagon-based search (AHBS) was introduced which considers the blocks size and block shape of each mode of a macroblock and can achieve least search points with a little degradation. In [6], initial search point prediction and early termination for fast motion estimation was suggested which can be cooperated with AHBS.

Meanwhile, there have been some works to speed up FSA itself without deterioration of the motion estimation error of FSA. The representative works were PDE (partial difference elimination algorithm), SEA (successive elimination algorithm), MSEA (multi-level SEA) and so on. Among them, we would like to pay attention to SEA where a test is performed for whether a search point can be or not a candidate of the optimal vector and thereafter the search points that fail in the test are excluded from the set of candidates for the optimal vector and they are not proceeded further [7]. MSEA can be considered as a generalized version of SEA [8][9]. It hierarchically applies the test done in SEA, varying the resolution of blocks from low resolution to high resolution.

V. Roca and F. Rousseau (Eds.): MIPS 2004, LNCS 3311, pp. 187–193, 2004.

In this paper, we introduce a new method of SEA effectively applicable to multi-reference frame applications such as H.264 [10]. Our method is not for the immediately previous image of a current image, namely the first reference image, but for the other reference images, namely second reference image, third one, and so on. We show that the computation complexity for motion estimation process for the reference images following the first reference image can be reduced using the relation between the first reference image and the other reference images.

2   Background

Prior to explaining our method, we need to introduce a conventional fast algorithm, SEA, for motion estimation. We consider that the size of a block is $N \times N$, the size of the search window is $(2M + 1) \times (2M + 1)$, and $f(p, q, t)$ represents the pixel value at position (p, q) in frame t. Then a block is expressed by As described in [7], the constraint in SEA for relieving the search process is as follows:

$$||\mathbf{M}| - |\mathbf{R}|| \leq SAD(m, n), \tag{1}$$

where $\mathbf{R}$ and $\mathbf{M}$ denote the N^2 dimensional column vectors corresponding to a current block and a reference block, whose elements are $r_{iN+j} = f(p+i, q+j, t)$ for $0 \leq i, j < N$ and with element $m_{iN+j} = f(p + i - x, q + j - y, t - 1)$ for $0 \leq i, j < N$ and $-M < x, y < M$, respectively, and $| \cdot |$ is the sum norm of a vector, for instance,

$$|\mathbf{R}| = \sum_{i=0}^{N-1} \sum_{j=0}^{N-1} |f(p + i, q + j, t)|, \tag{2}$$

and

$$SAD(m, n) = |\mathbf{R} - \mathbf{O}_{t-1}|, \tag{3}$$

where $\mathbf{O}_{t-1}$ is the N^2 dimensional column vector with element $o_{iN+j} = f(p+i-m, q+j-n, t-1)$. Assuming we have obtained $SAD(m, n)$ for an initial matching candidate block with the motion vector (m, n), Eq.(1) must be satisfied in order that (x, y) is a better matching candidate. Here is the idea of SEA, which is to perform the search process only on those blocks whose sum norms satisfy Eq. (1).

3   Proposed Method

In this section, as in SEA [7], we will drive an additional inequality to constrain the search process while preserving the optimal motion vector and then describe our algorithm. Since the multiple reference frame approach has been employed by video systems such as H.264 [10], motion estimation methods need to be efficiently extended instead of a straightforward extension. Thus, we introduce a computationally efficient extension of SEA for two or more reference frames. For simplicity, only two reference frames will be considered. Employing the notation in the previous section, we define a block located at (x, y) in frame $t - 2$ as a column vector $\mathbf{P}$ with element $p_{iN+j} = f(p+i-x, q+j-y, t-2)$ for $0 \leq i, j < N$. Then the constraint on the search process for frame $t - 2$ will be derived, while the constraint on frame $t - 1$ will follow the conventional SEA.

Prior to explaining the proposed method, we consider the following straightforward implementation of motion estimation for two reference frame: (1) apply SEA to frame $t-1$, and obtain the optimal motion vector (m^*, n^*) for the frame, and (2) in the same manner, apply SEA to frame $t - 2$, using $SAD(m^*, n^*)$ as an initial minimum SAD of the search process for the frame $t - 2$. This approach will be called SEA2 for use in the next section.

From Minkowski's inequality, $||\mathbf{A}| - |\mathbf{B}|| \leq |\mathbf{A} - \mathbf{B}|$, we have

$$||\mathbf{P} - \mathbf{M}| - |\mathbf{M} - \mathbf{R}|| \leq |\mathbf{P} - \mathbf{R}|, \tag{4}$$

where $|\mathbf{P} - \mathbf{R}|$ corresponds to the SAD between a current block in frame t and a reference block with displacement of (x, y) in frame $t - 2$. Like the derivation in SEA, let's assume we have obtained $SAD_2(m, n)$ for an initial matching candidate block with the motion vector (m, n) which is newly defined considering both of reference frames, i.e.,

$$SAD_2(m, n) = min\{|\mathbf{R} - \mathbf{O}_{t-2}|, SAD(m^*, n^*)\}. \tag{5}$$

For a current position (x, y) to be better matching candidate, the following has to be satisfied,

$$|\mathbf{P} - \mathbf{R}| \leq SAD_2(m, n). \tag{6}$$

Then,

$$||\mathbf{P} - \mathbf{M}| - |\mathbf{M} - \mathbf{R}|| \leq SAD_2(m, n). \tag{7}$$

This is the main result proposed in this paper. $|\mathbf{M} - \mathbf{R}|$ corresponds to the SAD between the current block, $\mathbf{R}$, and the reference block, $\mathbf{M}$, indicated by the motion vector (x, y). And $|\mathbf{P} - \mathbf{M}|$ is the SAD between the blocks, $\mathbf{P}$ and $\mathbf{M}$, located at the same position $(p + i - x, q + j - y)$ in two reference frames at $t - 1$ and $t - 2$.

Reminding of the procedure of SEA2, in step (1), if the inequality in Eq. (1) is satisfied, $|\mathbf{M} - \mathbf{R}|$ is computed because the vector is a candidate of the optimal vector. In this case, if $|\mathbf{P} - \mathbf{M}|$ is available by some means, the details of which will be given later, the inequality in Eq. (1) for frame t-2 as well as the inequality in Eq. (7) can be used to constrain the search process for $\mathbf{P}$. This is our key idea where the SAD between the blocks in frame t and frame $t - 1$ is used to eliminate the search process for the next previous frame $t - 2$, providing step (2) with another additional inequality, Eq. (7). On the other hand, if the inequality in Eq. (1) is not satisfied in step (1), our method will be not applied because $|\mathbf{M} - \mathbf{R}|$ should be additionally calculated which is not on the ordinary process of SEA.

Finally, Fig. 1 shows the pseudo-code of our method. It should be noted that our method has computational gain as much as the condition at line 13 is satisfied. That is, the search process for frame $t - 1$(line 2 to line 10) is the same as the conventional SEA, but the one for frame $t - 2$(line 11 to line 20) is different. Thus, we have gain in the search process for frame $t - 2$. For instance, assume that the condition at line 2 is not satisfied with about 70 percent of blocks and the condition at line 13 is satisfied with half of those blocks. Then, we can save the computation of about 35 percent, $0.7 \times 0.5 = 0.35$, in frame $t - 2$. Conclusively, we have gain in the search process for frame $t - 2$ depending on the conditions in Eq. (1) and Eq. (7).

Fast computation of $|\mathbf{P} - \mathbf{M}|$: $\mathbf{P}$ and $\mathbf{M}$ are associated to blocks located at the same position but in different frames $t - 1$ and $t - 2$. Thus, $|\mathbf{P} - \mathbf{M}|$ is the SAD between those blocks. To develop its fast computation, suppose that the size of an image is $W \times H$ and the representation of the column vectors, $\mathbf{P}$ and $\mathbf{M}$, is generalized to $\mathbf{P}_{p,q}$ and $\mathbf{M}_{p,q}$, where $\mathbf{P}_{i,j} = \{f(p+i, q+j, t-2), 0 \leq i, j < N\}$ and $\mathbf{M}_{i,j} = \{f(p+i, q+j, t-1), 0 \leq i, j < N\}$. Then the computation takes two following steps.

(1) For the whole frames, obtain the absolute difference frame, $d(p, q) = |f(p, q, t-1) - f(p, q, t-2)|$. This requires $W \times H$ sum operations and $W \times H$ absolute operations, or N^2 sum and absolute operations for each block.

(2) In this step, we assume to adopt the unrestricted motion estimation scheme, which removes the need of extraordinary treatment for the blocks near picture boundaries. Then, compute $|\mathbf{P}_{p+1,q} - \mathbf{M}_{p+1,q}|$, using $|\mathbf{P}_{p,q} - \mathbf{M}_{p,q}|$ that have been calculated for the immediately previous search point. $|\mathbf{P}_{p+1,q} - \mathbf{M}_{p+1,q}|$ requires only $2N$ sum operations for $2N$ pixels newly come in $\mathbf{P}_{p+1,q}$ and $\mathbf{M}_{p+1,q}$, and $2N$ subtraction operations for $2N$ pixels gone out from $\mathbf{P}_{p,q}$ and $\mathbf{M}_{p,q}$.

Assuming absolute operation being equivalent to sum operation, on average, the computation overhead for each block OH is

$$OH = 2N^2 + 4N, \tag{8}$$

where $2N^2$ and $4N$ are from step (1) and (2), respectively. Considering block matching at each search point that takes N^2 operations, the computation overhead corresponds to only about two search operations for a large value of N.

4   Experimental Results

Our algorithm is compared with SEA2 which is a straightforward extension of SEA for multiple reference applications. In the experiments, we used seven image sequences with CIF at 30Hz, each of which consists of 100 frames. The block size and the search range are 16×16 and ± 15, respectively. The number of the reference frames is selected as 2. The results of the proposed method and SEA2 are shown in Table 1, where $ANSP_{t-1}$ and $ANSP_{t-2}$ stand for the

```
1    M_R_available = 0; // flag for check if |M − R| is available
2    if(||M| − |R|| > SAD₂(m, n)) continue;
3    else{
4       SAD_{t−1} = |M − R|;
5       M_R_available = 1;
6       if(SAD_{t−1} < SAD₂(m, n)){
7          SAD₂(m, n) = SAD_{t−1};
8          (m, n) = (x, y); // update the candidate vector
9          Ref_index = 0; // update the reference frame index
10   }}
11   if(||P| − |R|| > SAD₂(m, n)) continue;
12   else{
13      if(M_R_available &&||P − R| − |M − R|| > SAD₂(m, n)) continue;
14      else{
15         SAD_{t−2} = |P − R|;
16         if(SAD_{t−2} < SAD₂(m, n)){
17            SAD₂(m, n) = SAD_{t−2};
18            (m, n) = (x, y);
19            Ref_index = 1;
20   }}}
```

Fig. 1. Pseudo-code of the proposed method

average number of search positions per block in the reference frame $t − 1$ and $t − 2$, respectively, and 'Mot & Dau' in the column of 'Image' means mother and daughter image sequence. Since two methods have the same complexity of motion estimation for frame $t − 1$, we should focus on the complexity for frame $t − 2$. For comparison, therefore, the column of $ANSP_{t−2}$ has two entries of the proposed method and SEA2. In $ANSP_{t−2}$, our method saves the computation of 39% to 86% compared with SEA2. It is interesting that our method is most effective for Container and Mobile sequences which have quite different characteristics. The ANSP for our method should include the block overhead computation. However, since the overhead amounts to just about 2 search points, it can be ignored in consideration of the quantities shown in Table 1. The details of $ANSP_{t−2}$ for each frame of Foreman sequence are shown in Fig. 2.

5   Conclusion

We proposed a computationally effective extension of SEA for multi-reference frames. It saves a considerable amount of computations in motion estimation for the reference frame temporally preceding the first reference frame, while it preserves the same estimation accuracy as FSA. It was realized by adding a new inequality to the inequality of SEA. Though we dealt with the case of considering two reference frames, it can be easily generalized to the case of more than two reference frames.

Table 1. Performance of the proposed method and SEA2

Image	ANSP of FSA	$ANSP_{t-1}$	$ANSP_{t-2}$		PSNR
			SEA2	Proposed	
Coastguard	961	429.6	423.0	122.7	30.41
Container	961	223.1	221.3	31.4	38.35
Foreman	961	188.8	179.0	98.3	33.95
Mobile	961	335.1	334.0	46.6	25.95
Mot & Dau	961	231.6	229.0	141.4	40.50
Stefan	961	322.9	311.5	174.6	26.73
Table tennis	961	469.9	458.5	280.3	30.66

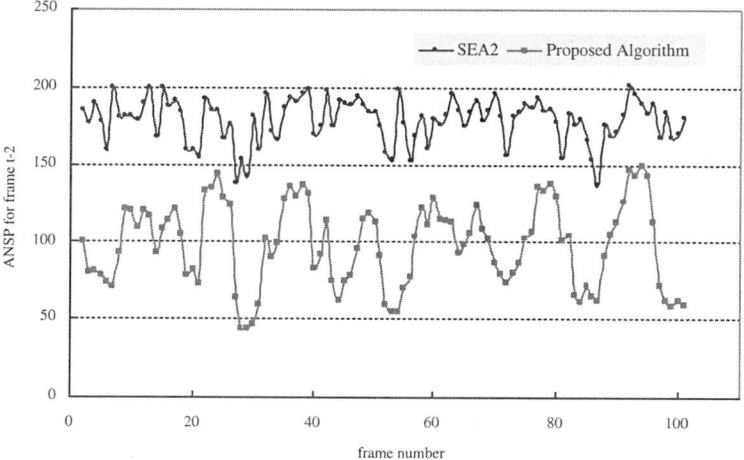

Fig. 2. $ANSP_{t-2}$ for each frame of Foreman sequence

Acknowledgements. This research was supported in part by the Ministry of Education, Korea, under the BK21 Project.

References

1. F.Dufaux and F. Moscheni, "Motion estimation techniques for digital TV: A review and a new contribution," Proc. IEEE, vol, 83, pp. 858-879, June 1995.
2. L.M. Po and W. C. Ma, "A novel four-step search algorithm for fast block motion estimation," IEEE Trans. Circuits Syst. Video Technol., vol. 6, pp. 313-317, June 1996.
3. L.K.Liu and E. Feig, "A block-based gradient descent search algorithm for block motion estimation in video coding," IEEE Trans. Circuits Syst. Video Technol., vol. 6, pp. 419-423, Aug. 1996.

4. S. Zhu and K.-K. Ma, "A new diamond search algorithm for fast block matching motion estimation," in Proc. Int. Conf. Inform., Comm., Signal Processing, Singapore, Sept. 9-12, 1997, pp. 292-296.

5. J. Zhang, Y. He, S. Yang, Y. Zhong, "Fast motion estimation method for MPEG-4 AVC," ISO/IEC JTC1/SC29/WG11, MPEG02/M8985, Shanghai, October, 2002.

6. Z. Chen, P. Zhou, Y. He, "Fast motion estimation for JVT," JVT-G016.doc, Joint Video Team (JVT) of ISO/IEC MPEG & ITU-T VCEG, Pattaya II, 7-14 March, 2003.

7. W. Li and E. Salari, "Successive elimination algorithm for motion estimation," IEEE Trans. on Image Processing, vol. 4, no. 1, pp. 105-107, Jan. 1995

8. X. Q. Gao, C. J. Duanmu, and C. R. Zou, "A multilevel successive elimination algorithm for block matching motion estimation," IEEE Trans. on Image Processing, vol. 9, no. 3, pp. 501-504, March 2000.

9. J. Y. Lu, K. S. Wu and J. C. Lin, "Fast full search in motion estimation by hierarchical use of Minkowski's inequality," Pattern Recognition, vol. 31, no. 7, pp. 945-952, pp. 945-952, 1998.

10. ITU-T recommendation H.264 — ISO/IEC 14496-10 AVC, "Draft text of final draft standard for advanced video coding," Mar. 2003.

Using Dynamic Replication to Manage Service Availability in a Multimedia Server Cluster

Jonathan Dukes and Jeremy Jones

Department of Computer Science, Trinity College Dublin, Ireland
{Jonathan.Dukes, Jeremy.Jones,}@cs.tcd.ie
http://www.cs.tcd.ie/CAG

Abstract. Dynamic replication policies assign non-disjoint subsets of multimedia presentations to nodes in a server cluster, replicating selected presentations to achieve load-balancing, while avoiding complete replication of the multimedia archive on every node. This paper presents a development of our existing *Dynamic RePacking* policy, which creates a configurable minimum number of replicas of selected presentations, increasing their availability. These additional replicas are assigned to nodes in a manner that allows load-balancing to be maintained when nodes fail. By separating replication to achieve load-balancing from replication to increase the availability of individual presentations, the trade-off between availability and storage cost can be controlled. This is illustrated by performance results from a prototype multimedia server cluster, which uses a group-communication service to implement inter-node communication.

1   Introduction

Commodity software solutions for providing on-demand multimedia streaming services, such as RealNetworks' Helix Universal Server or Microsoft's Windows Media Services, are easily deployed using stand-alone PCs. When the demand for an on-demand service exceeds the capacity of a single computer, the resources of multiple PCs can be aggregated in a server cluster.

The distribution of multimedia content among the nodes in a server cluster is of critical importance. One approach that may be adopted is to clone the multimedia server and all available content on each node. Client requests can be redirected to the least-loaded node in the cluster, maximizing the number of streams that can be supplied. Cloning large multimedia archives – which may be many terabytes in size – on large numbers of nodes becomes prohibitively expensive as the size of the archive or the number of nodes increases. While the cost-per-gigabyte of disk storage equipment is decreasing, the cost of managing such large archives, replicated many times, is an important consideration [7], as are environmental issues such as cooling and power consumption [2]. The time required to replicate large archives when adding new nodes or replacing failed ones must also be considered, since an increase in the mean-time-to-repair of a node will significantly reduce service availability.

V. Roca and F. Rousseau (Eds.): MIPS 2004, LNCS 3311, pp. 194–205, 2004.

Server striping [9] has been proposed in the past as a method of increasing the service capacity of a multimedia server cluster, while only storing a single copy of each presentation. This approach, however, has disadvantages relating to scalability, availability and the reconstruction of streams for playback by clients, and these disadvantages have been extensively discussed in the past, for example, by Lie et al [10].

Given the current trend towards the use of content distribution networks, in which clients receive content from nearby edge servers, we believe there is a need for low-cost cluster-based multimedia servers that are self-managing, cost effective to deploy and maintain, scale incrementally and degrade gracefully as nodes fail. The *HammerHead* architecture discussed in this paper has been designed with these requirements in mind.

HammerHead is based on the use of dynamic replication of content in a loosely-coupled cluster, since we believe this approach offers the best compromise between scalability, availability, performance and cost. Unlike the cloned server cluster model described above, a non-disjoint subset of the presentations in a multimedia archive is assigned to each cluster node, avoiding complete replication of the archive on every node. Client demand for each presentation is periodically evaluated and used to determine an assignment of presentations to nodes that will allow load-balancing to be achieved, based on expected demand. Some files may be replicated to facilitate load-balancing, increase performance (by increasing the potential to perform load-balancing) or to increase the availability of selected presentations.

The *Dynamic RePacking* content replication policy, developed for our prototype HammerHead cluster, is described in our previous work [6]. This paper describes a significant development to the basic Dynamic RePacking policy, which allows a configurable minimum number of replicas to be created for individual presentations, according to application requirements. In addition, replicas are assigned to nodes in a manner that allows load-balancing to be maintained when nodes fail. We argue that by separating replication to achieve load-balancing from replication to increase the availability of individual presentations, the trade-off between increased service availability and increased storage cost can be controlled. Unlike the policies that have been proposed in the past, which were evaluated using simulation studies, we present results obtained from a prototype HammerHead server cluster to illustrate this trade-off.

In the next section, we describe how the basic Dynamic RePacking policy has been enhanced to allow the availability requirements of individual presentations to be satisfied. In section 3, we briefly describe the HammerHead architecture. Detailed experimental results are presented in section 4. Finally, we discuss related work in section 5 and summarize our work in section 6.

2   Dynamic RePacking

The basic Dynamic RePacking algorithm is a significant development of the existing MMPacking algorithm proposed by Serpanos et al. [13]. Each node has an associated *target cumulative demand*, representing the proportion of a cluster's aggregate bandwidth contributed by that node. Similarly, each presentation has an associated *demand*, representing the proportion of the aggregate cluster bandwidth required to supply the average number of concurrent streams of that presentation. Dynamic RePacking assigns a non-disjoint subset of the K presentations in a multimedia archive to the N nodes in a cluster such that the cumulative demand for the presentations assigned to a node is equal to the node's target cumulative demand.

Dynamic RePacking creates a near-minimal number of replicas to achieve load-balancing, with all but the most demanding presentations assigned to just a single node. Using this basic policy, the loss of any single node will result in the unavailability of a significant proportion of the presentations. In practice, it may be desirable to create additional replicas of some or all of the presentations, reducing the number of presentations lost when nodes fail and thus reducing the impact that node failures have on achieved service load. We now describe how the basic Dynamic RePacking algorithm is applied iteratively to combinations of nodes, creating additional replicas of selected presentations in a manner that allows load-balancing to be maintained when nodes fail.

Each presentation is assigned a *minimum replica count* (MRC) according to application requirements. For example, an application might require that streams of each of the most demanding 50% of the presentations can be supplied in the presence of a single node failure, thus requiring that each of these presentations is stored on at least two nodes, corresponding to a MRC of two. After assigning a MRC to each presentation, presentations are assigned to nodes in phases. After phase 0, which is equivalent to the basic Dynamic RePacking algorithm, each presentation is assigned to at least one node, with a near-minimal number of replicas created to achieve load-balancing. After phase 1, those presentations whose MRC is two or more will be assigned to at least two nodes, such that load-balancing can be maintained after a single node failure. After phase 2, those presentations whose MRC is three or more will be assigned to at least three nodes, such that load-balancing can be maintained after any two nodes fail, and so on, until the MRC of each presentation has been satisfied.

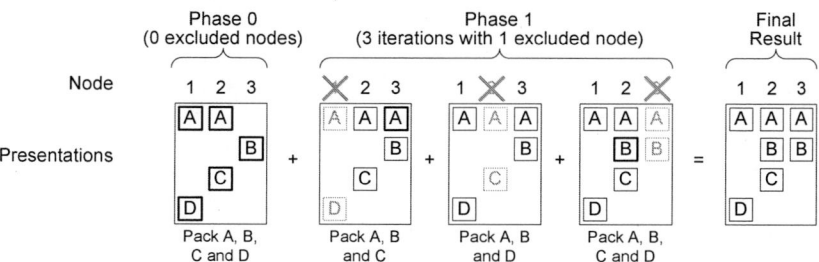

Fig. 1. Illustration of Dynamic RePacking over two phases with a total of four iterations. The MRC of presentations A and B is two while that of C and D is one.

In each phase, f, the Dynamic RePacking algorithm is used to assign presentations to nodes with f nodes excluded, as if those nodes had failed, as illustrated in Figure 1. Dynamic RePacking is executed $\binom{N}{f}$ times in each phase f and for each of these iterations, the presentations are packed as though a different combination of f nodes had failed. The effect of applying Dynamic RePacking iteratively in this manner will be to treat each combination of excluded and included nodes as a separate load-balancing problem, superimposing the assignment produced by each iteration over the assignments of previous iterations, as shown in Figure 1. Thus, in the first phase ($f = 0$), Dynamic RePacking will be applied $\binom{N}{0} = 1$ times, with no excluded nodes. In the second phase ($f = 1$), there will be $\binom{N}{1} = N$ iterations, each of which applies the Dynamic RePacking algorithm to a different combination of $N - f$ included nodes and f excluded nodes. The set of K' presentations assigned during each iteration is the union of *the set of presentations on the f excluded nodes that have not yet reached their MRC* and *the set of presentations on the $N - f$ included nodes, regardless of their MRC.* In other words, each iteration in phase f uses the basic Dynamic RePacking algorithm to assign presentations to nodes such that the availability of those presentations whose MRC is greater than f is guaranteed and, in addition, load-balancing can be maintained if the f excluded nodes were to actually fail. The number of phases must be at least equal to the maximum MRC assigned to any presentation.

Each iteration of the Dynamic RePacking policy proceeds as follows[1]. K' presentations are assigned to $N' = N - f$ nodes in round-robin order. Each node is assigned a *target shortfall* – the difference between the target cumulative demand of the node and the cumulative demand for the presentations assigned to the node during the current iteration. (At the beginning of each iteration, before any presentations have been assigned, the target shortfall of each node will be equal to its target cumulative demand. A node's target shortfall will decrease as presentations are assigned to it.) As the assignment proceeds in rounds, a heuristic approach is used to select a presentation to assign to the current node, in order to avoid excessive replication. The presentation selected for assignment to the current node is:

- the presentation with the lowest demand, which
  - if its MRC has been satisfied, has already been assigned to the current node by a previous iteration, or
  - if its MRC has not been satisfied, was previously stored on the current node, but has not yet been reassigned to the node

  or, if no such presentation exists,
- the presentation with the lowest demand, which was previously stored on the current node, but has not yet been reassigned to the node

  or, if no such presentation exists,
- the presentation with the lowest demand.

---

[1] Due to space limitations, we present only a brief description of the basic Dynamic RePacking algorithm. A more detailed explanation of the algorithm appears in our previous work [6]

The first criterion attempts to avoid creating additional replicas of those presentations whose MRC has already been satisfied by an earlier iteration, while giving equal priority to those presentations that have yet to satisfy their MRC. This will have the effect of sharing the workload of a failed node among the remaining nodes, were the current combination of f excluded nodes to actually fail. The second criterion reduces the cost (incurred by creating new replicas) of adapting to changes in the relative demand for presentations between successive reevaluations, if the first criterion cannot be satisfied. The final "fall-back" criterion causes a new replica to be created, requiring cluster resources to perform replication.

Assigning a presentation to a node reduces the target shortfall of the node by the presentation's estimated demand. If, after assigning a presentation to a node, the demand for the presentation is completely satisfied, then the presentation is removed from the list of unassigned presentations. If, however, the demand for the selected presentation exceeds the node's remaining target shortfall, the demand for the presentation is reduced by the remaining target shortfall and the presentation remains on the list for later assignment to another node, thus creating an additional replica of the presentation. After assigning a presentation to a node, assignment continues with the next node in round-robin order. Each round of assignment ends when either a presentation has been assigned to the last node or when, after assigning a presentation, the target shortfall of the current node is still greater than that of the next node. The nodes are sorted in descending order of target shortfall at the beginning of each round.

3   HammerHead Server Cluster

The HammerHead architecture implements a cluster-aware layer on top of a commodity multimedia streaming server, an instance of which runs on each cluster node. HammerHead consists of two main components, as illustrated in Figure 2. The HammerSource component provides an abstract view of the local commodity multimedia server on each node and is responsible for extracting state information from the server instance, capturing events of interest, communicating those events to the cluster-aware layer and initiating the download of new replicas from other cluster nodes or from remote locations. The HammerServer component is a stand-alone service that maintains the state information published by the HammerSource on each cluster node and updates this information in response to HammerSource events (such as the start and end of client streams or the creation of a new replica). An aggregated cluster state is formed from the state of each node and this information is used by HammerServer instances to perform load-balancing by redirecting client requests to suitable cluster nodes, and to implement the Dynamic RePacking policy described in the previous section. The aggregated cluster state is replicated at each cluster node, allowing the workload associated with client redirection to be shared and allowing the implementation of the dynamic replication policy to continue when nodes fail. The Ensemble [8] group communication toolkit is used to implement the communication channel between `HammerServer` and `HammerSource` instances.

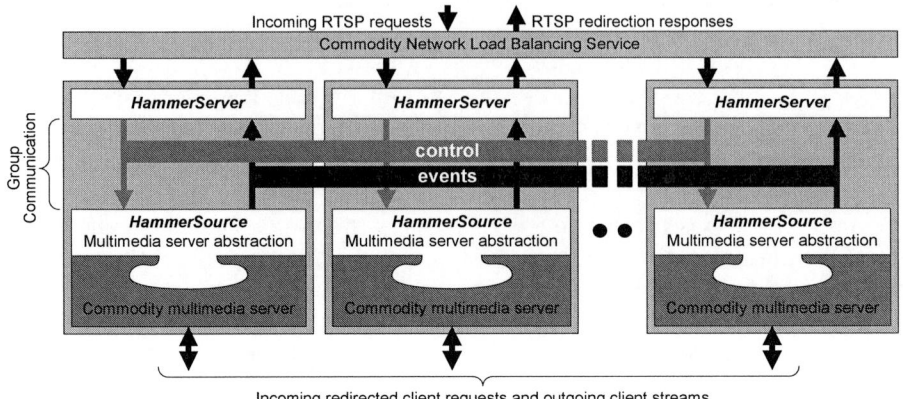

Fig. 2. HammerHead architecture

Client requests are distributed among `HammerServer` instances using a commodity network load-balancing service [11]. On receiving a request, a `HammerServer` instance examines the state of each node and redirects the client to the least-loaded commodity multimedia server instance. The client contacts the commodity multimedia server directly and the `HammerServer` layer plays no further active role in supplying the stream (although it will learn of the start and end of the stream through the events issued by the `HammerSource` plug-in and can use these events to update each node's published state and to estimate the demand for each presentation [6]).

Periodically (e.g. every five minutes, in the experiments described below), a coordinating `HammerServer` instance, designated by Ensemble, uses the estimated demand for each presentation to reevaluate the assignment of presentations to nodes, allowing the cluster to adapt to changes in client behaviour. This reevaluation may require new replicas to be created or existing replicas to be removed. The coordinating `HammerServer` instructs the relevant `HammerSource` plug-ins to implement the necessary changes. New replicas cannot be created instantaneously and must be downloaded from another cluster node, requiring resources that could otherwise be used to supply client streams. Similarly, the removal of a replica must be delayed until the streams serviced by the replica have ended and until any new replicas of the same presentation have been created. The coordinating `HammerServer` manages the creation and removal of replicas and coordinates the implementation of these changes with the redirection of client requests.

4   Performance Evaluation

A four-node HammerHead cluster was constructed and its performance evaluated using a set of hosts to generate RTSP [12] requests and consume the resulting multimedia streams. Although the pattern of generated requests was synthetic, the resulting multimedia streams were indistinguishable from requests generated by real users, from the perspective of the multimedia server. Each presentation's popularity was determined by a Zipf [17] distribution and requests were generated with exponentially distributed inter-arrival times. The mean request inter-arrival time was selected to achieve a target cluster load of 100%, corresponding to full utilization of the service capacity of the cluster. In the prototype cluster, Windows Media Services was used as the commodity multimedia server. Since Microsoft recommends limiting the outgoing stream bandwidth of a Windows Media Services instance to 50% of the bandwidth of the network adapter, and since each cluster node used a 100Mbps network adapter to supply client streams, the service capacity of each cluster node was limited to 50Mbps. For our experiments, using a four-node cluster, this corresponded to a cluster bandwidth of 200Mbps, or an average of 736 concurrent streams, each lasting 758 seconds. The inter-node replication streams used to create new replicas required the same network bandwidth as a client stream and were transported over the same network. Thus, the creation of new replicas temporarily reduced the resources available to supply client streams, at both the source and destination end of each replication stream. The demand for each presentation and the assignment of presentations to nodes was reevaluated using Dynamic RePacking every five minutes.

Four different Dynamic RePacking policy configurations have been evaluated. The first policy, referred to as *MinOne*, created only enough additional replicas of each presentation to facilitate load-balancing and to maintain load-balancing after a single node failure (two Dynamic RePacking phases). In other words, each presentation was assigned a MRC of one. The second configuration, *Top10*, assigned the most demanding 10% of the presentations to at least two cluster nodes, corresponding to a MRC of two, with the remaining presentations assigned a MRC of one. Similarly, the *Top50* configuration assigned the most demanding 50% of the presentations a MRC of two. Finally, the *MinTwo* configuration assigned every presentation to at least two nodes, corresponding to a MRC of two. A cloned server cluster configuration, in which every presentation was assigned to every node and Dynamic RePacking was disabled, was evaluated for comparison.

We begin by comparing the performance of each of these policy configurations in the absence of node failures. Figure 3 (a) shows the achieved service load, as a proportion of the target service load and Figure 3 (b) shows the storage capacity required by the policy, as a proportion of the minimum storage capacity required to assign each presentation to a single node. The performance achieved by the cloned configuration represents the optimal performance for this prototype cluster. This configuration, however, also incurs the highest storage capacity utilization (400% of the minimum capacity required to store each presentation on

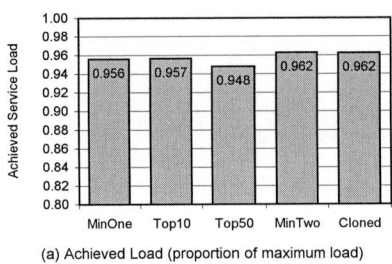

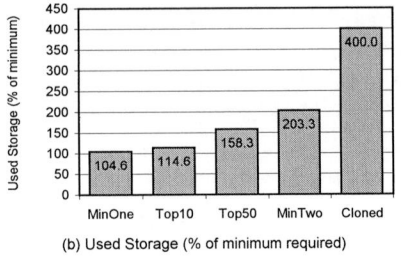

(a) Achieved Load (proportion of maximum load)

(b) Used Storage (% of minimum required)

Fig. 3. Dynamic RePacking policy

a single node). Periodic service capacity overloads, resulting from exponentially distributed interarrival times, and the latencies involved in the redirection of clients and updating the cluster state, prevented the cloned configuration from achieving 100% service load. The *MinOne* configuration significantly reduces storage capacity utilization to just 105% of the minimum value, with a reduction in achieved load of just 0.6%. The *MinTwo* policy achieved a service load which was approximately the same as the cloned configuration, while halving the storage capacity utilization. The *Top10* policy represents a compromise between the *MinOne* and *MinTwo* policies, both in terms of achieved load and storage utilization.

The *Top50* policy exhibits a small reduction in performance, compared with the *MinOne* policy, despite the creation of additional replicas of the most demanding presentations. This results from frequent changes in the relative ranking of presentations by estimated demand and, as a result, frequent changes in the MRC of those presentations. The resulting creation of new replicas reduces the bandwidth available to supply client streams. This suggests that, while evaluating demand over the short term is sufficient to achieve load-balancing, the evaluation of the MRC for each presentation should be based on statistics gathered over longer periods. (A further experiment, which ranked presentations according to their popularity measured over the duration of the experiment, rather than demand measured over half-hour periods, as was the case in these experiments, resulted in an increase in achieved service load to 0.963.)

To compare the performance of the five Dynamic RePacking configurations when failures occur, an experiment was conducted in which one node was removed from the cluster every hour. The achieved service load is plotted against time for target loads of 100% and 50% in Figure 4. Figure 5 shows the number of presentations still available after each node failure, as a proportion of the number of presentations originally in the archive. (We borrow the term *harvest* from Brewer's discussion of giant-scale services [3] for this metric.)

A cloned server cluster maintains 100% harvest despite any node failures. As a result, the cluster is able to achieve a service load approximately equal to the lower of the target service load and the remaining service capacity of the cluster. This is the optimal performance achievable when nodes fail. For a 100% target

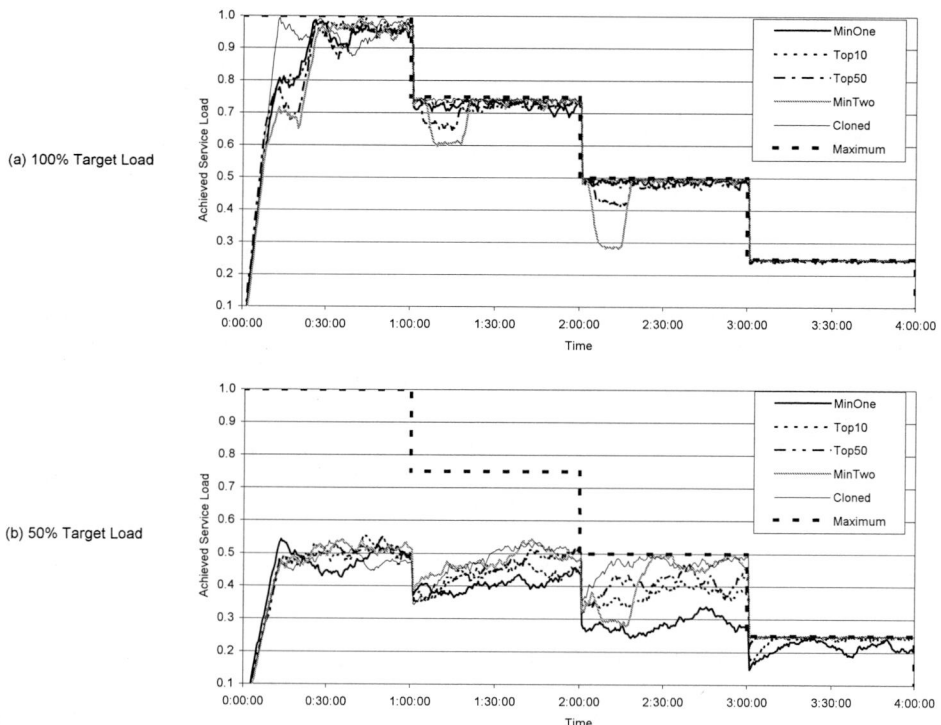

Fig. 4. Effect of node failure on achieved service load

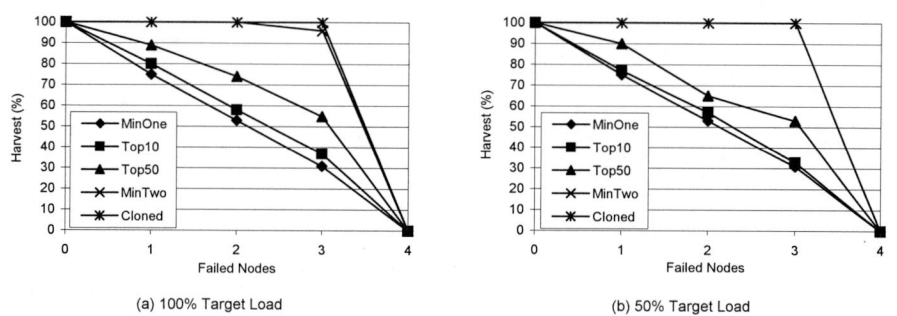

Fig. 5. Effect of node failure on harvest (% presentations available for streaming)

load, when a node failure occurs, the streams being serviced by the failed node all fail and, since the remaining nodes are already fully utilized, they cannot take on the workload of the failed node. For the 50% target load, however, after the first and second node failures, there is unused service capacity on the remaining nodes and the achieved service load is observed to recover after each failure.

For the 100% target service load, the characteristics of the *MinOne* and *Top10* configurations are similar to those of the cloned configuration, despite the reduction in harvest shown in Figure 5 (a). This behaviour occurs because, with a target load of 100%, there are sufficient requests for the remaining presentations to fully utilize the remaining nodes and because the assignment of presentations to nodes facilitated load-balancing, based on the estimated demand for each presentation. The *MinTwo* and *Top50* policies exhibit a temporary decrease in achieved load immediately after each node failure, since cluster resources that would otherwise be used to supply client streams are needed to re-replicate presentations up to their required MRC. This decrease in performance is present for the *Top10* configuration but is less noticeable.

For the 50% target service load, however, the reduction in harvest causes a reduction in achieved load, since a 50% target load does not generate a sufficient number of requests for each remaining presentation to fully utilize the remaining nodes. The impact of any reduction in harvest is reduced by creating additional replicas of the most demanding presentations. For example, using the *Top10* configuration, harvest decreases to 57% after two node failures but the achieved service load recovers to approximately 0.4, or 80% of the target service load.

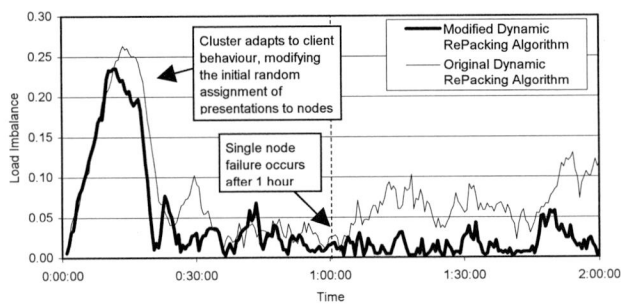

Fig. 6. Comparison of the basic and modified Dynamic RePacking algorithms

Finally, we illustrate how the modified Dynamic RePacking policy – presented in this paper – maintains load-balancing when node failures occur. For this experiment, the target service load was 75% and the the cluster was configured to disable Dynamic RePacking after the first node failure, preventing the cluster from recovering from the failure and demonstrating the behaviour to greater effect. In both cases, the MRC of each presentation was one. Figure 6 shows the load-imbalance of the cluster over a two hour period, with a single node failure occurring after one hour. The load-imbalance is expressed as the standard deviation between the load on each cluster node. The results clearly show that, while both configurations exhibit a similar level of load-imbalance up to the time when the node failure occurs, load-balancing is maintained using the modified algorithm, while there is a significant increase in load imbalance using the original algorithm. The initial peak in load imbalance during the first thirty minutes of operation illustrates how the Dynamic RePacking algorithm was used to modify the initial random assignment of presentations to nodes, adapting to the client request pattern.

5   Related Work

Although other dynamic replication policies have been proposed in the past [5, 4,10,14,15,16], all of the existing policies appear only to have been evaluated using simulation studies, whereas we present results from a prototype cluster implementation. None of the existing policies propose the separation of replication to achieve load-balancing from replication to increase availability, with the exception of [15], which does not consider load-balancing when assigning replicas to cluster nodes.

Anker et al. [1] propose a highly-available, fault-tolerant video-on-demand service, based on the use of group communication. The authors, however, do not address the dynamic replication policy used or its implementation and instead focus on rebalancing workload when nodes join or leave a cluster and on associated stream transport and buffering issues. Unlike this existing work, the HammerHead architecture described here restricts the use of group communication to inter-node communication, thereby allowing commodity multimedia players to access the service.

6   Conclusions

A modification of our basic Dynamic RePacking content replication policy has been proposed, which allows additional replicas of selected presentations to be created according to application-specific availability requirements. The modified policy assigns replicas to nodes in a manner that allows load-balancing to be maintained when nodes fail.

We have analysed the performance and behaviour of the modified Dynamic RePacking policy using a prototype HammerHead server cluster. The results presented demonstrate that selective dynamic replication can approach the performance achieved by replicating every presentation on every node, while significantly reducing the storage cost. The trade-off between increased availability and storage cost has been illustrated and it was shown that, by exploiting the highly skewed distribution of multimedia presentation popularity, a small increase in storage cost can significantly reduce the impact of node failures on performance. In the future, we intend investigating alternative methods for assigning a minimum replica count to each presentation.

References

1. T. Anker, D. Dolev, and I. Keidar. Fault tolerant video on demand services. In *Proceedings of the 19th International Conference on Distributed Computing Systems*, Austin, Texas, USA, June 1999.
2. L. A. Barroso, J. Dean, and U. Hölzle. Web search for a planet: The google cluster architecture. *IEEE Micro*, 23(2):22–28, March/April 2003.
3. E. A. Brewer. Lessons from giant-scale services. *IEEE Internet Computing*, 5(4):46–55, July/August 2001.
4. A. Dan and D. Sitaram. Dynamic policy of segment replication for load-balancing in video-on-demand servers. *ACM Multimedia Systems*, 3(3):93–103, July 1995.
5. A. Dan and D. Sitaram. An online video placement policy based on bandwidth to space ratio (BSR). In *Proceedings of the 1995 ACM SIGMOD International Conference on Management of Data*, pages 376–385, San Jose, California, USA, May 1995.
6. J. Dukes and J. Jones. Dynamic replication of content in the hammerhead multimedia server. In *Proceedings of EUROMEDIA 2003*, April 2003.
7. J. Gray and P. Shenoy. Rules of thumb in data engineering. In *Proceedings of the 16th International Conference on Data Engineering*, pages 3–12, San Diego, California, USA, February/March 2000.
8. M. Hayden. *The Ensemble System*. PhD thesis, Department of Computer Science, Cornell University, 1997.
9. J. Y. B. Lee. Parallel video servers: A tutorial. *IEEE Multimedia*, 5(2):20–28, April–June 1998.
10. P. W. K. Lie, J. C. S. Lui, and L. Golubchik. Threshold-based dynamic replication in large-scale video-on-demand systems. In *Proceedings of 8th International Workshop on Research Issues in Data Engineering (RIDE)*, Orlando, Florida, USA, February 1998.
11. Microsoft Corporation. Network load balancing technical overview. White paper, Microsoft Corporation, January 2000.
12. H. Schulzrinne, A. Rao, and R. Lanphier. Real time streaming protocol (RTSP). IETF RFC 2326 (proposed standard), April 1998.
13. D. N. Serpanos, L. Georgiadis, and T. Bouloutas. MMPacking: A load and storage balancing algorithm for distributed multimedia servers. *IEEE Transactions on Circuits and Systems for Video Technology*, 8(1):13–17, February 1998.
14. N. Venkatasubramanian and S. Ramanathan. Load management in distributed video servers. In *Proceedings of the International Conference on Distributed Computing Systems*, Baltimore, Maryland, USA, May 1997.
15. X. Wei and N. Venkatasubramanian. Predictive fault-tolerant placement in distributed video servers. In *IEEE International Conference on Multimedia and Expo 2001 (ICME 2001)*, Tokyo, Japan, August 2001.
16. J. L. Wolf, P. S. Yu, and H. Shachnai. DASD dancing: A disk load balancing optimization scheme for video-on-demand computer systems. In *Proceedings of the ACM SIGMETRICS Conference on Measurement and Modeling of Computer Systems (ACM SIGMETRICS '95)*, pages 157–166, Ottawa, Ontario, Canada, May 1995.
17. G. K. Zipf. *Human Behaviour and the Principle of Least Effort: an Introduction to Human Ecology*. Addison-Wesley, 1949.

Design of a Hybrid CDN

Karl-André Skevik, Vera Goebel, and Thomas Plagemann

Department of Informatics, University of Oslo
{karlas,goebel,plageman@ifi.uio.no}

Abstract. Peer-to-peer (P2P) based networks have several desirable features for content distribution, such as low costs, scalability, and fault tolerance. However, they usually fail to provide guarantees for content quality. In order to combine the desired features of classical Content Distribution Networks (CDNs) and P2P based networks, we use a hybrid CDN structure with a P2P streaming protocol. Based on an empirical analysis of BitTorrent and simulations, our design attempts to discourage freeloaders and reduces performance problems due to firewalls by incorporating a proxy based structure. This proxy based structure also makes it possible to incorporate caching, which has often been identified as lacking in P2P networks.

1 Introduction

Content Distribution Networks (CDNs) represent an area which has been subject for research in many years, but only recently the basis for widespread deployment has become available with the increasing availability of broadband Internet connections to private consumers. Powerful computers have also made efficient video compression techniques possible.

We regard Peer-to-peer (P2P) networks as an interesting basis for a streaming CDN system, because of the potential for scalability and fault tolerance in a P2P network. The use of caches in CDNs can significantly reduce network load, but caching is not often used by P2P networks. Content authenticity and quality is also often difficult to verify in typical P2P file distribution networks. A big problem with CDNs on the other hand, is their potentially high cost, which makes it desirable to spread the cost between users as it is done in P2P networks.

Our proposed hybrid solution aims to combine the best features of CDNs and P2P networks.

The P2P based file distribution application BitTorrent provided an opportunity to measure the effects of firewalls in P2P networks. We use caches with proxy functionality in order to avoid the performance issues which we identified and quantified through our empirical BitTorrent study. The hierarchical structure also implicitly creates clusters of nearby hosts, such as those using the same ISP. One of the contributions of this paper is the observation of how sensitive the performance of a P2P based streaming system is to firewalls. We also examine the effects of clients not contributing to the network, and find that the negative effects of these problems can be reduced with the use of caching

V. Roca and F. Rousseau (Eds.): MIPS 2004, LNCS 3311, pp. 206–217, 2004.

proxies. We describe our design of a P2P based hybrid CDN, which uses a traditional provider backbone and caching proxies. We have verified our design using simulations with real world workloads from the BitTorrent measurements and synthetic workloads.

There are several proposals for P2P based streaming systems, including [1], [2], [3], [4], [5], [6], [7], [8], but these do not address the consequences of firewalls for P2P based systems, which is an intrinsic motivation for our design of a hybrid CDN.

The remainder of this paper is structured as follows. Section 2 presents the results of our BitTorrent measurements. Our hybrid CDN is described in Section 3, with simulation results presented in Section 4. Finally, Section 5 gives our conclusions.

2   BitTorrent Measurements

BitTorrent [9] is a P2P based file distribution system, and not a streaming system, but the performance of both is affected by firewalls and freeloaders.

BitTorrent differs from similar P2P systems in that it does not implement functionality such as searching, which can be obtained through integration with the WWW. In BitTorrent, a group of clients cooperate in the distribution of a set of data, by spreading the load of serving among the participants. The protocol is designed in a way which discourages freeloaders, by having the nodes prefer peers from which data has been received. BitTorrent divides a file into *pieces* of $256KB$. Each host informs new clients of pieces it currently has, and notifies other clients upon receiving a new piece. A BitTorrent session starts with a single server and the file which is being *seeded*. A client which has downloaded the entire file also serves as a *seed* for the file until the user aborts the application. A list of active clients is maintained by the *tracker*, which is contacted by new clients. Having many seeds is beneficial, since these hosts increase the reliability of the session, and do not place any load upon it by downloading.

We modified the standard BitTorrent client to store the protocol interaction between clients to disk. No data was downloaded, but maintaining connections to as many clients as possible was attempted. This method does not give a perfect overview of all events, but provides an indication of the behavior of the clients in the session. The results also correspond to previously observed behavior. This data was combined with the output from a crawler [10], which regularly retrieved the list of peers from the tracker, and an application called *TorrentSniff*[1] which was used to get the number of clients and seeds.

When a client first connects to the tracker, the tracker appears to do a reverse connection to the client. If it succeeds, the new client is added to the list of potential peers. This ensures that the peer list consists only of hosts which should be able to accept incoming connections. A subset of this *peer list* is returned to clients upon request. Adding hosts which are unable to accept incoming connections would result in all clients wasting time trying to connect to closed hosts. While the actual operation of the tracker is unknown, the observed

---

[1] Available at `http://freshmeat.net/projects/torrentsniff/`

behavior matches the description above. The peer list can be used to determine if a given peer is behind a firewall or not; a host which has been returned in the peer list can be assumed to have been open at some point, otherwise it is most likely behind a firewall. The modified client additionally verified this by doing a reverse connection to the other hosts in the session.

We monitored several BitTorrent sessions, including RedHat torrents and MandrakeLinux torrents, over several months, from a machine in Oslo. Another machine at *Institut Eurecom* in France was used to verify our results. Both PCs were able to accept incoming connections. The results we present in this paper are based on examination of a *MandrakeLinux-10.0* torrent from the 31th of March 2004 to the 6th of May. The goal was to determine how long clients are connected to the network, and the effects of firewalls.

The relationship between seedtime and download speed is shown in Figure 1(a). The seedtime starts at zero once the entire file has been downloaded. One might expect a host which quickly downloads a file to leave early, since the user is still likely to be sitting by the machine. If downloading takes several days, the user might be elsewhere, leaving the application to run. The plot does to a certain extent fulfill these expectations; having an L shape. The majority of long seeding hosts achieved low average download speed.

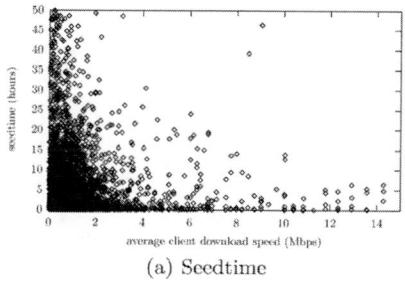

(a) Seedtime

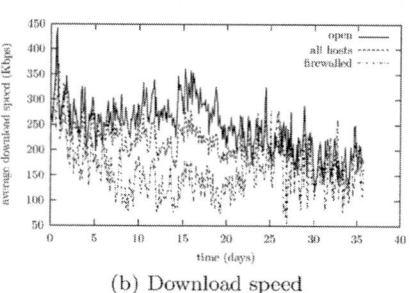

(b) Download speed

Fig. 1. BitTorrent measurement results

The effects of firewalls on download speed is shown in Figure 1(b). Hosts which are behind a firewall achieve significantly lower download rates, with an average of $0.2 - 0.3 Mbitps$ for open hosts. This value was seen to vary between BitTorrent sessions measured; in a similar session, [10] found average speeds in the range of $0.5 - 0.6 Mbitps$. The difference appears to be related to the number of clients in the session, which starts at 300, increases to 600 and then falls to around 200. Of the observed hosts, 40% are behind a firewall.

The BitTorrent protocol makes good use of a client while it is downloading, but provides no incentive for a client to seed afterwards. Especially, fast clients are quick to leave the network, as seen in Figure 1(a). The majority of clients stay for a long time, but they are often also the slowest.

Being able to accept incoming connections positively affects the download rate since a larger number of hosts can be communicated with. While this observation is obvious and not new, the difference appears to be significant and should be taken into consideration in a streaming network, where download speed is important. A more detailed presentation of our measurement results can be found in [11].

3   Protocol Operation

We propose a hybrid CDN which shares some properties with both BitTorrent and the WWW. File selection will likely be done via a traditional browser, and content presentation through a standard media player. Downloads are done using our P2P protocol. An overview of the different parts of the system is shown in Figure 2. The leftmost part is the client application which makes requests to the local host cache (LHC). How the data is retrieved from the network is not visible to the application, it only receives a sequential data stream, parts of which might be retrieved from many different peers. The LHC communicates with the proxy/site content cache (SCC), and receives the data stream from it, and the LHC of other clients on the same network. It also shares data in its cache to other local clients. Communication between the main server and SCCs is done in the same way; data can be retrieved from other SCCs, in addition to the main server.

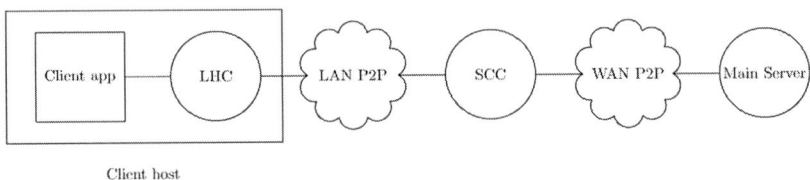

Fig. 2. Hybrid CDN overview

The protocol is similar to the HTTP protocol [12]. Three commands are used; *GET* to request data, *IGOT* to report the reception of data, and *HEAD* to retrieve information such as file size. Only the first command is used by the client application. To receive a stream of a given file, the application sends a *GET* request to the LHC. Seeking in the file can be done with an option to this command.

The P2P based communication, between different LHCs, SCCs, and the main server is more complicated and based on the transfer of fixed size blocks. Requests for specified blocks are made to the main server and results in either the block data itself, or a redirect message. The server maintains a list of SCCs and other hosts without a firewall which have successfully downloaded parts of a file. A redirect message includes a list of these alternate download locations. The block request is then made to hosts from this list, until one which can serve the block is found. Communication between a LHC and the SCC works in the same way. The SCC has the option of forwarding content to the LHC while it is receiving data from the main server.

The calculated checksum of a received block is reported to the SCC or server with the *IGOT* command. If it was downloaded after a redirection, the address of the host it downloaded the data from will also be reported. The block checksum does not protect against transmission errors, but is used to ensure that corrupt data is not propagated by caches, and to prevent a malicious host from claiming to have shared more data than it actually has. An *IGOT* command with a peer address and a correct checksum tells the server or SCC that this block can be found on at least two locations, and that the host specified as a peer shared a block. This information makes incentive creation in the server and SCC possible, by using a priority based on the amount of data shared by clients, when resources such as bandwidth are allocated.

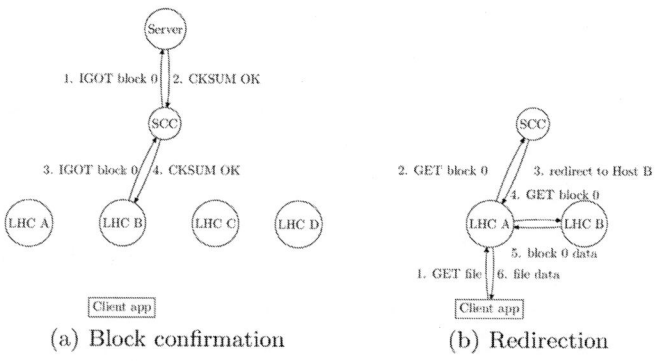

(a) Block confirmation   (b) Redirection

Fig. 3. Protocol operation example

A simple example will illustrate how the protocol is used. If for example the client using *LHC B* makes a request for *block 0* of a new file located on the *Server*, the request will be made to the site or ISP *SCC*, which will send a request to the *Server*. The data is returned in the opposite direction, from the *Server* to *LHC B*, via the *SCC*. When the entire block is received by the *SCC*, it sends an *IGOT* command to the *Server*, with the block checksum, as in Figure 3(a). It then receives a confirmation that the block was correctly received. The same is done by *LHC B*. The *Server* now knows that a copy of *block 0* has been successfully received by the *SCC*, which again knows that *LHC B* also has a copy.

At some point in time after this interaction, a client using *LHC A* makes a request for the same file, illustrated in Figure 3(b). *LHC A* makes a request to *SCC*, which redirects the client to *LHC B*. The same request is made to *LHC B*, which returns the data. After the block has been downloaded, the checksum is reported to *SCC* by *LHC A* in a way similar to step 3 and 4 in Figure 3(a), except that *LHC B* is specified as the source. *SCC* now knows that another copy of the block exists at *LHC A*, and that it still exists at *LHC B*, which has served a block to another node, possibly earning it a higher priority for future requests.

In summary, the SCC is placed outside the firewall of a site, under control of the site itself, or the ISP. This ensures that external clients have a host to connect to, eliminating the firewall problem. Clients make requests to this node as with a web cache, and cached data can be served directly from the node. The SCC and main server can refuse connections from clients with a low priority to create incentives for serving, making resource limits a tool to improve the performance of the network. The SCC has the option of requesting the missing parts of partially downloaded files. The LHC is responsible for retrieving blocks in a timely manner for the client, and monitoring connections to other peers.

Use of a proxy also automatically creates clusters which can share data directly at the lowest level of the distribution tree, reducing the load on the SCC. The administrator of the SCC can ensure that communication between local hosts is done in an efficient manner, without guesswork from the clients. For communication between different SCCs this will still be an issue, but local copies of popular files will likely exist at large sites.

The protocol and its operation is described in more detail in [11].

4   Simulations

In the previous sections, we have described a design for a hybrid CDN system that is based on the results of our BitTorrent measurements. This section describes simulations made to determine the potential for bandwidth savings with this design, compared to alternative approaches.

In the current state of this work, we are interested in the resource requirements for streaming, and the potential for reducing these, not the behavior of the system in a given network. Consequently, we use a simplified network model, and do not take factors such as varying link capacity and communication errors into consideration. All clients are assumed to fulfill the minimum resource requirements for participation. Blocks are requested sequentially, and a whole block is completed before the next is requested. The streamed content is $700MB$ in size and has a bit-rate of $1Mbitps$, for a total playtime of roughly 90 minutes. This might be somewhat higher than what is common among private Internet users today, but will likely be increasingly available in coming years. It will also result in a more demanding load on the system. In current networks, factors such as asymmetric home connections might limit peer uploads, but this is mainly a question of adapting content quality to the client base.

4.1   Simulation Load

The simulation loads are taken from two sources; our BitTorrent measurements, and a load generation tool. The first contains IP-addresses, connectivity duration and information about firewalls for a large number of geographically spread hosts, over a period of one month.

Connectivity times will likely be different for different P2P system types. The values in the BitTorrent based data is fixed, but we can compensate for this by varying connectivity times in the synthetic load. Comparing this real world load to the synthetic load should improve the realism of the simulation results.

The artificial load is generated using *MediSyn* [13], which uses a *generalized Zipf-like* distribution to create requests for a streaming server, but unfortunately does not support interactivity. A topology with nearly 10,000 end-node hosts was generated with the topology generation tool *Inet* [14] and requests were distributed randomly among these clients. The load from protocol processing is not data intensive and not considered here.

We are interested in the required network capacity at three points; the server, the client and the ISP. Network locality is not an issue at these three points, only between them (see Figure 2). We define an ISP here as a node through which traffic from a large number of clients passes, so we only look at ISPs with 50 or more clients. There is no difference between a site with many users, and what we here call an ISP. For the BitTorrent measurements, we used DNS information to group hosts together, while in the synthetic load we defined a node parent to 50 or more end-nodes as an ISP. In both cases, the ISP clients represent roughly half of the total number of clients.

The simulations are done by a specially written program, which takes the file requests as input, splits it into block requests, maintains state information, and produces information about the required capacity at different points in the network. The synthetic load uses a server with 1000 different files. The time clients stays connected is one of the simulation parameters. Cache size is taken into consideration; clients cache at most $1GB$ and ISPs $100GB$. A FIFO based cache replacement algorithm was used on the file blocks in the cache. This algorithm was chosen because it was easy to implement in the simulator, but it is likely that the performance of the system could be improved with a more sophisticated algorithm.

4.2   Simulation Models

To get a better idea of the potential efficiency of our system, we compare it to several design alternatives for content distribution; the client/server model, with and without caching, and a P2P based distribution model without caching.

Client/server Model: All files are stored on a single server, with multiple clients retrieving data directly from it. There is no client load from uploading in this case and firewalls are never a problem, since no incoming connections are made to any clients.

Caching is frequently used in the WWW, and in CDN systems to reduce the load on the server and should improve performance in this client/server scenario. Video files can be quite large, so ISPs or sites with many users have an interest in reducing the network load from these transfers. Caching by the ISP is one way to achieve this.

P2P Based Distribution: P2P based distribution can be viewed as having many caches, with varying degrees of locality; the ISP might still suffer multiple downloads of the same file. In this scenario, we let active clients offer downloaded blocks to other clients. Clients connect to the main server and are redirected to a client which has previously downloaded the requested part of the file. If no available client exists, the data is retrieved from the main server, which maintains a list of the last ten downloaders of each block. This list is updated with both uploader and downloader upon completion of a block download, since this was found to improve performance. The capacity of the clients is limited to one upload, so they contribute at least as much as they get.

P2P Based Distribution with ISP Caching: Elements of the two previous scenarios can be combined by having the ISP cache operate in a manner similar to the main server in the P2P scenario. It keeps track of which local clients have a given piece, and redirects local requests to these clients when possible. This reduces the load on the caches and eliminates duplicate downloads by local clients. The cache also serves external clients, existing as a permanent peer in the system. The server continues to work as in the previous scenario, by also redirecting requests when possible. ISP caches are expected to contribute a little more than private users, so an upload limit of five simultaneous external clients is used for ISP caches. Peer lists are updated with information about both uploader and downloader as in the P2P only case. This behavior corresponds to our proposed system, which is described in Section 3.

4.3   Load Comparison

The main part of the simulations are done using a synthetic simulation load, but we have also generated a load based on the real world data from the BitTorrent measurements. Only one file is distributed in the BitTorrent based load, so a synthetic load using only one file was generated for the purpose of comparing the two; the other simulations use multiple files.

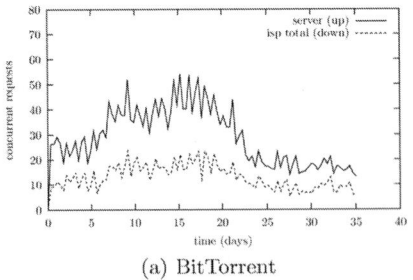

(a) BitTorrent

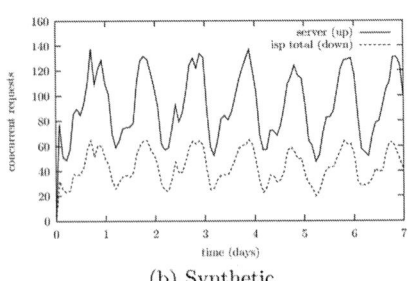

(b) Synthetic

Fig. 4. Requests

Figure 4(a) shows the request pattern in the BitTorrent based data, while the values for the synthetic load is shown in Figure 4(b). The two loads are similar, with daily spikes. The BitTorrent data is special in that it has a load increase spanning many days due to the easter holiday period.

Comparison of the simulation results from the BitTorrent based and synthetic loads showed no significant difference when the different simulation models were used.

4.4   Firewall Effects

Firewalls were found in the BitTorrent measurements to have significant impact on system performance. Since we have information about hosts with firewalls in the BitTorrent data, we can examine the consequences using simulations. For the synthetic load, we place firewalls at a similar portion of random clients. Hosts which are behind a firewall are not added to the list of possible peers since they never serve content. The client/server based model is not affected by firewalls; only the effects on the P2P based models are examined here.

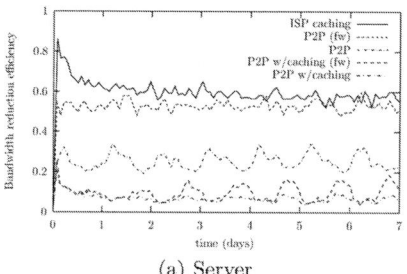 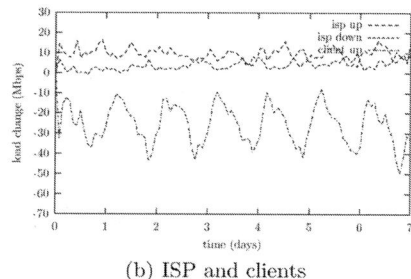

(a) Server   (b) ISP and clients

Fig. 5. Load changes due to firewalls

Figure 5(a) shows a detailed overview of the effects on the server load when firewalls are taken into consideration. Using P2P networking techniques eliminates most of the load on the server, but without caches, these benefits are significantly reduced, falling nearly to the level of ISP caching in a client/server network.

Figure 5(b) shows the load changes on the ISP and clients, when caches are added to the P2P networking model. Fewer clients contribute, so the client load falls, but apart from this, the changes are modest. The load on ISPs from uploads is increased.

4.5   Session Participation Duration

P2P networks utilize the resources of participating hosts to spread the load, and are affected by the time a host stays connected. In the synthetic load, we change this value to see the effects; clients stay for an extra 90 minutes after completing the download. In the P2P scenario, the result of the extra time is a $20-40 Mbitps$ reduction in the load on the server, with a similar increase in the load on the clients, as can be seen in Figure 6(b). With caches, there is a similar increase in the client load, and a small reduction in the load on the ISPs. The server load however is largely unchanged, but it is already very low. The P2P networking model is in other words sensitive to users leaving early, but caching reduces this sensitivity.

4.6   Simulation Result Summary

Figure 6(a) shows the simulation results for the different simulation models. P2P networks do not appear to provide a significant benefit over the client/server model with caching when firewalls are taken into consideration; the load on ISPs is clearly increased, and the server load only slightly reduced. Placing caches at the ISPs when P2P networks are used leads to a nearly eliminated server load, and reduced ISP load from downloads, at the cost of more outgoing traffic from the ISPs.

Using anti-freeloader techniques is desirable in P2P networks, to make clients contribute their resources for a longer time. Figure 6(b) shows the total bandwidth requirements for one week, first for the P2P networking model without caches, and then when they are used. Firewalls are included in both cases. The benefits are most significant in the former case, but there is also a small reduction in the outgoing ISP traffic in the latter.

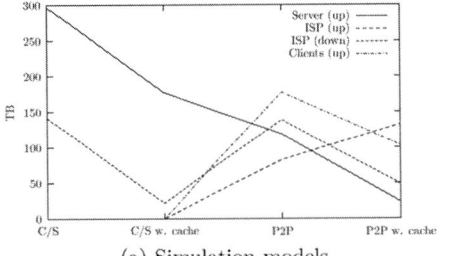
(a) Simulation models

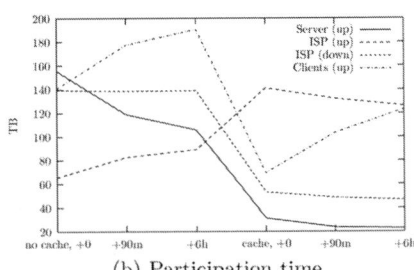
(b) Participation time

Fig. 6. Simulation results

4.7   Caveats

We have used measurements of the P2P file distribution application BitTorrent for validation of our design of a hybrid CDN. BitTorrent is a file distribution application, while we are designing a CDN streaming system. In spite of this difference, we benefit from the comparison, since our system faces many of the same challenges; firewalls and freeloaders are a problem for most P2P based networks.

A more important problem is the lack of interactivity in the simulation loads. Having measurements for a longer duration and for a higher number of clients would also be beneficial, but we do however not think that any of the fundamental observations we have made would change. A practical problem might be the installation of caches at ISPs and large sites, but the potential for bandwidth reduction and improved service for local users should make this attractive. The desire to do this has been expressed by ISPs [15], and significant potential for caching has also been observed in existing P2P networks [16], [17].

5   Conclusion

Broadband Internet connectivity in an increasing number of private households, and powerful compression techniques for high quality video, makes the deployment of streaming CDN systems increasingly realistic. In this paper, we have presented our design for a hybrid CDN.

In our measurements, we found that the performance of P2P based systems can be significantly affected by the presence of firewalls. Our simulations show that firewalls reduce benefits from load distribution also in P2P based streaming networks. P2P networks are also vulnerable to clients leaving early, which our measurements found not to be uncommon for well-connected clients. Our proposed system is based on proxies with cache functionality, placed at large ISPs and sites. Simulations show this system to still provide significant reduction in bandwidth requirements when firewalls are taken into consideration and clients leave the network early.

We have mainly focused on bandwidth usage as a metric for the behavior of our system. Other factors such as startup latency, QoS, and cache replacement strategies have not been evaluated. Other design options such as a suitable block size and transport protocol must be decided on. Many of these factors are difficult to simulate and easier to measure in the real world. For example, the efficiency of different incentives to stay connected is largely determined by human nature. Future work will be concentrated on implementing and examining parts of the system in the real world.

Acknowledgments. We would like to thank Ernst Biersack, Pascal Felber, Guillaume Urvoy-Keller and Mikel Izal Azcárate at Eurecom for fruitful discussions and provision of their crawler for our measurements.

References

1. Do, T.T., Hua, K.A., Tantaoui, M.A.: P2vod: Providing fault tolerant video-on-demand streaming in peer-to-peer environment. Technical report, School of Electrical Enginneering and Computer Science, University of Central Florida. (2003)

2. Loeser, C., Ditze, M., Altenbernd, P.: Architecture of an intelligent quality-of-service aware peer-to-peer multimedia network. In: Proc. of the 7th World of Multiconference on Systemics, Cybernetics and Informatics. (2003)

3. Jiang, X., Dong, Y., Xu, D., Bhargava, B.: Gnustream: a p2p media streaming prototype. In: Proceedings of IEEE International Conference on Multimedia & Expo. (2003)

4. Guo, Y., Suh, K., Kurose, J., Towsley, D.: P2cast: peer-to-peer patching scheme for vod service. In: Proceedings of the twelfth international conference on World Wide Web. (2003)

5. Guo, Y., Suh, K., Kurose, J., Towsley, D.: A peer-to-peer on-demand streaming service and its performance evaluation. In: Proceedings of 2003 IEEE International Conference on Multimedia & Expo. (2003)

6. Rejaie, R., Ortega, A.: Pals: peer-to-peer adaptive layered streaming. In: Proceedings of the 13th international workshop on NOSSDAV. (2003)

7. Hefeeda, M., Habib, A., Botev, B., Xu, D., Bhargava, D.B.: Promise: Peer-to-peer media streaming using collectcast. Proc. of ACM Multimedia 2003. (2003)

8. Hefeeda, M., Habib, A., Xu, D., Bhargava, B., Botev, B.: Collectcast: A peer-to-peer service for media streaming. ACM/Springer Multimedia Systems Journal. (2003)

9. Cohen, B.: Incentives build robustness in bittorrent. In: Workshop on Economics of Peer-to-Peer Systems. (2003)

10. Izal, M., Urvoy-Keller, G., Biersack, E., Felber, P., Hamra, A.A., L.Garces-Erice: Dissecting bittorrent: Five months in a torrent's lifetime. In: Passive and Active Measurements 2004. (2004)

11. Skevik, K.A., Goebel, V., Plagemann, T.: Analysis of bittorrent and its use for the design of a p2p based streaming protocol for a hybrid cdn. Technical Report 310, Department of Informatics, University of Oslo, http://www.ifi.uio.no/dmms/papers/129.pdf (2004)

12. Fielding, R., Gettys, J., Mogul, J., Frystyk, H., Masinter, L., Leach, P., Berners-Lee, T.: Hypertext transfer protocol – http/1.1. RFC2616. (1999)

13. Tang, W., Fu, Y., Cherkasova, L., Vahdat, A.: Medisyn: a synthetic streaming media service workload generator. In: Proceedings of the 13th international workshop on NOSSDAV, ACM Press. (2003)

14. Winick, J., Jamin, S.: Inet-3.0: Internet topology generator. Technical Report CSE-TR-456-02, University of Michigan. (2002)

15. Hasslinger, G.: Peer-to-peer networking from the view of internet service providers. Peer-to-Peer-systems and -Applications seminar. (2004)

16. Leibowitz, N., Ripeanu, M., Wierzbicki, A.: Deconstructing the kazaa network. In: 3rd IEEE Workshop on Internet Applications (WIAPP'03). (2003)

17. Gummadi, K.P., Dunn, R.J., Saroiu, S., Gribble, S.D., Levy, H.M., Zahorjan, J.: Measurement, modeling, and analysis of a peer-to-peer file-sharing workload. In: Proceedings of the 19th ACM Symposium on Operating Systems Principles (SOSP-19). (2003)

Using the Session Initiation Protocol for Connected Multimedia Services in a Ubiquitous Home Network Environment

Doo-Hyun Kim[1], Ji-Young Kwak[2], and Kyung-Hee Lee[2]

[1]Konkuk University, Seoul, Korea
doohyun@konkuk.ac.kr
[2]ETRI, Taejon, Korea
{jiyoung, kyunghee}@etri.re.kr

Abstract. This paper shows how connected multimedia services can be integrated within a networked embedded system. Such services can preserve a multimedia session even though the end-user moves from one computing environment to another. To benefit from immediate multimedia services, the user only needs to carry a small mobile pad, such as a PDA or a Cellular phone showing a list of I/O devices that can be used for an on-going multimedia session. It is not required to re-establish the session. When the user clicks and selects a suitable device from the list, the session is immediately migrated to the selected device. This paper focuses on the session preservation protocol and describes how we use SIP (Session Initiation Protocol) for session preservation. Our integrated architecture provides several features: registration and authentication of the mobile user and connected devices, detection of user's leave and presence, and session preservation. We also present an illustrative application: the connected video surveillance. We describe the design of a connected multimedia services system based on X.10 and SIP that supports both user mobility and multimedia session preservation in a ubiquitous wired/wireless home network environment.

Keywords: Home Networks, Smart Spaces, Connected Multimedia Services, Session Initiation Protocol, Session Preservation

1 Introduction

According to the familiar paradigm of ubiquitous computing [1-3], soon we will be able to access information and services virtually anywhere and at any time via any device, such as our phones, PDAs, laptops or even watches. The environment will be intelligent enough to pick up subtle user inputs like user movement, proximity, or temperature and supply any required service or information to the user through a mobile device by using voice/sound/lights or other media for communication.

In such ubiquitous computing environments, the user does not need to carry his computer that is crammed with a personal computing environment including his favorite applications and important data. Instead, the user can bring up these applications on any nearby computing devices such as a desk-top computer in the state they

V. Roca and F. Rousseau (Eds.): MIPS 2004, LNCS 3311, pp. 218–229, 2004.
© Springer-Verlag Berlin Heidelberg 2004

appeared when last invoked. As illustrated in Fig.1, the user just uses a computing device beside him as if he operated his own personal computer [4, 5].

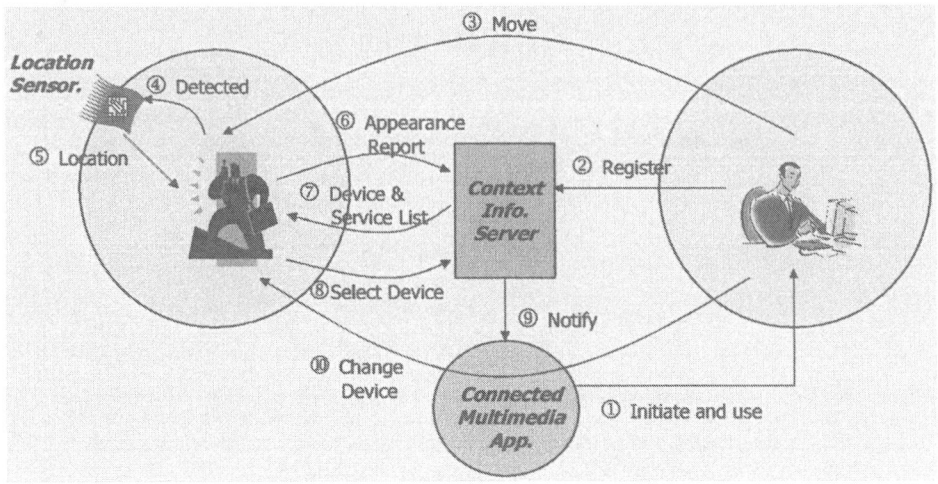

Fig. 1. Typical flow of a connected multimedia service.

In order to construct such intelligent environments, it is necessary to make provision for making the information of devices dynamically attached and detached to a home network, and for representing the capability of devices attached to a home network. Besides these basic environmental functionalities, it is necessary to provide a variety of features at the level of physical devices, the operating system, communication protocols, the middleware, man-machine interfaces, and so on. This paper focuses on the middleware level, especially on the session management protocol, which basically provides a means for creating, modifying, and terminating a multimedia session between devices including a user's handheld device. However, such a basic session initiation and management protocol needs to be extended to support locating the user's current IP address and re-establishing the session with the user's new nearby device for making the session created in the previous space to continue in a new space seamlessly.

1.1 Related Work

One of the significant examples of ubiquitous computing systems is the SmartSpaces [4] scenario that consists of intelligent services accessible to mobile users via handheld devices connected over short-range wireless links. These services will be integrated seamlessly into the environment that the user is familiar with, enabling easy and automatic usage.

In addition to the SmartSpaces scenario, there are several on-going projects. The MIT Oxygen project envisions the connected devices and intelligent space as well as the connected multimedia services [6]. The University of Washington is involved in the Portolano project seeking an invisible computing infrastructure and distributed services facilitated with the multiple interfaces in a coherent manner [7]. The CMU

(Carnegie Melon University) Aura project aims at minimizing distractions of the user's attention, creating an environment that adapts to the user's context and needs. One of key components of Prism, the Aura architecture, is the Context Observer providing reconfigurations to maximize the use of local resources subject to various resource utility functions specified by the tasks [8].

Compared to these on-going projects, we focus on the session preservation protocol and describe how to use SIP (Session Initiation Protocol) [9] for session preservation, the aspect not explored in these projects.

1.2 Our Contributions

Our integrated architecture provides several features: registration and authentication of the mobile user and connected devices, detection of user's leave and presence, and session preservation. We also describe the X.10 and SIP-based design of a connected multimedia services system supporting both user mobility and multimedia session preservation. SIP is an application-layer protocol that provides signaling for personal mobility whereas X.10 [10] is a communications protocol for remote control of electrical devices. These properties make the two protocols particularly suitable for environments with networked appliances.

We also needed a capability negotiation scheme to adapt the QoS of an application to available resources in a dynamic and autonomous way. This capability negotiation is performed by the body part of the SIP protocol: SIP describes the capability of terminal devices using SDP (Session Description Protocol) [11].

This paper is organized as follows. In the Section 2, we present the integrated architecture for the SmartSpace and the connection multimedia services system in the home network environment. In the Section 3, we explain how SIP is used for session preservation. Finally, in Section 4 we give an example of a connected home surveillance application as a preliminary implementation and conclude the paper with some suggestions for future work and research directions.

2   The Connected Multimedia Services System

In this section, we describe the proposed system architecture that provides the connected multimedia services through a networked embedded system. We also present the components of the connected multimedia service system and introduce two protocols, SIP and X.10 used for communication between components.

2.1   SIP Protocol and User Mobility

SIP is an application layer signaling protocol for creating, modifying, and terminating sessions with one or more participants. SIP defines several methods: INVITE, ACK, BYE, OPTIONS, CANCEL, REGISTER, MESSAGE, and others. Responses to methods indicate success or failure, distinguished by status codes, 1xx (100 to 199) for progress updates, 2xx for success, 3xx for redirection, and higher numbers for failure [12].

Entities in SIP are user agents, proxy servers, and redirect servers. The SIP user agent listens for incoming SIP messages and sends SIP messages upon user actions or incoming messages. The SIP proxy server relays SIP messages and the SIP redirect server returns the location of a host. Both the redirect and proxy server accepts registrations from users containing the current location of the user. The location can be stored either locally at the SIP server or in a dedicated location server. Deployment of such SIP servers enables personal mobility, since a user can register with the server independently of location and thus be found even if the user has changed location or the communication device [12].

2.2 Capability Representation

SIP relies on the Session Description Protocol (SDP) for carrying out the negotiation for codec identification. Thus, SIP supports session descriptions that allow participants to agree on a set of compatible media types. It can also use XML (Extensible Markup Language) as its body. So, SIP can deliver descriptions of devices based on XML. These features make SIP particularly suitable for the home environment.

2.3 Device Control

X.10 is a communication protocol that allows compatible products to talk to each other using the existing electrical wiring in the home. Control signals are transmitted between devices over the electrical wiring of the user's home using a modulated carrier signal overlaid on the normal AC power supply. This makes X.10 particularly suitable for the home environment.

The X.10 home network consists of X.10 transmitters and X.10 receivers. X.10 communicates between transmitter and receiver modules using RF (Radio Frequency) signals transmitted over the existing AC wiring that carry digital information [13].

2.4 Integrated Architecture

As shown in Fig. 2, the proposed system architecture consists of four sub-systems: the Service Manager, the Context Information Provider, SIP User Agent running on Static Embedded Devices, Mobile Clients, and Master. These four sub-systems form the *SmartSpace* as illustrated in the left or right part of Fig. 2. The *SmartSpace* provides intelligent services accessible to mobile users via handheld devices. The session of on-going multimedia services is preserved in the *SmartSpace* even though the user moves from one *SmartSpace* to another.

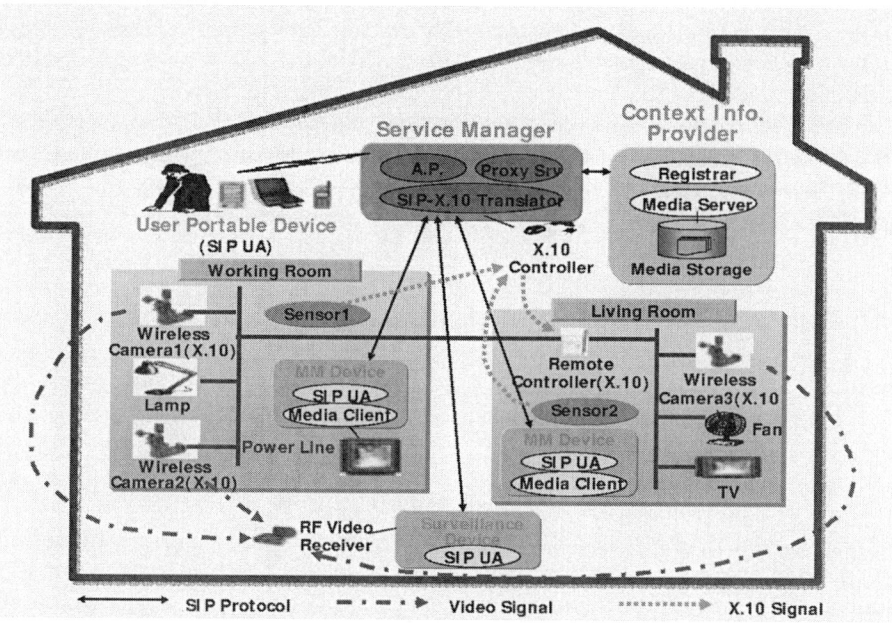

Fig. 2. Architecture of the connected multimedia services system.

The Service Manager manages devices that are dynamically attached to or detached from a *SmartSpace*. All the devices are assumed to be capable of dealing with SIP User Agent. The X.10 devices, which are not SIP-based devices natively, are encapsulated by SIP-X.10 Translator. Therefore, the Service Manager is composed of three modules: a SIP-X.10 Translator, an Access Point, and a Proxy Server. The primary goal of the SIP-X.10 Translator is to provide protocol conversion between SIP and X.10. The Access Point supports the communication of mobile users by assigning a dynamic IP address to mobile devices connected over wireless links. The Proxy Server is a SIP application server that accepts requests in order to service user requests and sends back responses to these requests. The Proxy Server accepts registration requests from devices and services, and relays these requests to the Context Information Provider.

The Context Information Provider manages the device list and sends the updated list of devices to the user's hand-held device, whenever the device list is modified. The registration information of networked devices and user's device forms <Identity, Status, Service-Type, Location, IP Address, Occupying-User> and <Authentication, Location, IP Address> respectively. When the Status is "On-Going," the corresponding device is in service at this moment. Thus the device list also contains the list of services. The Context Information Provider offers the function of two modules, a Registrar and a Media Server. The Registrar is a SIP location server that accepts the SIP REGISTER requests and offers location services. The Media Server provides media services to Media Clients through the Media File Storage.

The Master, which is not shown in Fig. 2, plays a role of a global server for the distributed Context Information Providers. The Master is to support the session preservation for inter-home connected multimedia services, and is assumed to be located outside of a home.

The SIP User Agent, a SIP application terminal that initiates the SIP request or returns a response to the received SIP request on behalf of the user is loaded and runs on every static embedded device and mobile device of the user. It receives the service list from the Service Manager.

3   Operation of the Connected Multimedia Services System

A *SmartSpace* is a space that contains myriad devices working together to provide users with access to information and services. The user with a portable device who enters a *SmartSpace* is detected by a sensor and registers the location information at the Service Manager. If the authentication fails during the registration process, the fact of the unauthorized access request may be reported to the Master. Once registered, the portable device periodically receives the updated list of services. If the user is in the middle of the service at a previous space, the service is offered to the user continuously on the basis of the information about the on-going services in the previous space stored at the Service Manager. The user is able to choose a service, select a function, and execute the function among listed available services. Thus, a session of on-going multimedia services is preserved even though the user moves from one *SmartSpace* to another. In the following, we explain the operation of connected multimedia services in the view of the call flow of SIP messages exchanged between interacting SIP entities.

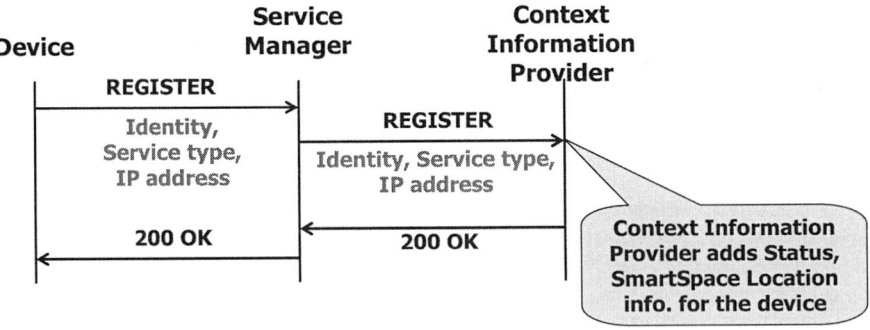

Fig. 3. Local device registration process

3.1   Registration and Authentication

Fig. 3 shows the registration process of a device located in a space statically. When a local device is turned on, the description information of the device is conveyed in the

SIP REGISTER message to the Context Information Provider. The registration process is completed when the Context Information Provider sends a response message to the received request through the Service Manager. In this process, the description information of a device that consists of identity, status, service type and *SmartSpace* location information is also added to the service list of a Service Manager. The description information and IP address of the device is added to the database of the Context Information Provider.

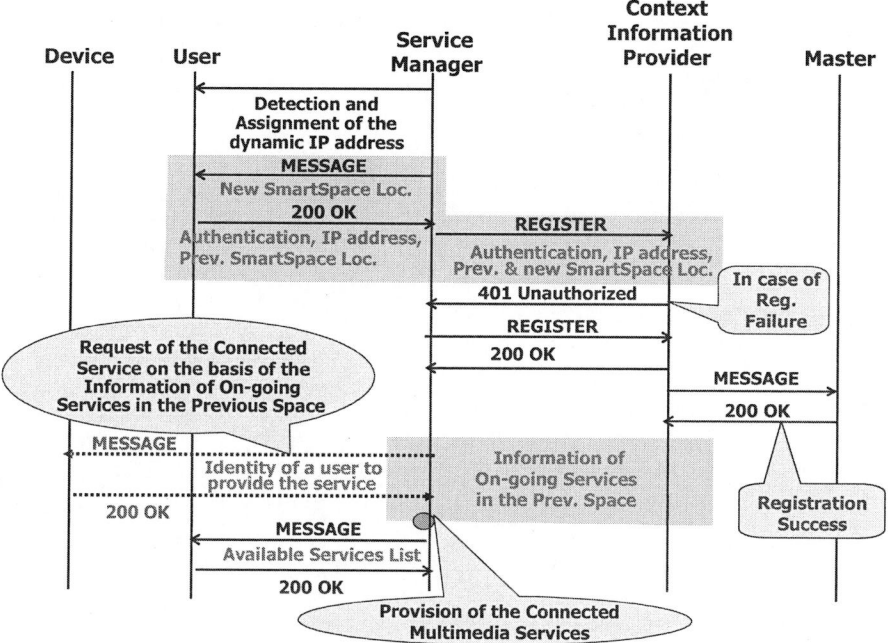

Fig. 4. User registration and authentication process.

Fig. 4 illustrates the registration and authentication process of users. When the user with a PDA enters the *SmartSpace*, the Service Manager detects the user appearance by means of a sensor and assigns a dynamic IP address to the PDA by the Access Point. This location information is automatically provided by the system (e.g. by sensor1 in Fig. 2) and not by the user himself who does not need to care where he is.

It is necessary that the user should be authenticated to user the service ater moving into a new space, because an unauthorized user could take over the device from the original user. The Service Manager requests the authentication and location information of the user by using the SIP MESSAGE and initiates the registration and authentication of the user. When the user sends a response message to the received request to the Service Manager, the registration and authentication process of the user makes progress. The list of available services is delivered to the user upon completion of the user registration. If the user is in the middle of a service in the previous space, this service is offered to him continuously.

The connected multimedia services are provided by the Service Manager on the basis of the information about on-going services in the previous space, that is, the information contained in the response message received from the Context Information Provider. In other words, the Service Manager requests the connected service to the corresponding device in the space to which the user has moved on the basis of the information about on-going services in the previous space. The request is done by sending the SIP MESSAGE to the corresponding device. On the contrary, if the user registration fails, the fact of the unauthorized access request is reported in the SIP MESSAGE to the Master.

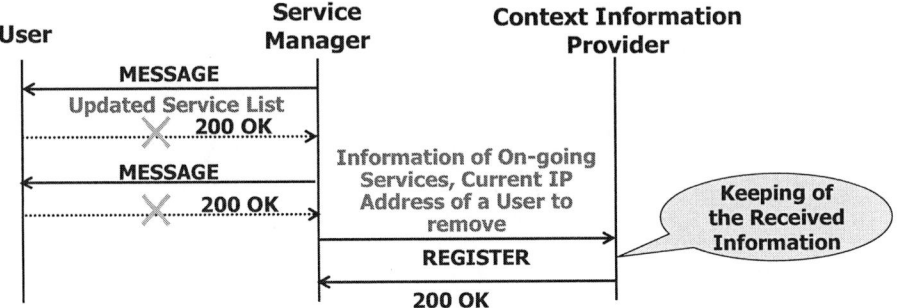

Fig. 5. Detection process of the user that leaves the current space.

Whenever the service list is updated, the Service Manager sends this list to all the users that exist in the current space. Then, the user must reply with a response message to the Service Manager. If the user does not send the response to successive requests, the Service Manager considers the user as having been moved. Fig. 5 shows the Service Manager that stores the information about on-going services of the user who has moved by sending this information in the SIP REGISTER message to the Context Information Provider. Keeping the information about on-going services is necessary for providing connected multimedia services.

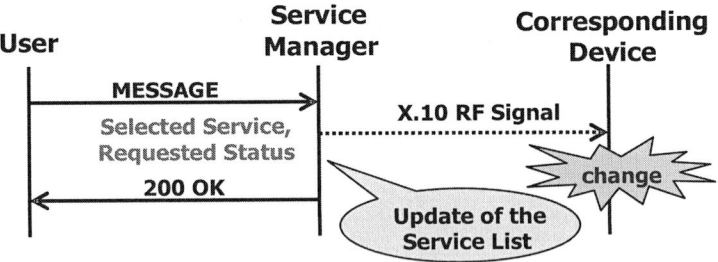

Fig. 6. Remote control of household appliances.

3.2   Request and Provision of Services

A *SmartSpace* offers various services such as remote control of networked household appliances, streaming of media files, and surveillance. Services in the *SmartSpace* are described by means of a simple, open XML schema that is the body part of the SIP protocol. In addition to that, the *SmartSpace* makes use of the X.10 protocol as a means for remote control of household appliances.

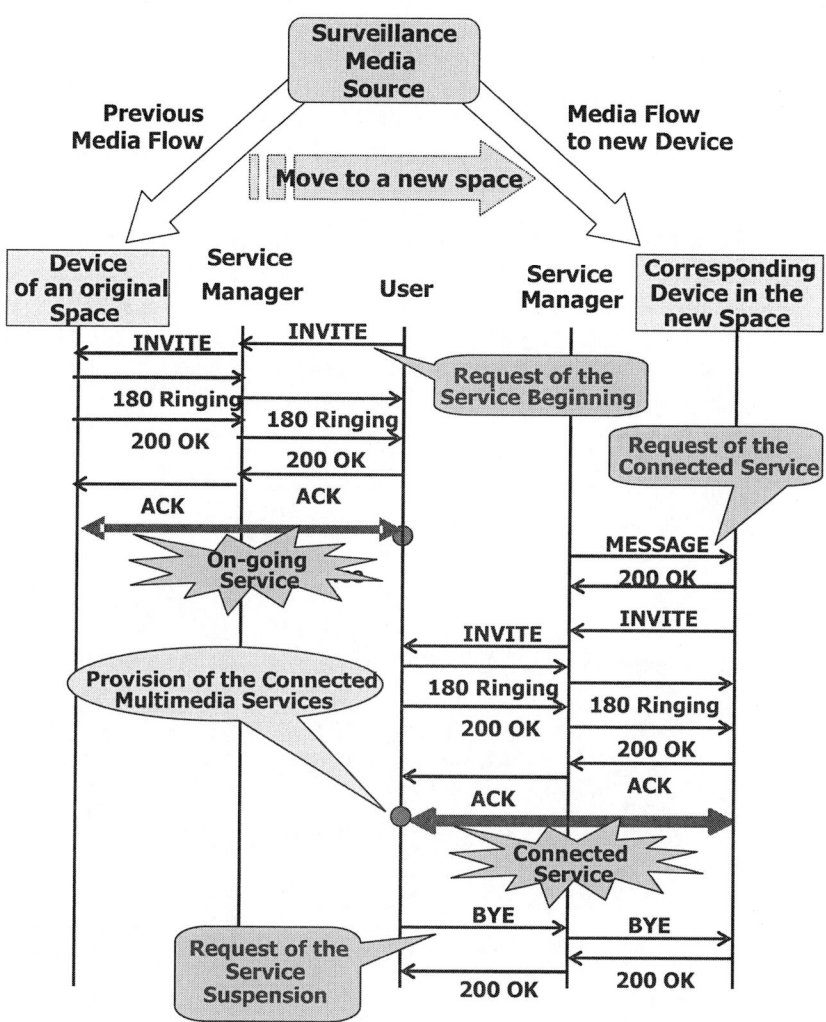

Fig. 7. The preservation process of a service session.

The process shown in Fig. 6 offers remote control of electrical devices such as lights, appliances, and other devices at home. With a PDA, the user selects the remote control of household appliances among the listed available services. Then, the Service Manager receives the information about the selected service in the SIP MESSAGE from the user (or the User Agent) and sends the X.10 RF signal to the corresponding device though the SIP-X.10 Translator. Thus, the state of the corresponding device is changed according to the request of the user and the service list of the Service Manager is updated. In order to inform the user that the selected service is executing, the Service Manager sends the response message to the user.

There are many services such as streaming of media files, surveillance, and others that must continue even if the user moves between spaces. Fig. 7 shows the session preservation process of these application services in the view of the call flow of SIP messages. In this process, the request for services is conveyed in the SIP INVITE message from the user (or the User Agent) to the Service Manager. The capability negotiation is performed by the body part of the SIP protocol, as SIP describes capability of terminal devices and available services in SDP or XML in its body. When a service session between the user and a corresponding device is established, the device offers the requested service. If the user leaves in the middle of the session, the service must be offered to him continuously in the new space. To do so, the Service Manager sends the information about the on-going services in the previous space in the SIP MESSAGE to the corresponding devices in the new space, if the preservation of the on-going services is possible there. Then, the corresponding devices in the new space offer the on-going services by continuing the established session with the user that has moved.

This continuous operation is guaranteed before and after the migration of the user by detecting his movements and keeping the information about on-going services. The detection of the user that has moved is done on the basis of the response (or its absence) to the delivered service list.

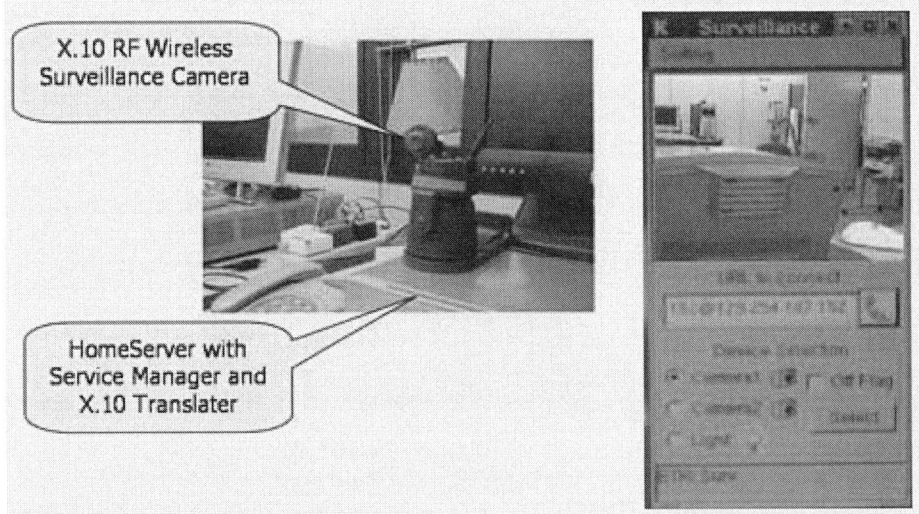

(a) Wireless Surveillance Camera and HomeServer with Service Manager

(b) Snapshot of the Connected Surveillance Application running on Linux Terminals

Fig. 8. Connected Surveillance.

4 Conclusions

In this paper, we have presented a novel software architecture providing connected multimedia services through a networked embedded system. The proposed system architecture was designed to support ubiquitous multimedia services by coupling the SIP and X.10 protocols. The properties such as the user mobility of SIP and the appliances control of X.10 make the two protocols particularly suitable for the provision of ubiquitous multimedia services.

The integrated architecture proposed in this paper is now under implementation. As an example application, Fig. 8 presents snapshots of the connected surveillance service that uses X.10 RF wireless cameras and the HomeServer [14]. When the user wants to leave his home, he can activate this service and leave with his hand-held device like a PDA equipped with WLAN or Bluetooth. Later on, when he wants to check the status of his home, he opens the PDA and selects the on-going surveillance session from the list of the available services. At the same time, he chooses the nearest desk-top from the list of available display devices next to him. Then, the stream of images from his home starts displaying on the desk-top. Fig. 8(a) shows the wireless camera and the HomeServer located at his home while Fig. 8(b) presents the GUI and the displayed stream of images from the wireless surveillance camera.

This example illustrates an initial implementation stage of our integrated architecture for the SmartSpace. As the future work, we have to integrate highly advanced schemes such as capability negotiation, transcoding, and QoS management [15, 16] to take advantage of heterogeneous capabilities of devices. Moreover, a highly reliable authentication framework and more sophisticated real-life applications will be studied and implemented on top of the currently implemented basic functions.

Acknowledgement. This work was supported in part by ETRI, Korea

References

1. Mark Weiser, "Some Computer Science Problems in Ubiquitous Computing." Communications of the ACM July 1993
2. IBM Pervasive Computing, http://www.ibm.com
3. M. Satyanarayanan, "Pervasive Computing Vision and Challengs", IEEE Personal Comm, vol.6, no.8, Aug.2001, pp.10-17.
4. L. Kagal, V. Korolev, S. Avancha, A. Joshi and T. Finin, "A SmartSpace Approach to Service Management in Pervasive System," Grace Hopper Celebration of Women in Computing 2002.
5. K. Takashio, G. Soeda, H. Tokuda, "A Mobile Agent Framework for Follow-Me Applications in Ubiquitous Computing Environment," 21st International Conference on Distributed Computing Systems Workshops (ICDCSW '01), April 16-19, 2001.
6. MIT, "Oxygen Project," http://oxygen.lcs.mit.edu/.
7. University of Washington, "Portolano - An expedition into invisible computing," http://portolano.cs.washington.edu/
8. CMU, "Aura Project : Pervasive invisible computing." http://www-2.cs.cmu.edu/~aura/

9.   M. Handly, H. Schulzrinne, E. Schooler, and J. Rosenberg, "SIP: session initiation proto-col," Request for Comments (Proposed Standard) 2543, Internet Engineering Task Force, Mar. 1999.
10. http://www.x10.com
11. Handley, M. and V. Jacobson, "SDP: session description protocol," RFC 2327, April 1998.
12. E. Wedlund, H. Schulzrinne, "Mobility Support using SIP," Second ACM/IEEE Interna-tional Conference on Wireless and Mobile Multimedia (WoWMoM'99), Seattle, Wash-ington, August 1999.
13. http://www.homeautomationzone.com/
14. Kyunghee Lee, D.H. Kim, J.Y. Kim, D.M. Sul, S.H. Ahn, "Requirements and Referential Software Architecture for Home Server based Inter-Home Multimedia Collaboration Services," IEEE Transactions on Consumer Electronics, Vol. 49, No. 1, Feb. 2004, pp. 145-150.
15. D. Wu, Y.T. Hou, and Y.Q. Zhang, "Transporting Real-Time Video over the Internet: Challenges and Approaches", Proceedings of the IEEE, Vol. 88, No. 12, 2000.
16. H. Vin, "Heterogeneous Networking", IEEE Multimedia, Vol. 7, 1995, pp. 84-88.

The Digital Call Assistant:
Determine Optimal Time Slots for Calls

Manuel Görtz, Ralf Ackermann, and Ralf Steinmetz

Multimedia Communications Lab
Department of Electrical Engineering and Information Technology
Darmstadt University of Technology
Merckstr. 25, D-64283 Darmstadt, Germany
{Manuel.Goertz, Ralf.Ackermann, Ralf.Steinmetz}@KOM.tu-darmstadt.de
http://www.kom.tu-darmstadt.de

Abstract. Communication plays a key role in today's businesses. Reaching a communication partner often has become a time consuming task. A multitude of potential communication channels with individual addresses forces a callee to guess an appropriate device at the right time. Under these circumstances additional information about a high probability to reach the calling target at a specific point in time enhances efficiency in communication. The decision when to call and the choice of the communication channel can be based on these information. This paper presents a *Digital Call Assistant* to determine an optimal time slot to place a call. The proposed approach combines calendar events and context information. The combination of these two information sources allows the creation of call plans which provide a list of possible time slots for communication with another user. A trust concept will assure that these sensible data will only be shared among trusted peers. Pending call requests and open call slots are presented to the user. The proposed planning application is going to form a novel part in our context-aware communication service framework.

1 Introduction

The current working requirements in companies have lead to a change in the daily schedule of employees. In many cases working behavior has become nomadic. The working environment is no longer bound to a specific place and time. Nevertheless, employees are typically still integrated in the company by means of communication over networks. A variety of electronic appliances supports the demands for this kind of mobility. This has lead to the observation that an individual usually carries a multitude of electronic helpers, such as cell phones or PDAs. These companion devices are typically designed for a specific purpose. At each point in time the user has access to a certain set of communicating devices. Additionally, the choice of active communication entities depends on the user's current context. Context is in this case determined by a superset of aspects such as location, situation or environment.

Communication is an essential part of our culture and a requirement for today's business models. Thus, communication must be as efficient as possible for all involved communication partners. However, the caller often faces the problem of how and when

V. Roca and F. Rousseau (Eds.): MIPS 2004, LNCS 3311, pp. 230–241, 2004.
© Springer-Verlag Berlin Heidelberg 2004

to optimally reach the callee. Based on assumptions about the communication partners current context the most suited communication type can be chosen. The callee's context is guessed from typical daily routines and history information.

However, this approach is characterized by a high uncertainty. There is no exact a priori knowledge of the called party's current situation, condition or mood. In many case multiple attempts have to be made to reach the communication partner at an adequate point in time. Finding an appropriate moment will increase the efficiency of the caller and additionally decrease the possibility of disturbing the communication partner.

A typical example is an ongoing meeting which the communication partner B attends. The caller (A), who is not aware of the meeting and its duration, will need to re-try the call attempts. A call on the cell phone might disturb B if the phone is not switched off. Some telephone systems provide a call-completion service. A caller can activate this service. When the callee becomes available again this is signaled to the requester. However, if A is not available then, B might now activate a call-completion. This would lead to a loop of mutual activated call-completion requests and unsuccessful call attempts.

The Digital Call Assistant proposed in this paper tries to provide a solution to this *reachability problem*. Both entities, the caller and the callee, negotiate a point in time which is promising for a call. The negotiation is done between agents that act on behalf of the users. The planning algorithm can be parameterized with a number of requirements. Context information is used to react on dynamic changes in the user's schedules. Thus, the assistant saves time for both communication partners.

The rest of the paper is structured as follows. Section 2 introduces necessary concepts and provides definitions. The proposed planning algorithm is shown in Section 3. The overall system design of the Digital Call Assistant is covered by Section 4. The paper is finally concluded with a summary and an outlook.

2   Concept and Components

The approach proposed in this paper relies on the utilization of context information and calendar events. This section provides the basic concepts and definitions of both information types in order to establish a common understanding.

2.1   Context

Context forms an important concept in the proposed planning algorithm. Individual users are surrounded by their context. This is expressed by the very generic definition that can be found in [1]:

> *Context*: That which surrounds, and gives meaning to, something else.

Actions of users are performed in a specific context. Contexts are often *rich* objects like situations and cannot be completely described [2]. McCarthy states that the main question of *what context is* cannot be answered as a result of a unique conclusion [3]. Several definitions exist in the area of computer science, too. The following definition adapted from [4] has been chosen as notion for context in this paper.

Context is any information that can be used to characterize the situation of a subject and its interaction with optional objects. Objects are persons, places, or applications that are considered relevant to the subject.

In general applications are considered *context-aware* if they use context to provide relevant information and/or services to the user. The relevancy depends on the user's task [5]. Main usage scenarios of context in such applications have been identified in [6]. These applications automatically *adapt* their behavior according to discovered context (using active context), or *present* the context to the user using reductions of all possible information and/or *store* the context for the user for later retrieval (using passive context).

Context Usage. The main prerequisite of a context-aware system is the process of context sensing. This process spans several processing steps. The following steps have been identified: The phases of *acquisition, synthesis, dissemination* and *use*. The principle procedure is described in the *context cycle* shown in Figure 1. The Context Cycle model follows the principle of the Omnibus Model for multi-sensor fusion [7].

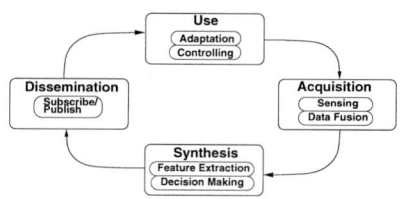

Fig. 1. The Context Cycle

The automatic acquisition of context information is a prerequisite to appropriately model real world situations. A common way is the utilization of a multitude of sensors. Sensors are used to observe the physical world.

Two types of sensors can be distinguished [8]. *Physical sensors* are hardware components that measure parameters in the environment. They provide the information on electronic level, typically as analog output or as digital signals. *Logical sensors* are components that provide information that is not directly sensed from the environment but represents aggregated information about the observed world. Information sources can be a clock as a sensor that offers time and a server offering the current exchange rate. Logical sensors most often supply information as digital signal over a common interface such as a serial data connection or an HTTP-connection.

Each sensor S can be seen as a time dependent function that provides the system with a set of values which give a description of the context at a specific time. The function $S: t \rightarrow X$ returns a scalar, vector, or a symbolic value (X) [9]. The output of a single sensor may often not produce sufficient information for the desired purpose. The concept of *sensor fusion* describes the combination of sensory data or data derived from sensory data. The assumptions is that the resulting information is in some sense better than it would be possible when sources were used individually.

The context synthesis process assesses significant features of the context. This process uses the sensor information as an input and creates an abstract representation of the captured situation. The combination of several context values provides a very powerful mechanism to determine the current situation. Location, entity activity and time are typical context sources and form the *primary context*.

Knowledge of the current location and time together with a user's calendar gives a good estimation of the user's current social situation. It is preferable that a user's context is detected automatically. Finally, context information has to be disseminated to a context consumer which stores or uses the information. Applications use the context information as an implicit input for e.g. parametrization of functional blocks.

2.2   Calendar Information

Usage of an electronic calendar is a common procedure that structures daily routines and reminds of important dates. Employees usually have access to a company-wide group calendar in addition to a private calendar. Calendar entries are commonly composed of a descriptive text, a start and end time, a categorization and a location (e.g. room number) information. Enterprise business has become considerably dependent of calendaring and scheduling.

Inter-enterprise scheduling can be achieved by central group calendars or by inviting employees to dates. However, the calendaring applications often use proprietary formats. Sharing information across the Internet thus may raise interoperability problems. The Internet Calendaring (iCal) [10] approach tries to provide a common format to exchange information between dissimilar calendaring and scheduling applications. Therefore, it forms an adequate basis for the purposes discussed in this paper.

3   Approach

In this section the proposed approach to tackle the addressed reachability problem is shown. The *planning algorithm* is a first step towards a Digital Call Assistant for communication processes. The solution approach uses user calendar events to determine the best point in time to call someone. Additionally, context information is used to reflect the dynamics of user's activities.

3.1   Components

The basic components of the algorithm are explained next. Each user possess a number of resources, such as communicating devices. The set of communication devices varies with the time and the context of the user. A specific type of such a device (A) is capable of establishing a communication with a certain set of other device types.

$$A \multimap \{A,C,D\}$$

Each user controls two scheduling lists. The *call-in plan* contains a schedule of what resources for communication are available at what time. The complementary *call-out plan* comprise the time slots for possible call request with the according communication resources. The format of the list follows the iCalendar specification. The core definition has been extended by additional necessary information.

3.2   Planning Algorithm

The principle sequence of the planning procedure is shown in Figure 2. The scenario comprises Context Servers and Calendar applications that are co-located with the planning applications. The sequence can be divided into the subsequences *initialization, call, planning, updating* and *running*.

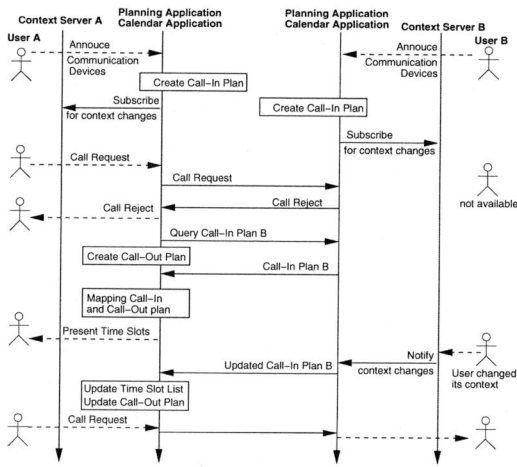

Fig. 2. Sequence of the Planning Algorithm

Initialization phase. The users subscribe to the planning application. During the subscription the users announce their communication resources to the application. The communication resources can be categorized regarding their capabilities. These facilities are taken from a compatibility list that describes what device is able to communicate with which other device.

The planning application queries an iCalendar compliant calendar application about the current calendar entries of the subscribed users. The number of entries can be limited by obtaining an arbitrary time horizon, e.g. a time range of 12 hours. Next, the calendar events are analyzed and a static call-in plan is generated. This plan serves as a coarse structure.

The call-in plan entries are derived from the predefined preference list associating calendar categories such as *meeting, travel* or *lunch* with possible available communication entities. Each calendar event is augmented with the available communication channels. Additionally, the free-times of the user are also associated with the available communication devices. The entries are supplementary associated with a context. This association allows the dynamic adaption of the time range of the call plan entries.

The call-in and call-out plans are stored as a directed acyclic graph (DAG). Each entity E controls a set of resources R. Each of the resources has a list of time slots where communication with this resource is possible. These time slots are denoted v_n. A time slot must at least contain its start time t_s and end time t_e, denoted as $v_n = (t_{s_n}, t_{e_n})$.

In order to support the dynamics of the user's situation and the resulting changes in their daily schedule context sensing techniques are used. The planning application binds the calendar events with the according context whenever it is possible. This allows the application to adapt to the current user's context. To be aware of context changes the planning application subscribes at the Context Server for the required contexts.

Call phase. After the initialization phase the user specifies its call intension. Therefore, the user provides the planning application the parameters such as the callee's address, importance of the conversation and the type of communication. A number of communication types have been defined for the proposed solution. These types are used to classify the communication events. Additionally, the attributes can be used to express the intention of the communication. An event can be augmented by more than one attribute. These information allow the called party to prepare scheduled calls. The following attributes are distinguished.

asynchronous. Describes communication via e-mail or facsimile.
synchronous. Attribute for real-time interactive communication such as phone calls.
uni-directional. A one-way exchange of information that do no require any direct feedback, such as calls to answer machines.
bi-directional. The transmitted messages is answered by the callee.
informative. The content is purely informative. This kind of message is often transmitted in a uni-directional or asynchronous fashion.
urgent. The communication request has to be processed with the highest priority.
private. Classifies the conversation as private.
public. The content of the message is declared as public.

The planning application interacts with a communication instance that initiates the call on behalf of the user. Such a 3rd party call control entity could e.g. be a back-to-back user agent (B2BUA) in a SIP environment. In case that the communication request was not successful, e.g. the called party was busy or not available, the planning application will start its algorithm. The algorithm will try to find a suitable time slot for the next call attempts.

Planning phase. The planning phase is the core of the algorithm. In the first step a static call-out plan of the calling user is created. This step could also be done in the initialization phase. The planning application then queries on request the calling target for a current call-in plan. The corresponding planning assistant responds with the call-in plan. The exchange is similar to the sharing of context information shown in [11].

A suitable trust and privacy concept is needed to secure this very sensible information. The concept of trust and context has been investigated in [12] Additionally, the shared information should be available in a hierarchical structure. Such that persons belonging to a specific trust group will only receive the according information detail level.

Figure 3(a) shows the call-out plan of the caller on the left side. On the right side of Figure 3(a) the call-in plan of the callee is depicted. Both partners possess three communication devices shown as resources in the figure. After receiving the other party's plan the *mapping algorithm* is executed. The algorithm determines possible time slots for efficient communication. A temporal ordered list of these time slot is returned. These time slots are shown as overlapping areas in Figure 3(b). The beginning of the next possible time slot is marked as t_1. The example assumes exclusively disjunct resources.

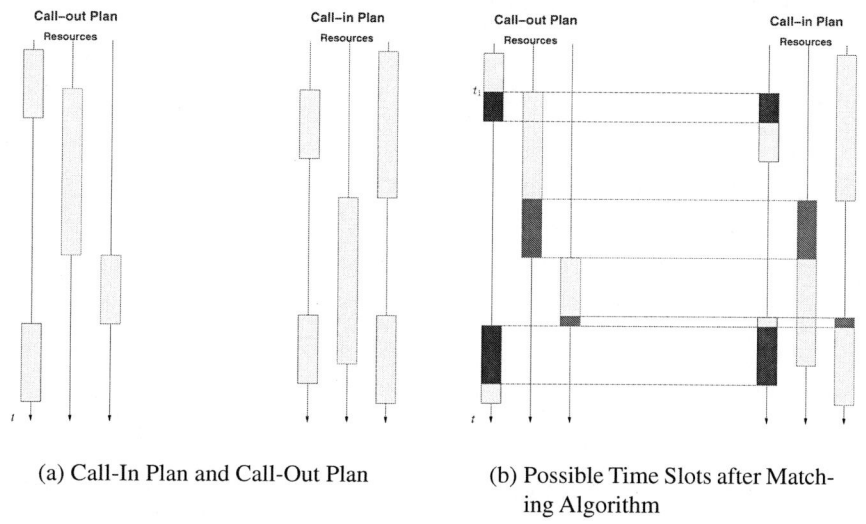

(a) Call-In Plan and Call-Out Plan

(b) Possible Time Slots after Matching Algorithm

Fig. 3. Matching of Resources in Call-In Plan and Call-Out Plan

Updating phase. During pending call attempts the schedule of the call-in and call-out plan can change. When the user's context changes this information is signaled to the Context Server. If the context of the user and the context information bound to the calendar events are not in consistence the planning application is notified. A mismatch might occur if a scheduled meeting lasts longer than previously expected or the user is back earlier from lunch.

Upon receiving such a change the call-in plan and call-out plan of the user are updated. Thereafter, the update is signaled to all subscribed planning applications. The utilization of a context sensing infrastructure is an important concept since users are usually not diligent in updating their calendar entries manually. The context-awareness of the application allows for the automatic adaption of the scheduled plans.

Running phase. The caller has a list of pending call requests. If the planning algorithm has marked a call requests as *promising* the user can initiate this call request. If the call is still unsuccessful the algorithm will be run again. The phases *call* and *updating* are usually executed multiple times during the use of the application.

3.3   Matching Algorithm

Two algorithms are shown in more detail. The *matching algorithm* (Alg.1) compares two nodes v_n, v'_n of the graphs representing the call-in and call-out plans. If the two nodes have an overlapping region, the algorithm will return the maximum overlapping region. Otherwise an empty set is returned.

Algorithm 1 The Matching Algorithm

1: **function** MATCH(v_1, v_2)
2:     $s \leftarrow \min(t_{s_1}, t_{s_2})$
3:     $e \leftarrow \max(t_{e_1}, t_{e_2})$
4:     **if** $e > s$ **then**
5:         **return** (s, e)
6:     **else**
7:         **return** null
8:     **end if**
9: **end function**

Algorithm 2 describes the building of a time slot list, that contains all identified overlapping regions in temporal order. The resource graphs are compared node by node. A drawback of the algorithm is the exponential runtime of $O(n^2)$. It occurs since all entries of the call-in plan have to be matched with all entries in the call-out plan.

However, an exhaustive search is not necessarily required. Each first match of $v \in R$ with $v' \in R'$ is saved as a pivot element. The next node of R does not need to examine all nodes prior to the pivot element. Additionally, no element has to be checked after a missed matched since the nodes in R are ordered regarding their time. This strategy significantly decreases the number of operations.

4   The Digital Call Assistant System Design

The planning algorithm introduced in Section 3 will be part of our context-aware communication infrastructure. The communication platform is based on an Open Source IP Telephony system [13] using the Session Initiation Protocol (SIP) [14]. However, the proposed approach is in no way limited to just SIP as communication protocol. The overall system design is depicted in Figure 4. It comprises the components Context Server, Calendar Server, Planning Application and a Communication Server.

Algorithm 2 Building Time-Slot List

Require: R and R' as ordered list regarding start time
 1: **function** MAPPING(R, R')
 2:     $i \leftarrow 0$
 3:     **for all** $v \in R$ **do**
 4:         **for all** $v' \in R'$ **do**
 5:             **if** $t_s \geq t'_e$ **then**
 6:                 next
 7:             **else if** $t_e \leq t'_s$ **then**
 8:                 next
 9:             **else**
10:                 $T[i] \leftarrow$ MATCH(v, v')
11:                 $i \leftarrow i + 1$
12:             **end if**
13:         **end for**
14:     **end for**
15:     **return** T
16: **end function**

4.1   The Digital Call Assistant System Setup

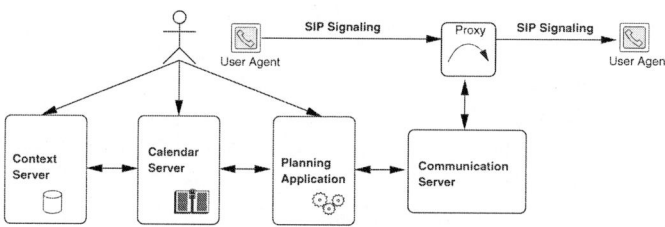

Fig. 4. The Digital Call Assistant System

4.2   Context Provision Infrastructure

The prototype system uses a *Context Server* as an integration component and context source for other context-aware applications. The planning application is a consumer of the context. The planning algorithm requires the current context to parameterize or modify the schedules of the participants. The context server is shown in Figure 5. A multitude of context information sources transmit their data to the server. The data will be encoded in an extended syntax of the Presence Information Data Format (PIDF) [15].

Low-level information sources such as temperature or light sensors are encapsulated by *virtual sensors* (VS). The purpose of the these virtual sensors is the provision of an abstraction to the vendor specific data type and communication. Typical context information sources are devices, such as Bluetooth sender, RF/IR-Badges or iCalendar-compliant applications.

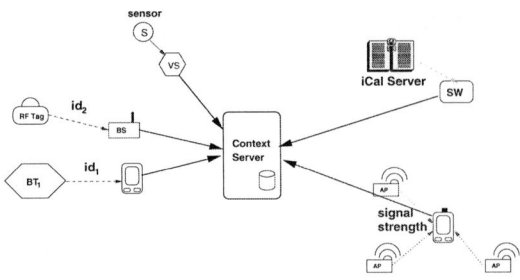

Fig. 5. A Context-aware System Using a Multitude of Context Information Sources

The context server provides the context information via a Web Service interface. The planning application can address the requests for context data using the Simple Object Access Protocol (SOAP). The client application can use a request/response mechanism to query the necessary information. Additionally, a subscribe/notify mechanism can be used to be informed when a context change has taken place.

Location Information. In order to test the proposed call planning approach context information needs to be available. Therefore, an indoor location sensing system has been implemented. Different indoor location sensing systems have been built, installed and evaluated to show the feasibility of our approach. The investigated concepts in our lab cover location sensing using a Wireless LAN infrastructure, the utilization of Bluetooth and a location system with infra-red beacons and radio frequency tags.

A location sensing application based on a IEEE 802.11 Wireless LAN infrastructure has been implemented. This approach follows the core ideas of RADAR [16]. The intended granularity of the system was the detection of the room a user is currently in. The detection was achieved by comparing the measured signal strength values with an a-priori prepared signal strength map of the floor. The evaluation [17] resulted in a detection rate of approximately 85% if three or more Access Points (APs) are visible for the measuring device. The rate drops to roughly 60% if only two APs were available. A PDA running Linux and patched drivers for the WLAN cards were used as the prototype device.

A similar approach was undertaken using a Bluetooth environment [18]. A PDA with a Bluetooth interface receives the beacons send out by the base stations. The 48 bit address of each base station provides a tag for identification. The actual location is obtained from a lookup-table that holds a corresponding symbolic location information for every identifier. An advantage of both approaches is that they use commodity wireless network technologies. WLAN and Bluetooth already exist in most office-like environments. Device such as PDAs and notebooks are often equipped with Wireless LAN and Bluetooth interfaces. Therefore, no additional hardware cost for location sensing on the client side is needed.

4.3   Extending and Using the iCalendar Format

The core specification [10] of the iCalendar defines a set of data fields for shared and distributed electronic calendaring. A MIME content type was chosen as calendar format. The following calendar components are defined: VEVENT, VTODO, VJOURNAL, VFREEBUSY and VTIMEZONE. For the purposes of the presented Digital Call Assistant the existing fields are used whenever possible. Where needed the content is interpreted according to our needs. The MIME content is suitable to be transmitted with a variety of protocols such as SMTP or HTTP.

The transport protocol of the events is derived from the iCalendar Transport-Independent Interoperability Protocol (iTIP) [19]. The protocol provides basic methods to negotiate free time slots. However, the negotiation does not provide the necessary functionality needed for the proposed planning algorithm and adaption through context information.

The Calendar User Type (CUTYPE) provides a mechanism to specify the type of calendar user. The parameter can e.g. contain the types ROOM and RESOURCE. These parameters are used to convey information about the location of the user and the available communication resources. To mark a time slot as free or busy the FBTYPE parameter is used. It allows to set the time slot to FREE or various BUSY types. Additionally, individual notations can be defined. For interoperability these new notions have to be registered with the IANA. The following listing shows an exemplary data set for the Digital Call Assistant.

Listing 1.1. iCalendar data used for the Digital Call Assistant

```
BEGIN:VEVENT
DTSTAMP:20040613T220546Z
ORGANIZER:MAILTO:mgoertz@kom.tu-darmstadt.de
CREATED:20040613T215649Z
UID:DigitalCallAssistant
LAST-MODIFIED:20040613T220437Z
SUMMARY:Meeting
LOCATION:S3106/348
CLASS:PUBLIC
PRIORITY:3
CATEGORIES:Conference
CUTYPE:RESOURCE: sip-phone, mobile-phone
FBTYPE=BUSY:20040616T109000Z/20040616T103000Z
DTSTART:20040616T090000Z
DTEND:20040616T100000Z
TRANSP:OPAQUE
END:VEVENT
```

5   Conclusion

Efficient communication promises savings in time and money. The loop of unsuccessful calls and unsuccessful call-backs has to be avoided. Knowledge about time slots when the communication partner can be optimally reached is the key information for this purpose. The decision when to place the call can be based on this information. In this paper a Digital Call Assistant has been introduced. The assistant's planning algorithm determines time slots which are appropriate for communication. Call-in and call-out plans are the basis for the algorithm. These plans are derived from user's electronic calendar (iCalendar). Context information is used to continuously adapt the static schedule to the user's current situation. The call assistant is part of our context-aware communication service infrastructure. The presented algorithm is identifying the time slots straight forward. Enhancements (e.g. trust concept) to this system are left for further work.

References

1. Imperial College Department of Computing: Foldoc: On-line dictionary. (http://foldoc.doc.ic.ac.uk/foldoc/))
2. Guha, R.: Contexts: A Formalization and Some Applications. PhD thesis, Stanford (1991)
3. McCarthy, J., Buvač, S.: Formalizing Context (Expanded Notes). Computing Natural Language (1997)
4. Dey, A.K.: Providing Architectural Support for Building Context-Aware Applications. PhD thesis, Georgia Institute of Technology (2000)
5. Dey, A.K.: Context-aware computing: The cyberdesk project. In: AAAI 1998 Spring Symposium on Intelligent Environments, Palo Alto, CA, AAAI Press (1998) 51–54 http://www.cc.gatech.edu/fce/cyberdesk/pubs/AAAI98/AAAI98.html.
6. Chen, G., Kotz, D.: A survey of context-aware mobile computing research. Technical Report TR2000-381, Dept. of Computer Science, Dartmouth College (2000)
7. Bedworth, M., O'Brien, J.: The omnibus model: A new architecture for data fusion? In: Proceedings of the 2nd International Conference on Information Fusion (FUSION'99), Helsinki, Finnland (1999)
8. Schmidt, A., Beigl, M., Gellersen, H.W.: There is more to context than location. Computers and Graphics **23** (1999) 893–901
9. Brooks, R.R., Iyengar, S.: Multi-Sensor Fusion: Fundamentals and Applications. Prentice Hall, New Jersey (1998)
10. Dawson, F., Stenerson, D.: Internet Calendaring and Scheduling Core Object Specification (iCalendar). RFC 2445 (1998)
11. Goertz, M., Ackermann, R., Steinmetz, R.: Enhanced SIP Communication Services by Context Sharing. In: 30th EUROMICRO Conference 2004. (2004)
12. Martinovic, I., Goertz, M., Ackermann, R., Mauthe, A., Steinmetz, R.: Trust and context: Two complementary concepts for creating spontaneous collaborative networks and intelligent applications. In: Proceedings of SoftCOM'04, International Conference on Software, Telecommunications and Computer Networks, Croatia/Italy (2004)
13. Vovida Networks, Inc.: Vovida vocal system. (http://www.vovida.org/)
14. Rosenberg, J., Schulzrinne, H., Camarillo, G., Johnston, A., Peterson, J., Sparks, R., Handley, M., Schooler, E.: SIP: Session Initiation Protocol. RFC 3261 (2002)
15. Sugano, H., Fujimoto, S., Klyne, G., Bateman, A., Carr, W., Peterson, J.: Presence Information Data Format (PIDF). Internet Draft draft-ietf-impp-cpim-pidf-08.txt (2003)
16. Bahl, P., Padmanabhan, V.N.: RADAR: An in-building RF-based user location and tracking system. In: IEEE INFOCOM, Tel-Aviv, Israel, IEEE Computer Society Press (2000) 775–784
17. Goertz, M., Perez, A., Ackermann, R., Mauthe, A., Steinmetz, R.: Location Sensing using RADAR. Technical Report TR-KOM-2003-09, Multimedia Communications Lab, Darmstadt University of Technology (2003)
18. Goertz, M., Ackermann, R., Mauthe, A., Steinmetz, R.: A Protype Setup for Location-Aware Personal Communication Services. In: Evolute Workshop 2003. (2003)
19. Silverberg, S., Mansour, S., Dawson, F., Hopson, R.: iCalendar Transport-Independent Interoperability Protocol (iTIP) Scheduling Events, BusyTime, To-dos and Journal Entries. RFC 2446 (1998)

Augmented E-commerce: Making Augmented Reality Usable in Everyday E-commerce with Laser Projection Tracking

Jun Park[1] and Woohun Lee[2]

[1] Dept. of Computer Science, Hongik University, Seoul 121-791, Korea
jpark@cs.hongik.ac.kr, http://id.kaist.ac.kr
[2] Dept. of Industrial Design, Korea Advanced Institute of Science and Technology, Daejon 305-701, Korea

Abstract. Advances in network and mobile communication technologies enabled transformations of markets to the forms of E-commerce. However, current E-commerce technologies cannot provide enough information on the physical dimension, material color, and tactile impression of the products. There exists fundamental discrepancy between the internet-based cyber world and the user's real environment. By superimposing 3D virtual products on the user's real environment, Augmented Reality (AR) technologies might resolve this discrepancy. However, AR implementations often require bulky infra-structures and/or laborious installations, and hence are not easily available. In this paper, we present a marker-free, easy-to-use, and stand-alone AR system based on laser projection tracking (AR Pointer). A prototype of Augmented E-commerce system was developed based on AR Pointer technology. Using this system, untrained users can just "Put & feel a product" in order to find the match of the virtual model of a product in the real environment.

1 Introduction

Worldwide spread of Internet technologies and services enabled web-based E-Commerce services since the middle of 1990's. Up to present, B2B(Business to Business) and B2C(Business to Client) E-Commerce markets have grown steadily, and more and more people are utilizing E-commerce services. Recently, in addition to Internet-based online retailing, E-Commerce has been expanded to mobile environments and living rooms by use of mobile phones and Digital TV sets. The major advantage of E-commerce is convenience in commercial transactions: a user can easily search for a product and compare it with other products in a short amount of time simply by mouse-clicking. However, E-Commerce has deprived users of the reality. Two-dimensional images and texts on the screen are not sufficient to provide information on the physical dimension, material color, texture, tactile impression, and manipulation feedback of a product. Most of those who have on-line shopping experience would be persuaded that we might have unpleasant surprise of receiving products different from our expectations.

V. Roca and F. Rousseau (Eds.): MIPS 2004, LNCS 3311, pp. 242–251, 2004.
© Springer-Verlag Berlin Heidelberg 2004

The difference between the impressions on the images and the actual products is due to the fundamental discrepancy between the internet-based cyber world and the real environment. To resolve the discrepancy, 3D virtual products can be provided using web 3D tools such as Viewpoint or Cult3D. They may provide rich experience to the users [1], but still 3D virtual products are not in the same context as the users' real environment. To correctly resolve the discrepancy, the user's real environments (user's office or home) and the virtual object (the product) should be seamlessly mingled. By superimposing 3D virtual environment on the user's real environment, Augmented Reality (AR) technologies are used to merge real and virtual environments [2]: the size and other physical properties of the products can be expressed in the user's real environment context. However, there have been only a few AR researches applicable to E-Commerce. ARIS(Augmented Reality Image Synthesis) is one of such systems. In ARIS project, an E-commerce system for interior decoration is under development that may improve the shopping experience of visitors (potential customers) using Augmented Reality methods [3]. However, ARIS requires various infrastructures, difficult to be practically utilized in the usual life. There have been few practical AR systems practically usable and available to the untrained users. Many technical limitations of AR need to be overcome so that AR technologies can be applied in E-Commerce services.

2   Limitations of Augmented Reality Technologies for Everyday Use

Augmented Reality, providing magical feeling of immersion, has been one of the most attractive research topics in the human computer interaction and computer graphics fields since early 1990's. Thanks to advances in hardware devices, tracking technologies, and display technologies, AR implementations are available on commercial off-the-shelf desktop and notebook computer systems. However, AR has not been successfully utilized in many practical applications, especially in E-commerce applications, because of its limitations and technical difficulties as summarized in the following.

2.1   Bulky Infrastructures and Complicated Configuration for Using AR

Because of technical similarities, AR systems have been built on existing Virtual Reality (VR) technologies. In most research and development, AR employed VR tracking and display devices that are often bulky and heavy, and/or require laboratory installations of sourcing devices (e.g., for magnetic and ultrasonic trackers). Vision-based tracking technologies, such as ARToolKit system [4], often use artificial markers, and hence do not require hardware installation in the environment. However, markers need to be installed and calibrated in advance. There are also (partial) occlusion problems of the markers, which may result in tracking failure.

2.2   Use of Unwieldy Head-Worn Displays (HWD/HMD)

In AR applications, video see-through displays are preferred to optical see-through displays for higher registration accuracy. However, video see-through displays are often unwieldy and fatiguing, hindering user's natural view of and interaction with the real environment. Thanks to recent technical advances, AR has been successfully ported on hand-held devices such as PDA's and mobile phones [6][7][8], demonstrating the potential of producing practical applications. Hand-held AR systems may be used in many applications including collaboration [5] and navigation [13].

2.3   Lack of Stand-Alone Systems

Because AR integrates many advanced technologies, such as tracking, display, video, and computation technologies, it has been difficult to produce a stand-alone (or all-in-one type) AR system. For tracking systems, many technologies require installation of sourcing devices, on which tracking devices physically depend. Although vision-based tracking systems do not require sourcing devices, they rely on pre-installed markers [4][14]. Inertial trackers work as a stand-alone tracking system, but the tracking information they provide cannot be used to construct geometrical relations between the real and the virtual environment, which is required for image overlays. GPS and digital compass also do not depend on other devices, but because of low accuracy, they are not useful in many AR applications other than navigation and guidance. To summarize, existing AR systems are dependent on tracking infrastructures or accessories such as markers. For non-experts, a plug-and-play stand-alone AR system is desired.

To produce practical applications including E-commerce applications, portable, marker-free, easy-to-use, and stand-alone AR system is desirable. Such an AR system can be used in everyday E-Commerce life of untrained users.

In this paper, we introduce a stand-alone, plug-and-play type AR Pointer system to be used for Augmented E-Commerce applications. The proposed AR system is easy to use and marker-free, and hence more general-purpose in applications. Instead of traditional tracking systems that require laborious installations, a small laser and a video camera are used as a tracking unit in our system.

3   AR Pointer System

Diffractive optical elements (DOE's) transform a single laser beam into various structured light patterns as shown in Fig. 1. These patterns, projected on a planar structure (wall, table, floor, etc.), change their shapes as the distance and the orientation between the laser and the plane change. Through calculating the geometrical relation (distance and orientation) between the plane and the laser, the laser patterns can be used in place of markers that are widely used in vision-based tracking systems. In other words, a projected DOE pattern may play a role of a marker to provide tracking information.

Fig. 1. Laser light patterns

3.1 Use of Laser Devices

Laser pointer has been often used in Human Computer Interaction (HCI) applications in order to support interactions among computer systems, visual environments, and human. It has been used as a 2D pointing device in conferences or presentations in place of traditional pointing devices such as mice [9][10]. Laser pointer has also been used as a travel aid device for blind people [11]. In this system, a stereo camera was used to calculate the depth to the pointed surface. A red-pass filter was also used for robustness of image analysis. In collaborative AR environments, lasers have been used as an interaction device to select, move, and rotate visual elements [12]. However, to the best of our knowledge, lasers have not been used as a real-time tracking device up to present.

3.2 AR Pointer System Components

AR Pointer system is composed of a laser with a DOE, a video camera, and an optional LCD monitor. A laser and a video camera are combined into one unit to be used as a tracking and video capture device. An optional LCD monitor can be used as a display unit, to compose a hand-held AR system together with laser-camera unit. For registration information calculation and virtual object overlay, the laser-camera unit can be connected to any type of computing systems (Fig. 2). Depending on the computing system, AR Pointer system could be constructed as a desktop system (Fig. 3) or a mobile system (notebook PC / tablet PC / PDA / mobile phone). As a mobile unit, AR Pointer can be used also as a hand-held AR system as in TransVision[5] or NaviCam[13] (Fig. 4).

　　AR Pointer system is distinguished from the previous AR systems, showing many advantages. Firstly, AR Pointer is a stand-alone system: there is no requirement of laborious installation process, bulky and unwieldy tracking devices, or marker installations and calibrations. Secondly, it is also highly mobile and easy-to-use as in plug-and-play peripheral devices. Thirdly, pointing with the laser patterns, AR Pointer system can overlay virtual products on any planar surface of user's real environment. These advantages enable AR Pointer system to be used in everyday E-commerce life of untrained users.

　　Most tracking systems are exocentric in terms of tracking coordinate systems (the origin of the coordinate system is located outside the tracking unit). It is because tracking is dependent on external tracking devices or installed artificial markers. However, AR Pointer is egocentric because the sourcing device (laser) and sensing device (video camera) are of one unit. AR Pointer system can be used to "Point a spot & Register an object", which is an intuitive interaction approach. In other words, a

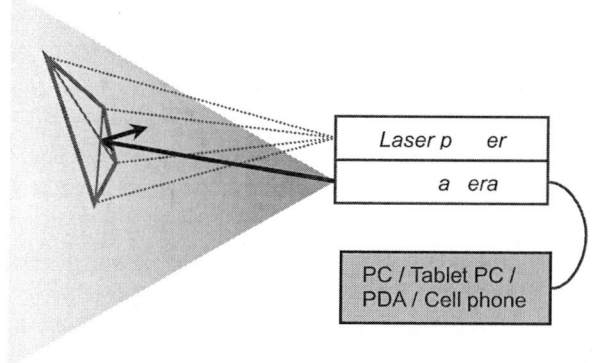

Fig. 2. AR Pointer system components

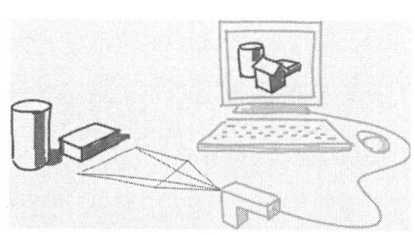 

Fig. 3. Desktop AR Pointer system   **Fig. 4.** Mobile AR Pointer system

virtual object can be placed in the user-specified position just by pointing at using the laser: AR Pointer system is an AR pointing device.

AR Pointer system is similar to vision-based tracking systems in that it is also based on "seeing" visual elements. However, AR pointer system does not require marker installations and calibrations. Therefore, AR Pointer can be used to overlay on the environments where marker installation is difficult or impossible. All the more, AR Pointer system can be used on the other side of a glass window or a show-case: it is capable of performing non-contact and/or intrusive operations.

4 Pose Computation and Transformations

There are three major coordinate systems in AR Pointer system (Fig. 5). The laser coordinate system (L) is used to describe DOE pattern feature positions. The tabletop coordinate system (T) is used to represent the position and orientation of the laser patterns projected on the planar surface. Lastly, the camera coordinate system (C) is used to represent image coordinates of measurements (the laser patterns) and to overlay the virtual products in augmented views.

There are two important coordinate transformations, each of which should be calculated on-line or off-line.

T_{CL}: from the laser coordinate to the camera coordinate transformation

T_{CT}: from the tabletop coordinate to the camera coordinate transformation

Parameters of laser coordinate system are the laser DOE starting point (O_L) and the DOE pattern features (Xi). Parameters of camera coordinate system are the focal point (Oc), field of view (FOV: horizontal(θh) and vertical(θv)), and image resolution (Sx, Sy). These parameters can be calculated off-line except for the image resolution (user may specify the image resolution in run-time).

4.1 Coordinate Transformation Calculation

In order to correctly superimpose visual overlays on the tabletop, it is necessary to determine the coordinate transformation from the camera to the tabletop coordinate system (or the inverse) The input measurements are the coordinates of laser pattern features on the camera image. To determine the tabletop coordinate system, in our approach we calculated the feature 3D positions of the laser patterns projected on the tabletop.

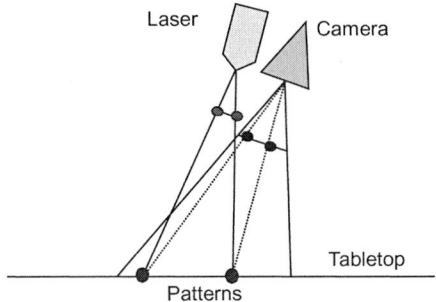

Fig. 5. AR pointer Coordinate Systems

As can be seen in Fig. 5, these 3D positions are the intersections of the rays from the laser DOE starting point to the DOE pattern features, and the rays from the camera focal point to the image features [14]. To calculate the intersection (more precisely, the closest point between the rays), the rays should be represented in a common coordinate system (in our implementation, camera coordinate system was used). To represent rays in a common coordinate system, coordinate transformation between the laser and the camera coordinate systems should be known. This transformation was calculated off-line as explained in the following section. Detailed steps for computing ray intersections are described in the following.

(1) Laser feature patterns are transformed into camera coordinate system as follow.

$$X_i^C = T_{CL} \cdot X_i^L \qquad \text{(DOE pattern features)}$$

$$O_L^C = T_{CL} \cdot O_L^L \qquad \text{(DOE pattern start point)}$$

(2) Ray from laser focal point to the feature points represented in camera coordinate system are obtained (there is no unknown parameter).

$$l_{Li}^C = O_L^C + t_i (X_i^C - O_L^C)$$

(3) Rays from camera focal point to the image feature points are calculated using the camera FOV.

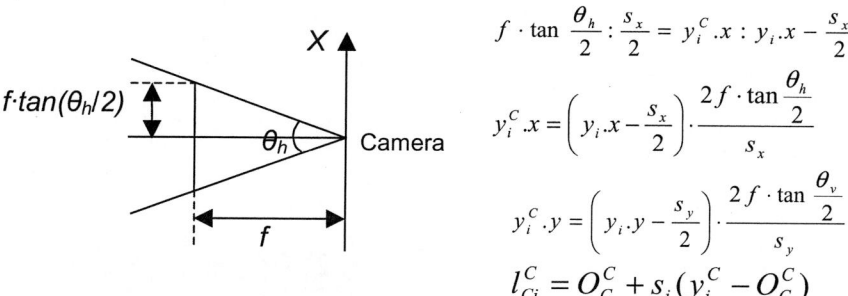

$$f \cdot \tan \frac{\theta_h}{2} : \frac{s_x}{2} = y_i^C.x : y_i.x - \frac{s_x}{2}$$

$$y_i^C.x = \left(y_i.x - \frac{s_x}{2} \right) \cdot \frac{2f \cdot \tan \dfrac{\theta_h}{2}}{s_x}$$

$$y_i^C.y = \left(y_i.y - \frac{s_y}{2} \right) \cdot \frac{2f \cdot \tan \dfrac{\theta_v}{2}}{s_y}$$

$$l_{Ci}^C = O_C^C + s_i (y_i^C - O_C^C)$$

(4) The intersections (the closest point between the rays) of the rays are computed.

The intersections of these pairs of rays are the positions of the laser pattern features projected on the tabletop, represented in the camera coordinate system. With three or more non-collinear points, the tabletop coordinate system can be constructed, and the transformation between the camera coordinate system and the tabletop coordinate system can be calculated.

4.2 Laser to Camera Coordinate Transformation

The transformation between the laser coordinate system and the camera coordinate system should be known in order to represent the rays in a common coordinate system. We used an off-line calibration method, which is described in the following.

(1) The laser pattern was aligned perpendicular to a visual marker, and the distance between the laser and the marker was measured (Fig. 6-top).

(2) The camera image of the visual marker was taken with the laser switched off (Fig. 6-bottom).

(3) The 3D visual marker coordinates were measured.

(4) The camera pose was computed using an optimization method based on the 3D marker positions and the 2D image coordinates.

In step (1), the transformation from the laser to the marker coordinate system (T_{ML}) was obtained (simple translation on Z axis). In step (4), the transformation from the marker to the camera coordinate system (T_{CM}) was calculated. Transformation from the laser to the camera coordinate system (T_{CL}) was then calculated using T_{CM} and T_{ML}.

$$T_{CL} = T_{CM} \cdot T_{ML}$$

This calibration process was performed once after system assembly: it does not have to be repeated. The images used for "from the laser coordinate system to the camera coordinate system" transformation calculation are shown in Fig. 6.

4.3 The Prototype of AR Pointer System

Our prototype of AR Pointer system is composed of a laser with a 5x5 grid pattern DOE, an IEEE-1394 video camera, and an LCD monitor (Fig. 7). The laser and the camera were placed parallel and apart. As the distance between the laser and the camera increases, the tracking accuracy might be improved.

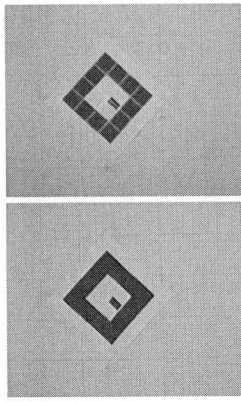

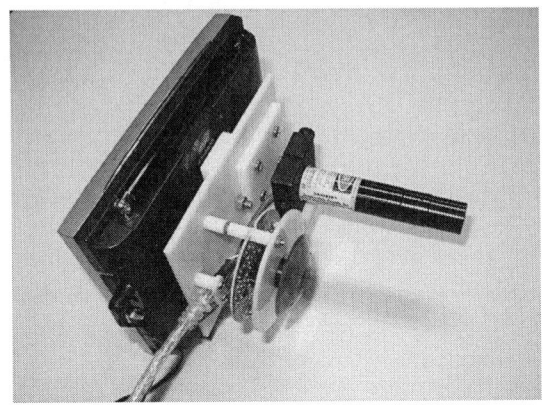

Fig. 6. Images used for laser to camera transformation calculation with laser on (top) and off (bottom)

Fig. 7. AR Pointer system hardware configuration

For robust detection of bright laser patterns, the automatic camera exposure control was disabled (manually controlled to obtain dark images). The video images were captured in VGA resolution at about 15fps. The depth range of the system was from 15cm to 4 meters depending on the illumination condition.

5 Prototyping Augmented E-commerce Web Sites

We built an Augmented E-commerce system based on our noble stand-alone AR Pointer system. Fig. 8 shows the Augmented E-commerce prototype embedded to an E-Commerce web site. The scenario is very intuitive. A user searches for a product on-line and drag-and-drops the 3D model of an interesting product onto the AR window. AR Pointer system, then, registers the virtual object on the user pointed spot in the real environment. The user interacts with the virtual objects by moving and rotating in order to estimate the dimension and shape of the product, and to find the matches with the room environment. In other words, users can "Put & feel a product" in reality.

Using Augmented E-commerce system, users' shopping experience can be greatly improved. Users can perceive the products in his/her office or home environment through LCD monitor without wearing unwieldy HMD's or HUD's.

We have performed usability tests of the system. Untrained users could place the virtual products with no difficulty, and perceive the size of the bookshelf visually comparing with papers of A4 size (Fig. 9). In another experiment, users placed a virtual toy truck beside toy robots. Users estimated the size of the truck easily by comparing with other reference objects such as a soda can and a sports watch (Fig. 10). As a stand-alone system, users also could utilize AR pointer to place virtual products on various places (on the wall and on the desktop in our experiments).

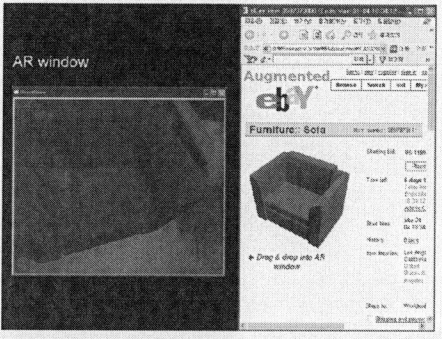

Fig. 8. Augmented E-commerce

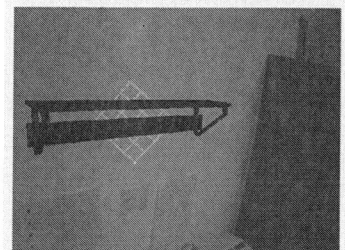

Fig. 9. Augmented view of a shelf

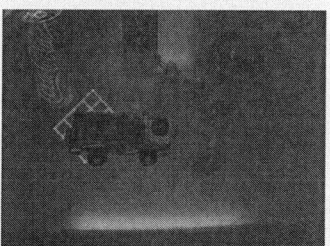

Fig. 10. Augmented view of a toy truck

6 Conclusions

We introduced a laser-projection based AR Pointer system and Augmented E-Commerce prototype that combines web contents and AR technologies. AR Pointer is a stand-alone, marker-free, and easy-to-use AR system that can be used in everyday E-Commerce life of untrained users. AR Pointer can be used to overlay virtual objects on the environments where marker installation is difficult or impossible. Owing to its non-contact and intrusive nature, AR Pointer system can be also used on the other side of a glass window or a showcase. AR Pointer system is less influenced by visual noise because the tracking is based on bright laser patterns: it is less dependent on illumination conditions and less influenced by occlusions.

Inertial trackers can be integrated with AR Pointer system to avoid tracking failures when laser patterns cannot be correctly detected. An inertial sensor can be also used as a control device to rotate the virtual objects in the augmented views. AR Pointer system can be improved by enabling tracking on non-planar surfaces. If executed in real-time, 3D shapes of the surface may be estimated for advanced interaction with the real environments.

References

1. H. Li, T. Daugherty & F. Biocca, Characteristics of virtual Experience in electronic commerce: A protocol analysis, Journal of Interactive Marketing, 15 (3), 13-30, 2001
2. Azuma, Ron; Baillot, Yohan; Behringer, Reinhold; Feiner, Steven; Julier, Simon and MacIntyre, Blair. "Recent Advances in Augmented Reality." In IEEE Computer Graphics and Applications, 25(6):24-35, Nov-Dec 2001
3. http://aris-ist.intranet.gr/documents/ARIS_D8.2.doc
4. M. Billinghurst and H. Kato, "Collaborative Mixed Reality", In Proceedings of International Symposium on Mixed Reality (ISMR '99). Mixed Reality-Merging Real and Virtual Worlds, pp. 261-284, 1999
5. Jun Rekimoto, "TransVision: A hand-held augmented reality system for collaborative design", International Conference on Virtual Systems and Multimedia (VSMM'96), pp. 85-90, 1996
6. Virtual Mosquito Hunt-Cebit2003(http://w4.siemens.de/en2/html/press/newsdesk_archive /2003 /e_0311 _d.html)
7. Daniel Wagner and Dieter Schmalstieg, "First Steps Towards Handheld Augmented Reality", Proceedings of the 7th International Conference on Wearable Computers (ISWC03), 2003
8. Oliver Bimber, Video See-Through AR and Optical Tracking with Consumer Cell Phones (Unpublished, http://www.uni-weimar.de/~bimber /research. php)
9. Carsten Kirstein and Heinrich Müller, "Interaction with a Projection Screen Using a Camera-Tracked Laser Pointer", Proceedings of the International Conference on Multimedia Modeling, IEEE Computer Society Press, pp.191-192, 1998
10. D.R. Olsen Jr. and T. Nielsen, "Laser pointer interaction", Proceedings of the SIGCHI conference on Human Factors in Computing Systems, pp.17-22, 2001
11. F. Fontana, A. Fusiello, M. Gobbi, V. Murino, D. Rocchesso, L. Sartor, and A. Panuccio. "A Cross-Modal Electronic Travel Aid Device", In F. Patern`o (Ed.): Mobile HCI 2002, LNCS 2411, pp. 393–397, 2002
12. Jun Rekimoto and Masanori Saitoh, "Augmented Surfaces: A Spatially Continuous Workspace for Hybrid Computing Environments", Proceedings of CHI'99, pp.378-385, 1999
13. Jun Rekimoto and Katashi Nagao, "The World through the Computer: Computer Augmented Interaction with Real World Environments", Proceedings of UIST'95, pp.29-36, 1995
14. Ulrich Neumann and Jun Park, "Tracking for Augmented Reality on Wearable Computers", Virtual Reality Journal, No.3, pp.167-175, Springer-Verlag, London Ltd. 1998

Architecture and Protocols for the Protection and Management of Multimedia Information

Víctor Torres, Eva Rodríguez, Silvia Llorente, and Jaime Delgado

Universitat Pompeu Fabra, Passeig de Circumval·lació, 8,
08003 Barcelona, Spain
{victor.torres, eva.rodriguez, silvia.llorente,
jaime.delgado}@upf.edu
http://dmag.upf.edu

Abstract. The management of multimedia information is a key issue when permissions, protection issues and rights are involved. The appearance of new multimedia description formats, new distribution channels and diverse business models makes it difficult to provide a general solution for performing such management. In this scenario, the description of architectures and protocols appears as a valid approach in the way to obtain results that could be adapted to the changing scenario present nowadays in the multimedia arena. MPEG-21, in its aim to provide a multimedia framework, is a good starting point for describing candidate modules for implementing a system that works inside the framework. Based on some of its parts, as currently 16 parts are being considered within the standard, we propose a modular architecture to cope with the complexity of the rights management and protection of multimedia information. This architecture is by no means complete, but is intended to map a standard to something more usable. The use of the modules defined in the architecture is illustrated with some examples, which show how protected multimedia information can be processed in the context of a real system based on an international standard, such as MPEG-21.

1 Introduction

According to the MPEG-21 standard, digital Items (DI) are the unit of representation of multimedia information. Inside a DI we can find different kinds of data about a multimedia content, specifically those related to the representation of the information specified in the different MPEG-21 parts. Furthermore, we can have information in other formats (standardised or not), representing the multimedia object itself, for instance, a video or an image.

Within these premises, our aim is to define a possible architecture for the management of the different kinds of data that could be represented inside a DI, focusing on the MPEG-21 data, understanding MPEG-21 data as those defined in the different parts of the standard, but keeping an eye on non-MPEG-21 data, too.

The strategy to accomplish this objective can be decomposed into phases. The first one is to identify each part present in the DI. Then, the mechanisms for managing each part should be described and implemented. The final result will be the descrip-

V. Roca and F. Rousseau (Eds.): MIPS 2004, LNCS 3311, pp. 252–263, 2004.

tion of the architecture for managing the different components of a DI, using the corresponding protocols when needed.

This paper presents the digital item structure and some of the MPEG-21 associated data it can carry. Moreover, it describes some preliminary ideas for the architecture needed to support such data as well as the protocols involved in its management. It also presents a use case in order to clarify the processing of a digital item with the proposed architecture.

2   Multimedia Information Representation Using MPEG-21

A Digital Item is defined in [1] as a structured digital object, including a standard representation and identification, and metadata within the MPEG-21 framework. It is the fundamental unit of distribution and transaction within this framework. The rest of this section presents some parts of the MPEG-21 standard.

2.1   Digital Item Declaration (DID)

The two major goals of the Digital Item Declaration part within MPEG-21 are first to establish a uniform and flexible abstraction and interoperable schema for declaring Digital Items and also to be as general and flexible as possible, providing hooks to enable higher level functionality and interoperability. The Digital Item Declaration (DID) technology [2] is defined in three normative parts: DID Model, Representation and Schema. DID corresponds to part 2 of the MPEG-21 standard.

Whereas the DID Model describes a set of abstract terms and concepts to build the model for defining Digital Items, the Representation part describes the DID Language (DIDL), that is, the syntax and semantics of each of the DID elements defined in the Model, as represented in XML. The Representation part specifies in a textual manner the semantics and construction rules of the elements and element-based objects, while uses XML Schema to describe the syntax of each element. Finally, these individual schemas are grouped together in the Schema part to provide the entire grammar of the DID representation in XML.

Within this model, a Digital Item is the digital representation of a work, and as such, it is the thing that is acted upon.

Figure 1 shows the relationship among some of the terms of the defined model. The Container is a potentially hierarchical structure that allows Items to be grouped together with their descriptors, components and resources. The term "Item" should be understood as a declarative representation of a Digital Item.

A DIDL document consists of a DIDL root element with an Item child element or a Container child element. An Item element is a grouping of possible sub-Items and/or Components, bound to a set of relevant Descriptors containing descriptive information about the item. A Component groups a Resource element with a set of Descriptors that contain descriptive information about the resource, plus a set of Anchors that specify points or regions of interest in the resource. It is intended to be the basic building block of digital content within a DIDL document. A Resource is an individually identifiable asset such as a video or audio clip, an image, an electronic ticket, a textual work or even also a physical object. A Reference element is used to link the contents of an element inside another element. A Statement element, which defines a piece of information pertaining to the parent element, can contain any kind of information, including descriptive, control, revision, tracking or identifying information. It can also contain any data format, including plain text and various machine-readable formats such as well-formed XML.

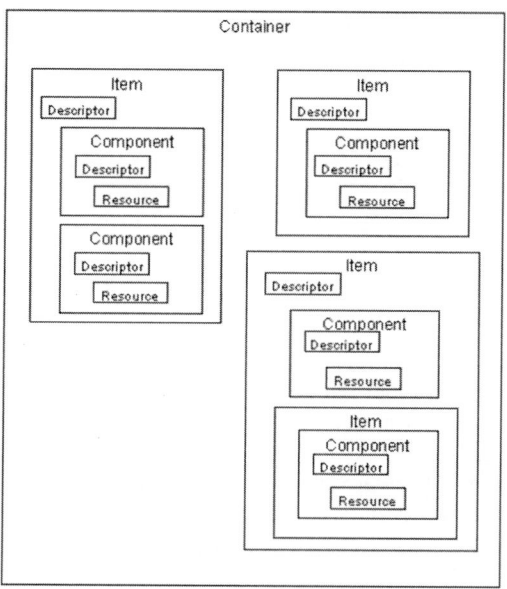

Fig. 1. Example of digital item declaration

The possibility of inserting data in any kind of format, specially well-formed XML, within a Statement provides a wide field for including information used to protect and process multimedia data, such as rights expressions, intellectual property management and protection information or adaptation descriptors.

2.2 Rights Expression Language (REL)

Rights expression languages (RELs) have been proposed to express rights and conditions of use of digital content. RELs can be used for example to describe an agreement between a content provider and a distributor, or between a distributor and an end user. Moreover, RELs can be used to express the copyright associated to a given digital content by specifying under which conditions the user is allowed to exercise a right.

Part 5 of the MPEG-21 standard [3] specifies the syntax and semantics of a Rights Expression Language (REL). In particular, it specifies the syntax and semantics of the language for issuing rights for users to act on Digital Items. One important concept in

REL is the License. A License is a container of grants that are formed by a principal that has the permission to exercise a right against a resource under some conditions that must be previously fulfilled. Figure 2 shows the structure of a REL License.

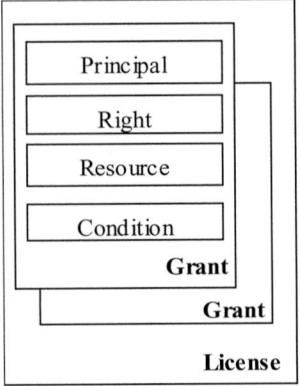

Fig. 2. REL License

Inside a REL license, the most important element is the Grant. A Grant is an XML structure that is formed by four elements:
-   Principal represents the unique identification of an entity involved in the granting or exercising of Rights.
-   Right specifies an action or activity that a Principal may perform on, or using, some associated Resource.
-   Resource represents the object against which the Principal of a Grant has the Right to perform.
-   Condition represents grammatical terms, conditions and obligations that a Principal must satisfy before it may take advantage of an authorisation conveyed to it in a Grant.

A Grant expresses that some Principal may exercise some Right against some Resource, subject, possibly, to some Condition.

MPEG-21 REL makes use of the Rights Data Dictionary [4], part 6 of the MPEG-21 standard, that comprises a set of clear, consistent, structured, integrated and uniquely identified terms. The structure of the RDD is designed to provide a set of well-defined terms for use in rights expressions.

2.3 Digital Item Processing (DIP)

Digital Item Processing (DIP) [5] specifies the syntax and semantics of tools that may be used to process Digital Items. The objective of DIP is to provide a normative set of tools for specifying the processing of a Digital Item in a predefined manner. In this way, it will be possible to extend the DIDL [2] in order to add user specific functionality inside the Digital Item, but maintaining the interoperability at the processing level.

A key component of Digital Item Processing is the Digital Item Method (DIM), that is, the mechanism that enables Digital Item Users to include sequences of instructions for adding predefined functionality to a Digital Item. Such a sequence of instructions is not intended to implement the processing of media resources, but might be used for adaptations of the Digital Item Declaration at the DID level. A DIM is expressed using the Digital Item Method Language (DIML), which includes a binding for Digital Item Base Operations (DIBO). The method definition may be referenced from or embedded in the DID. For example, a Digital Item representing a music album may contain a DIM to add a new music track to the album. Such a DIM can be used to ensure that the new music track is added to the Digital Item while maintaining a suggested format (i.e. elements added in the correct place in the DID structure, correct Descriptors are included, etc.).

Digital Item Base Operations are functional building blocks utilised by a Digital Item Method. They can be considered somewhat analogous to the standard library of functions of a programming language. A DIBO is described by a normatively defined interface and semantics, while the DIBO implementation will depend on the peculiarities of the terminal in which it is to be developed.

Digital Item extension Operations (DIXOs) specify a normative mechanism for enabling extended functionality in an efficient way. Their operation is like the DIBOs one except that these are not normatively defined. As well as DIMs, DIXOs may be referenced from or embedded in the DID.

2.4   Intellectual Property Management and Protection (IPMP)

Part 4 of the MPEG-21 standard, Intellectual Property Management and Protection [6], has concluded the requirements phase. A Call for Proposals has been done and as a result a first draft [7] has been published.

In this part of the standard an interoperable framework for Intellectual Property Management and Protection will be defined. MPEG-21 will provide a framework that encourages the creation of new services that can be used to support new business models. These services should meet the needs of the different members of the networks associated with the distribution of digital items.

An architecture that facilitates the management of trust relationships will be defined, too. This means that it must describe content usage requirements, peer capabilities and requirements associated with making trustworthy guarantees about MPEG-21 peer capabilities.

In IPMP it will be included the expression and enforcement of rights that are associated with digital item distribution, management and usage by all members of the value chains.

We made a contribution [8] to the Call for Proposals and it has been considered together with other proposals in the final draft. Our contribution defines an XML Schema to describe IPMP information.

3 Proposed Architecture

As MPEG-21 has not yet defined an architecture for a system implementing the functionality needed to process a digital item containing information associated to different parts of the MPEG-21 standard, we propose here the definition of a Multimedia Information Protection And Management System (MIPAMS), the DMAG-MIPAMS.

The system architecture consists on several modules, each of them providing a part of the functionality. The use of a modular approach allows the addition of new modules as needed.

The architecture distinguishes between the final user and the servers needed to provide the underlying infrastructure. In the middle, we have the "Intermediary". The functionality of this intermediary element could be also located either on the user side, on the server side or on both, depending on the characteristics of the equipment or terminal used by the final user to access to the multimedia content. Its main functionality is to hide the complexity of the system from a final user perspective. This will allow us to describe the architecture in a very general way, taking advantage of the equipment capabilities independence. For instance, if the final user accesses to the system using a personal computer, then the intermediary functionality could easily be located into his side. On the contrary, if the final user accesses the system through a PDA or a mobile phone, the intermediary functionality should be better located on the server side, possibly having a new server acting as the user in front of the different servers and providing the user the desired content. Another important functionality of the intermediary could be to provide access to different systems (with different content servers, license server, etc), hiding this issue to the final user.

Figure 3 shows the basic modules that could be present in the proposed architecture.

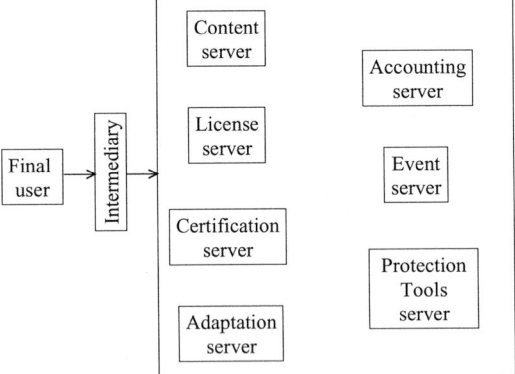

Fig. 3. Proposed architecture for the DMAG-MIPAMS

Briefly, the functionality of each module represented in the architecture is the following:

- Content server: It provides the content that final users may request. It can be internally decomposed into several modules, for instance, if we want to separate digital items describing resources from the resource itself or if the content is stored in an external system.

- License server: It provides licensing functionality needed to access the content. It includes license creation and license validation.
- Certification server: It certifies the entities present in the system, including other modules and final users. It includes registration, authentication or key delivery.
- Adaptation server: It performs the adaptation of the content depending on the characteristics of the final user terminal.
- Accounting server: Keeps track of what happens in the system, including statistics and traces.
- Event server: Receives events information associated to content usage in order to advise the author or distributor of the content, if needed.
- Protection tools server: It stores the tools needed for the protection of content.

The event server and the adaptation server relate to other two parts of the MPEG-21 standard, Digital Item Adaptation (DIA) [9] and Event Reporting (ER) [10]. We do not describe them here, as we want to focus on the rights expression and protection aspects of the system, which will be further developed in next sections.

The protocols appearing in the architecture presented are explained in some more detail in the next section and an example of use is provided in section 4.2 Use case.

3.1   Protocols Involved in the Architecture

An important issue in a distributed system is the secure communication between the different entities within it. In the system that we present, it is important to ensure the secure interchange of multimedia content, digital items, licenses and protection or adaptation tools between the different entities, servers and users. These entities communicate using the channel security provided by the SSL/TLS set of protocols [11]. The SSL protocol provides privacy and reliability between two communicating entities. One advantage of SSL is that it is application protocol independent. SSL is a commonly used protocol for managing the security of a message transmission over insecure networks.

In the DMAG-MIPAMS, the transport layer provides a secure authenticated channel, while the application layer provides a message protocol layer that permits the interchange of messages through the secure channel provided by the transport protocol.

The messages exchanged between the different entities in the system have a common structure. The main fields of these messages are an identifier, the content of the message and a digital signature. The content field of the message is defined in a different way depending on the purpose of the message. For example, in messages where the terminal or the intermediary request a license, the information related to the user, content and action is placed in this field. In a message that requests a protection or an adaptation tool, the data that identifies the required tool is placed in this field.

In the use case presented below, in section 4.2, we describe some of the messages interchanged between the Intermediary and the License Server or Protection Tools Server.

4 Multimedia Content Processing

We have introduced a possible architecture and a first approach for protocols in order to build a system to manage multimedia content in a protected environment, emphasising the case where we want to manage content that has associated digital rights.

A next step could be to develop mechanisms for processing digital content in this context.

MPEG-21 has also started work to describe how multimedia content can be processed. This is the MPEG-21 DIP (Digital Item Processing) [5].

In the next subsections we introduce our approach to include protection in the DIP.

4.1 Protected Multimedia Content Processing

Currently, the semantics of the standardised DIBOs (see 2.3) that form the basis of DIP [5] do not take into account some important concepts in multimedia content distribution such as the protection of multimedia content or the digital rights related to protected or unprotected content.

In this section we present the PlayResource DIBO, which processes protected multimedia content. We also specify its syntax and semantics.

PlayResource DIBO plays the specified protected and governed media resource. As the resource has associated rights expressions, a license-based authorisation is done in order to check if the user has permissions to perform the requested operation. Afterwards, the resource is unprotected with the appropriate IPMP tool, only if the user has been previously authorised. It has as inputs a resource node that represents the DIDL resource element describing the media resource, an MpegDIPResourceChangeObject containing the changes to be applied to the resource before playing it and a Boolean value indicating if the resource will be played asynchronously or not. The return value will be an MpegDIPResourceStatus object identifying the playing resource.

We propose the definition of a set of subroutines that can be used by the implementers of DIBOs that process protected multimedia content that can also have associated rights expressions, such as the presented DIBO or other ones like *ExecuteResource*, *PrintResource* or *StoreDIDNew*. A sample set of proposals for subroutines could be:

- *GetLicenses* obtains the licenses related to the operation that a user wants to exercise against a specific resource. It has as inputs the right, resource and the principal. It returns an array of Nodes that contains the REL licenses. The licenses can be retrieved from the License Server or from the DI.
- *AuthoriseUser* checks if a user has the permissions to perform the requested operation. It has as input the array of licenses returned by the *GetLicenses* subroutine. The returned value is a Boolean that is true if the user is authorised or false otherwise.
- *GetIPMPInfo* obtains the IPMP information related to the resource against which the user wants to perform the requested operation. This information is obtained from the Digital Item. Relevant IPMP information can be the Tool List that includes the list of IPMP tools used in order to consume the content or the Tool

Holder that is where the binary IPMP Tool can be placed if the Digital Item carries the IPMP Tool
- *UnprotectContent* obtains the IPMP Tool from the IPMPTools Server and unprotects the multimedia content the user wants to play.

Figure 4 shows how the *PlayResource* DIBO can use the proposed subroutines when a user wants to play a protected resource.

```
PlayResource(resourceNode, changes, async){
    Node[] licenses=GetLicenses(right,resource,principal);
    boolean isAuth=AuthoriseUser(licenses);
    if(isAuth){
        Node IPMPInfo=GetIPMPInfo(resourceNode);
        if(IPMPInfo != null){
            GetIPMPTools(IPMPInfo);
            UnprotectContent();
        }
        for(x = 0; x < resource.width; x++) {
            for(y = 0; y < resource.height; y++) {
                Show Pixel(x, y, resource[x][y]);
            }
        }
    }
    else
        Alert("Not Authorized");
}
```

Fig. 4. PlayResource DIBO

The DIBO described in this section is the result of a core experiment [12] presented by DMAG in the 69th MPEG meeting held in July 2004. This core experiment has made use of existing REL tools [13] [14] that we previously developed and contributed as MPEG-21 Reference Software [15].

4.2 Use Case

In this section we present a scenario to illustrate how the proposed architecture and protocols and the processing of protected multimedia content are related. Moreover, with this example we try to clarify the relationship among the different MPEG-21 parts presented, i.e. DID, DIP, REL, RDD and IPMP.

The scenario we propose is about a digital library. Imagine a student is subscribed to an electronic journal and wants to view an article. In the architecture we propose, first the user receives an MPEG-21 digital item with the protected multimedia content or a reference to it. Then, when the user accesses the DI, it is parsed and the user can choose the operation or method he wants to perform (view the article, print the article, etc.). Finally, if the user has the appropriate permissions he is authorised to perform the chosen operation.

The workflow of these actions is as specified below and graphically shown in figure 5:
- The user receives the Digital Item with the physical resource or a reference to the multimedia Content Server where it is placed. Moreover, the Digital Item has the associated Digital Item Processing information that will allow the user to choose the Digital Item Method that he wants to perform.

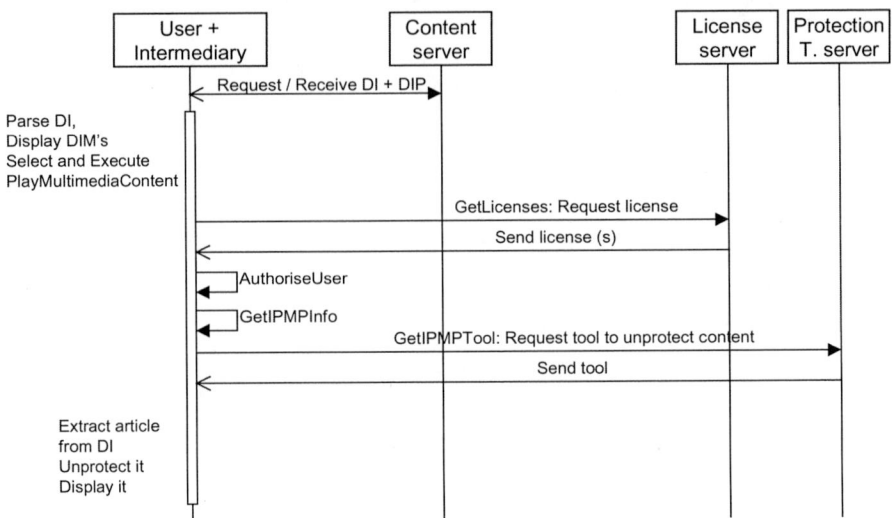

Fig. 5. Use case workflow diagram

- The user loads the DI in the appropriate application, which displays all the Digital Item Methods.
- The user chooses a PlayMultimediaContent DIM.
- PlayMultimediaContent DIM is executed and the PlayProtectedResource DIBO is called. This DIBO performs the following steps:
    - GetLicenses subroutine calls the Intermediary in order to obtain the licenses associated with the protected multimedia content, the action (play) and the user. Once a secure channel is established between the Intermediary and the License Server, a message is sent to request the license or licenses related to the action. In the content field of the message the user data, the resource and the right are indicated. Then, the License Server sends a message to the Intermediary and in its content field it places the appropriate REL licenses.
    - AuthoriseUser subroutine performs a license-based authorisation with the REL related information. As the user has a REL license with the appropriate permissions, he is authorised.
    - GetIPMPInfo subroutine obtains the Intellectual Property Management and Protection related information from the Digital Item. It obtains the IPMP Tool information needed to consume the protected resource.
    - GetIPMPTools subroutine calls the Intermediary in order to obtain the IPMP Tool from the Protection Tools Server where it is placed. First, a secure channel is established between the Intermediary and the Protection Tools Server. Then the Intermediary builds a message with the identifier of the IPMP Tool and sends it to the Protection Tools Server. Finally, the requested tool is sent to the Intermediary.
    - The content is obtained from the DI or the Content Server and it is unprotected with the IPMP Tool.
    - The unprotected multimedia content is played.

5 Conclusions

This paper has presented some preliminary ideas for the implementation of a distributed system for distribution and management of protected multimedia content supporting different parts of the MPEG-21 standard. These parts are mainly Digital Item Declaration (DID), Rights Expression Language (REL), Digital Item Processing (DIP) and Intellectual Property and Management Protection (IPMP). Also other parts of the standard have been considered, like Event Reporting (ER) and Digital Item Adaptation (DIA), although they have not been developed yet.

To be precise, we have presented an architecture that contains some of the modules needed to process digital items together with their associated rights and permissions as well as the tools needed to protect them.

Moreover, we have introduced the concept of processing protected multimedia content, by means of DIP, describing some DIBOs and subroutines that would be useful in order to cope with the processing of multimedia information contained or referenced in digital items that have associated rights expressions and require protection tools to access them.

The architecture proposed allows the inclusion or removal of modules and provides the needed flexibility depending on the system implemented. This permits offering protection and licensing functionalities to systems not following MPEG-21 standard at all, for instance, by using other rights expression languages, like Open Digital Rights Language (ODRL) [16] or external content servers.

We have already developed some tools [14], such as those needed to handle REL, and we are working on implementing some DIMs.

The architecture and protocols proposed will be the starting point for a complex multimedia system we are going to develop in the near future.

Other architectures exist for the management of protected multimedia content. Nevertheless, they correspond to proprietary systems and it is difficult to compare our architecture with them, as they do not follow any standard. Among projects funded by public administrations, the MOSES project [17] developed the Open SDRM [18] architecture, with which DMAG-MIPAMS could have some common points.

Acknowledgements. This work has been partly supported by the Spanish administration (AgentWeb project, TIC 2002-01336) and is being developed within VISNET, a European Network of Excellence (http://www.visnet-noe.org), funded under the European Commission IST FP6 program.

References

1. ISO/IEC, ISO/IEC TR 21000-1 Information Technology - Multimedia Framework (MPEG-21) - Part 1: Vision, Technologies and Strategy. December 2001.
2. ISO/IEC, ISO/IEC 21000-2 DID 2nd Edition CD. ISO/IEC JTC 1/SC 29/WG 11/N6409. March 2004.
3. ISO/IEC, ISO/IEC FDIS 21000-5 - Rights Expression Language. ISO/IEC JTC 1/SC 29/WG 11/N5839. July 2003.

4. ISO/IEC, ISO/IEC FDIS 21000-6 - Rights Data Dictionary. ISO/IEC JTC 1/SC 29/WG 11/N5842. July 2003.
5. ISO/IEC, ISO/IEC 21000-10 Study of CD "Digital Item Processing". ISO/IEC JTC 1/SC 29/WG 11/N6415. March 2004.
6. MPEG-21, http://www.chiariglione.org/mpeg/standards/mpeg-21/mpeg-21.htm. 2004.
7. ISO/IEC, ISO/IEC 21000-4 Intellectual Property Management and Protection Working Draft v.1. ISO/IEC JTC 1/SC 29/WG 11/N6644. July 2004.
8. Rodríguez, E.; Llorente, S.; Delgado,J. DMAG answer to MPEG-21 Intellectual Property Management and Protection Call for Proposals. ISO/IEC JTC 1/SC 29/WG 11/M10832. July 2004. http://hayek.upf.es/dmag/DMAGMPEG21Tools/m10832.pdf
9. ISO/IEC, ISO/IEC Digital Item Adaptation Amendment 1 Working Draft 2.0. ISO/IEC JTC 1/SC 29/WG 11/N6413. March 2004.
10. ISO/IEC, ISO/IEC MPEG-21 Event Reporting WD (v1.0). ISO/IEC JTC 1/SC 29/WG 11/N6419. March 2004.
11. Dierks, T. and C. Allen: The TLS Protocol Version 1.0. RFC 2246. January 1999.
12. ISO/IEC, ISO/IEC Work plan for Core Experiment on DIBOs for REL. ISO/IEC JTC 1/SC 29/WG 11/N6418. March 2004.
13. Delgado, J.; Gallego, I.; Rodriguez, E. Use of the MPEG 21 rights expression language for music distribution. Proceedings of the Third International Conference WEB Delivering of Music (WEDELMUSIC'03). Page(s): 16- 19. IEEE Computer Society. ISBN 0-7695-1935-0. September 2003.
14. DMAG MPEG-21 Tools, http://dmag.upf.edu/DMAGRELTools/Index.htm. 2004.
15. De Keukelaere F. et al, Text of ISO/IEC 21000-8 CD MPEG-21 Reference Software. ISO/IECJTC1/SC29/WG11/N6470. March 2004.
16. Iannella, R.: Open Digital Right Language (ODRL) Version 1.1. http://odrl.net/1.1/ODRL-11.pdf. August 2002.
17. MPEG Open Security for Embedded Systems (MOSES) project. http://www.crl.co.uk/projects/moses/
18. Serrao, C. et al. Open SDRM – An open and secure digital rights management solution. November 2003. http://www.crl.co.uk/projects/moses/Public/docs/IADIS03No74.pdf

Application of 'Attack Trees' Technique to Copyright Protection Protocols Using Watermarking and Definition of a New Transactions Protocol SecDP (Secure Distribution Protocol)

Mª Victoria Higuero[1], Juan José Unzilla[1], Eduardo Jacob[1], Purificación Sáiz[1], and David Luengo[2]

[1] University of the Basque Country, Faculty of Engineering of Bilbao,
48013 Bilbao, Spain
{jtphiapm, jtpungaj, jtpjatae, jtppusaa}@bi,ehu.es
http://det.bi.ehu.es
[2] University of the Basque Country, Faculty of Engineering of Bilbao,
48013 Bilbao, Spain
{jtbluvid}@bipt106.bi.ehu.es

Abstract. The ease and convenience that electronic commerce provides, especially when multimedia material is involved, is helping the growth in the number and volume of electronic transactions through the Internet. But this new type of trading operations has highlighted new problems related to copyright protection. Watermarking seems to be an interesting approach to solve these problems. However, watermarks by themselves do not provide enough protection, but they must be used in certain scenarios fulfilling a number of requirements. In this paper we present the results of the application of a risk analysis technique (specifically 'attack trees' technique) to the most demanding schemes developed so far, and a new protocol or business model derived from the main conclusions of this process, which we have called SecDP (Secure Distribution Protocol).

1 Introduction

Watermarking is seen in recent times as the most interesting technique to provide means of protecting Intellectual Property Rights in electronic commerce with digital contents environments. This technique consists in embedding a piece of information in the digital material, containing some data about such material (author, document's owner, and so on). However, watermarking by itself does not resolve the problem of proving ownership (even with the application of cryptographic tools [1]), or identifying the authors of fraudulent actions [2] [3] [4]. In this paper we describe the most important scenarios developed to try and achieve reliable copyright protection. But, considering that sooner or later any system can be put into attack, and that these frameworks can be seen as general information systems, we should analyse several aspects related to the security level that they provide, by trying to answer questions such as who the attackers might be in our environment, which objectives they have,

V. Roca and F. Rousseau (Eds.): MIPS 2004, LNCS 3311, pp. 264–275, 2004.

which type of attack is most likely to happen, and so on. In order to solve these questions is important to apply a formal and systematic method. We have selected 'attack trees' developed by Bruce Schneier [5] [6] as a practical way to analyse the security provided by these systems.

This paper presents some interesting issues regarding the process of applying 'attack trees' to the ECMS (Electronic Copyright Management System) protocol and a number of important conclusions, that will be used later on to design a new transaction model, which we also include in this paper.

The remainder of this paper is organised as follows: In Section 2, the basic properties of the three most important scenarios developed so far are discussed. In Section 3 a review of the main aspects of the application of the 'attack tree' technique to the ECMS model is presented. In Section 4 we introduce the new business model developed, which enhances the security provided by previous scenarios. Some conclusions and other important issues to be taken into account are summarised in Section 5.

2 Transactions Protocols

As mentioned before, watermarks must be used in structured scenarios or business models fulfilling several requirements [7] to provide adequate rights protection to all participants in the transactions. We describe in this section the basics of three of the most important models or protocols defined so far: TTP Protocol [3], Buyer-Seller Protocol [4], and ECMS system [2].

2.1 TTP Protocol

The main characteristic of this protocol, proposed by Qiao and Nahrstedt [3], is the use of a Trusted Third Party (TTP), that works as an intermediary between a buyer and the distributor of the multimedia material. The agents considered in this model are: a Seller or Distributor, a Buyer, and a TTP.

The global process goes as follows: when the Distributor wants to sell a digital content to a Buyer, he sends the original material (V) to the TTP. Then, the TTP embeds a mark generated for the Buyer and, finally, sends the marked document to the Buyer. Besides, the TTP will store the watermark and any other information that could be used to help to solve a possible dispute at a later stage. Sendings of contents are carried out using cryptography. The steps involved in the protocol are the following ones:

$$\text{Distributor} \rightarrow \{V\}K_{\text{TTP-Dist}} \tag{1}$$

$$\text{TTP: } W_B \tag{2}$$

$$\text{TTP: } V_w = V \oplus W_B \tag{3}$$

$$\text{TTP} \rightarrow \{V_w\}\, K_{\text{TTP-B}} \rightarrow \text{Buyer} \tag{4}$$

The main problem of this protocol is that the process depends entirely on the TTP. Because of that, if the TTP goes wrong, the whole system will go wrong, or if anyone compromises the security of the TTP, then the security of the whole system will be compromised. Besides, the performance of the system depends mainly on the performance of the TTP because it performs the most important tasks. And on the other hand this system does not provide support for author rights protection.

2.2 Buyer-Seller Protocol

The Buyer-Seller protocol was proposed by Memon and Wong [4]. This protocol makes use as well of a TTP entity, but in this case, the interchange of marked multimedia material takes place directly between the Distributor and the Buyer. The TTP gets only involved in the generation of the watermark used in every transaction, and later only if there is a dispute to be solved. This scenario uses cryptography for sendings, and privacy homomorphism for preventing the Buyer and the Distributor having the same unencrypted copy of the marked material. The steps involved in the basic process of the Buyer-Seller protocol are summarized below:

$$\text{TTP} \rightarrow \{\{W_B\}Kpu_B\}Kpr_{TTP} \rightarrow \text{Buyer} \tag{5}$$

$$\text{Buyer} \rightarrow \{\{W_B\}Kpu_B\}Kpr_{TTP} \rightarrow \text{Distributor} \tag{6}$$

$$\text{Distributor: } V'_w = V \oplus W_D \tag{7}$$

$$\text{Distributor: } \sigma(\{W_B\}Kpu_B) = \{\sigma(W_B)Kpu_B\} \tag{8}$$

$$\text{Distributor: } \{V''_w\}Kpu_B = \{V'_w\}Kpu_B \oplus \{\sigma(W_B)\}Kpu_B = \{V'_w \oplus \sigma(W_B)\}Kpu_B \tag{9}$$

$$\text{Distributor: Tabla: ID Buyer, } \{W_B\}Kpu_B, W_D, \{\{W_B\}Kpu_B\}Kpr_{TTP}, \sigma \tag{10}$$

$$\text{Distributor} \rightarrow \{V''_w\}Kpu_B \rightarrow \text{Buyer} \tag{11}$$

In the case of appearance of unauthorized copies, the Distributor will be able to identify the authors of the fraudulent actions, by using the data stored in the Distributor's database.

The main drawback of this method is the necessity to rely on a distributor to identify the author of infractions, which may not be too reliable. And, furthermore, buyers must have certain knowledge of the system operation, which is not desirable.

2.3 ECMS Protocol

The Electronic Copyright Management System (ECMS) model, developed by Piva, Bartolini and Barni [2], introduces a slight difference with respect to the previous methods, by considering the Author and Distributor as independent agents. This model is closer to reality, as authors and distributors are usually different entities; an Author creates contents and transfers them to a Distributor, for sale. On the other hand, this protocol allows all participants in e-commerce transactions to verify by themselves their ownership rights. The steps involved in this protocol are summarized as shown below:

First, the Author generates a CUN (Create Unique Number), identifying himself as the Author of the content, and uses CUN to create the first mark (W_A), embeds it in the document, encrypted with his private key, and sends the document with the first watermark to the TTP.

$$\text{Author: CUN} \tag{12}$$

$$\text{Author: } W_A = \{CUN\} \, Kpr_A \tag{13}$$

$$\text{Author: } V'_w = V \oplus W_A \rightarrow TTP \tag{14}$$

Afterwards, the content is registered at the Distributor. When the Author allows a distributor to sell copies of his creations, it creates and embeds a second mark identifying the Distributor as a legal seller of the contents. Then, the Distributor sends his PIN to the Author. The Author creates then a second watermark using Distributor's PIN and document's CUN, and embeds it into the marked document (previously watermarked with the first mark).

$$\text{Distributor} \rightarrow PIN_D \rightarrow \text{Author} \tag{15}$$

$$\text{Author: } W_D = \{ PIN_D \oplus CUN\} \, Kpr_A \tag{16}$$

$$\text{Author: } V''_w = V'_w \oplus W_D \tag{17}$$

$$\text{Author} \rightarrow [V''_w , W_D] \rightarrow \text{Distributor} \tag{18}$$

The purpose of the following steps is to prove the Buyer's participation in the transaction. To that end, the content is marked with the Buyer's PIN. The Distributor sends then the Buyer's PIN, the CUN and the second watermark to the TTP. The process consists of the following steps:

$$\text{Buyer} \rightarrow PIN_B \rightarrow \text{Distributor} \tag{19}$$

$$\text{Distributor} \rightarrow [PIN_B , CUN, W_D] \rightarrow TTP \tag{20}$$

$$\text{TTP: } W_B = \{ PIN_B \oplus CUN\} \, Kpr_{TTP} \tag{21}$$

$$\text{TTP: } V'''w = V'w \oplus W_D \oplus W_B \tag{22}$$

$$\text{TTP: Sign} = \{HASH(V'''w)\} \, Kpr_{TTP} \tag{23}$$

$$\text{TTP} \rightarrow [\text{Sign}, W_D] \rightarrow \text{Distributor} \tag{24}$$

$$\text{Distributor: } V'''w = V''w \oplus W_B \tag{25}$$

$$\text{Distributor} \rightarrow [V'''w , \text{Sign}, W_B] \rightarrow \text{Buyer} \tag{26}$$

This protocol is more complex than the previous ones, but enhances the level of security. It relies heavily on the services of a TTP, but transactions are very simple for buyers. This is normally a requirement in these scenarios, so as not to reduce people's willingness to buy contents.

3   Risk Analysis Application

Historically, risks analysis techniques have been applied to the study of catastrophes (fires, earthquakes, attacks and so on), and more recently they have started to be applied to the analysis and evaluation of the security level of information systems. Because of that, it is also interesting to apply this approach to the identification of weak points or vulnerabilities of the most demanding scenarios developed to provide property rights protection in electronic commerce with digital contents using watermarking. So far, researchers have focused their work mainly on the watermarks insertion process itself, working with different materials [8] [9] [10], on the protocols defining the transactions processes [2] [3] [4], and on the study and classification of the attacks against this type of systems [11] [12] [13] [14]. However, these pieces of work are focused on specific characteristics of the infrastructure, and do not usually analyse the system as a whole.

In this section we present the main aspects of the *'attack trees'* technique and the most relevant issues derived from the application of it to the ECMS protocol. We have selected the 'attack tree' technique because it provides the necessary tools to analyse the security of copyright protection systems, as well as answering basic security questions in a similar way as any other risk analysis technique. Additional advantages of this approach are its simple and practical character, and the possibility of reuse its results easily at a later stage.

On the other hand we present the results of the *'attack trees'* technique applied to the ECMS model because we consider this one as the most complex and secure scenario, as well as the most interesting process. Please do not forget that a big part of the process is similar in the other two protocols.

3.1 *'Attack Trees'*

'Attack trees' is a technique proposed by Bruce Schneier [6]. It is a very simple and intuitive technique, based on the use of expansion trees to show graphically the different threats that could affect any system. An *'attack tree'* is a way of thinking and describing the security of a system. Also, it is a way of building an automatic database describing the security of a system, allowing in this way to build possible countermeasures against the attacks. Besides, it is a way of capturing experience and reusing it.

An *'attack tree'* represents the possible attacks a system can suffer, and the corresponding countermeasures, in a tree expansion structure. The main node, or root node, represents the target of the attack. Logically, in a complex system there will be several main nodes, each one representing a possible threat. The leaf nodes represent different attacks in a hierarchical way, and these can be used to reach the above mentioned threat-goals. By repeating this process, a tree is generated, which describes how the system can be attacked. The next figure (Fig. 1) shows an example of an 'attack tree'.

In the example shown in the figure (Fig. 1), the goal is "Obtain CS's secret key" (CS represents the TTP). Attackers can reach this goal by different strategies: OR by "obtaining the secret key's keystore and its password", OR "by obtaining the secret key when it is decrypted in PC's memory", OR by "using any other way" (threat, blackmail, bribe and so on). AND nodes and OR nodes can appear in the tree. OR nodes represent alternatives, while AND nodes show different steps, all of them necessary to achieve the goal. For example, in the figure, the node labelled "Obtain encrypted secret key" is reached through "obtaining the keystore" AND "obtaining its password" nodes; it is an AND node as, to achieve it, it is required both, "obtain the keystore file" and "obtain the keystore's password".

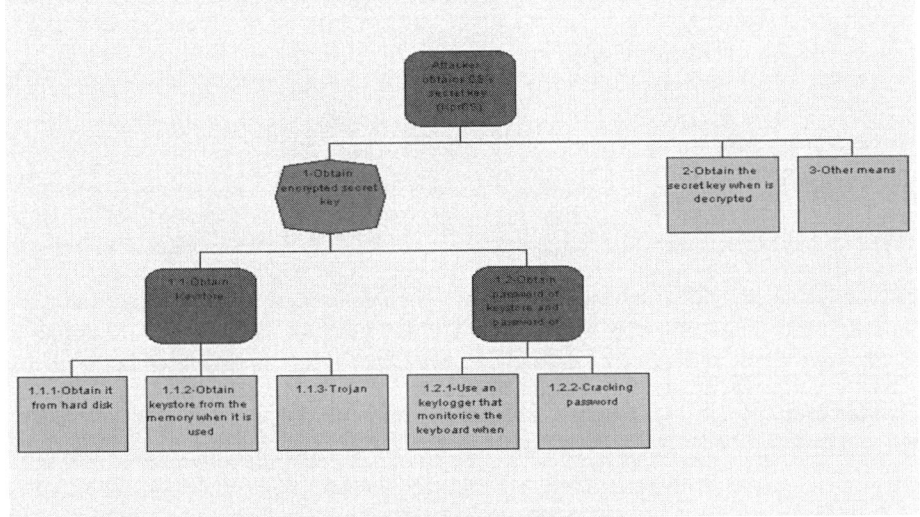

Fig. 1. Example of an 'attack tree'

Besides, every node can have associated some values characterising the threat. Examples of values are: cost of the attack, risk of the attack, technical skill required, possibility or impossibility, and so on.

It is very easy to combine values from nodes to obtain security conclusions, such as: the cheapest attack, best low-skilled attack, cheapest attack with low-risk attack, etc.

3.2   Application of 'Attack Trees' to the ECMS Protocol

In this section, a way to improve the security of the ECMS scenario is presented, enhancing the copyright protection. To do so, we have defined the main possible threats that the system is exposed to, due to its vulnerabilities. The impact of those threats has been valued, and countermeasures necessary to diminish their impact, or the risk of realisation of them, proposed. The analysed threats are:

- An attacker steals a marked image belonging to a registered system user.
- An attacker obtains an illegal copy and fraud cannot be proved, or an attacker proves (falsely) his ownership rights of a document.
- An attacker steals the authorship of an image from a legitimate Author.
- An illegitimate distributor distributes an image and fraud cannot be proved.
- A legitimate buyer cannot prove his rights (and the buyer notices it).
- A legitimate buyer cannot prove its rights to the Verification Authority (and he does not notice it).
- A legitimate distributor cannot prove his rights.
- There is ambiguity in the ownership of a document, among several agents.

In order to show the application of the 'attack trees' to the ECMS protocol, an example will be used, where an attacker has obtained a private key. The 'attack tree' for that threat is shown in the following figure (Fig. 2), where numbers and letters have been used, instead of drawings:

```
0 < OR > Attacker  obtains CS's secret key (KprCS)
   1 < AND > Obtain encrypted secret key
    1.1 < OR > Obtain Keystore
       1.1.1 Obtain it from hard disk
       1.1.2 Obtain keystore from the memory when it is used
       1.1.3 Troyano
    1.2 < OR > Obtain password of keystore and  password of
secret key
       1.2.1 Use a keylogger that monitors the keyboard when
password of keystore is written
       1.2.2 Cracking password
   2 Obtain the secret key when is decrypted
   3 Another means
```

Fig. 2. 'Attack tree' for the threat of obtaining CS's secret key (KprCS)

Applying the 'attack trees' methodology, we assign values to the final leaves in the tree, while the values for the rest of the nodes are obtained by combining the values of the nodes below, until reaching the root node. The values used to characterise these threats are:

- **Cost of attack:** It reflects the economic cost to be assumed by an attacker to carry out the attack. Its value ranges from 0 to infinite.
- **Assumed risk** by an attacker. It is an estimate of the probability of detection or identification of the attack. Its value ranges from 0 to 1.
- **Technical skill** required. It is an estimate of the technical skill level required to execute the attack.. Its value ranges from 0 to 100.

- **Accessibility by external attackers.** It express the dependence of the access type of the attacker on the attack execution. Its values can be TRUE or FALSE.
- **Impact of the attack.** It ranges from 0 to infinite, and reflects the cost implied with the execution of the attack.

The following table (Table 1) shows the values assigned to the final leaf nodes in the previous tree:

Table 1. Values of final leaf nodes in the threat "obtaining CS's secret key (KprCS)"

Threat Id.	Cost of attack	Assumed risk	Technical skill	Accessibility by external attackers	Impact
1.1.1	0	0.1	10	False	0
1.1.2	100000	0.3	50	True	0
1.1.3	10000	0.4	50	True	0
1.2.1	1000	0.3	50	True	0
1.2.2	200	0	20	True	0
2	50000	0.1	1	True	*
3	1000	0.2	50	False	0

The values for other nodes upwards, until reaching the root node, can be calculated from these figures, by means of simple logic or arithmetic operations (AND, OR, minimum value, maximum value, and so on), as shown in Table 2.

Table 2. Some values obtained for threat "obtaining CS's secret key (KprCS)" with AND and OR nodes

Threat Id.	Cost of attack	Assumed risk	Technical skill	Accessibility by external attackers	Impact
1.1	0	0.1	1	True	0
1.2	200	0	10	True	0
1	200	0.1	10	True	0
0	200	0.1	10	True	100000

And, afterwards, having obtained values for every node, we get the main characteristics for all possible attack scenarios, as shown in Table 3. Each line shows a possible attack to materialize the '0' threat. All the values in a file must occur simultaneously for this attack success.

Table 3. Attack scenarios

Scenario Id.	Attack scenario	Cost of attack	Accepted risk	Technical skill	Reachable by external	Impact
1	{1.1.1, 1.2.1}	1000	0.37	50	False	100000
2	{1.1.1, 1.2.2}	200	0.1	20	False	100000
3	{1.1.2, 1.2.1}	11000	0.51	50	True	100000
4	{1.1.2, 1.2.2}	10200	0.3	50	True	100000
5	{1.1.3, 1.2.1}	11000	0.58	50	True	100000
6	{1.1.3, 1.2.2}	10200	0.4	50	True	100000
7	2	50000	0.1	1	True	100000
8	3	1000	0.2	50	False	100000

The next step consists in characterising all possible attackers who could carry out the defined threats against our system. We have chosen the following attacker profiles:

Table 4. Attack profiles

Attacker	Description	Available resources	Accepted risk	Access
A	Usual Internet user	0-100€	<0.3	External
B	Rival distributor	10000-50000€	<0.05-0.1	External
C	Hacker	1000€	<0.2	External
D	Dissatisfied worker	1000€	<0.1	Internal

Finally, we have obtained attack profiles for different attackers. This is just an *'attack tree'* containing only achievable attacks for each attacker. In order to build the profiles, nodes not fulfilling attacker's features are eliminated from the *'attack tree'* associated to the threat. For example, to create the attack profile for a type C attacker, all attacks requiring a cost higher than 1000 € will be eliminated. The same is made for the rest of indicators.

3.3 Results of the Analysis

From the complete process of applying the *'attack trees'* methodology to the ECMS model, we have obtained the main attack scenarios for some defined threats, and the main attacker profiles to be considered in our environment. We have use the SecurI-Tree [16] tool for the study and analysis of the designed trees. And, amongst the most important conclusions obtained as a result of this process, we would like to emphasise the threats capable of causing the greatest impact in the system, which are the following ones:

- The ambiguity that appears when several organizations have the same marked document regarding the identity of the offender.
- MITM (Man-In-the-Middle) attacks against a client's registry phase.

Considering these issues, we have developed a new business model introducing countermeasures to avoid them. In order to solve the threat of ambiguity, we devised our own protocol (Secure Distribution Protocol: SecDP) starting from the basics of the ECMS protocol. And to avoid MITM attacks, we have modified the registry operation in the system, by relieving the user of security decisions such as manually accepting X.509 certificates during the establishment of a SSL connection.

4   SecDP: Secure Distribution Protocol

Our approach, which we have called SecDP, is based on some basic aspects of the ECMS protocol, introducing some modifications to try and avoid the ambiguities and problems mentioned in the previous section.

The steps involved in the SecDP protocol are the following ones: First, the Author generates a CUN (Create Unique Number) to identify himself as the Author of the content, and uses it to create the first mark (WA). Then he embeds the mark, encrypted with the Author's private key, in the digital content and sends it to the TTP.

$$\text{Author: CUN} \tag{27}$$

$$\text{Author: } W_A = \{CUN\} \, Kpr_A \tag{28}$$

$$\text{Author: } V'w = V \oplus W_A \rightarrow TTP \tag{29}$$

To avoid the possibility of the Author and Distributor having the same marked version, the second watermark is created and embedded by TTP.

$$\text{Distributor} \rightarrow PIN_D \rightarrow TTP \tag{30}$$

$$\text{TTP: } W_D = \{ PIN_D \oplus CUN\} \, Kpr_{TTP} \tag{31}$$

$$\text{TTP: } V''_w = V'_w \oplus W_D \tag{32}$$

$$\text{TTP} \rightarrow [V''_w , W_D] \rightarrow \text{Distributor} \tag{33}$$

It can be seen that steps 30 to 33 are rather similar to those in the ECMS protocol, the difference being that here the TTP creates and inserts the second mark, instead of the Author, as in the ECMS model. Thus, the TTP uses his private key to create the mark.

In order to avoid fraud between the Buyer and the Distributor, we propose the use of asymmetric cryptography; after marking the document with the third mark, the TTP encrypts the marked document with the Buyer's public key and sends it to the Distributor, so he cannot decrypt it (because the Distributor does not know the Buyer's private key). The steps involved are:

$$\text{Buyer} \rightarrow PIN_B \rightarrow \text{Distributor} \tag{34}$$

$$\text{Distributor} \rightarrow [PIN_B , CUN] \rightarrow TTP \tag{35}$$

$$\text{TTP: } W_B = \{ PIN_B \oplus CUN\} \, Kpr_{TTP} \tag{36}$$

$$\text{TTP: } V'''w = V''w \oplus W_B \tag{37}$$

$$\text{TTP: Firm} = \{HASH(V'''w)\} \, Kpr_{TTP} \tag{38}$$

$$\text{TTP: } \{ V'''w \} \, Kpu_B \tag{39}$$

$$\text{TTP} \rightarrow [\{ V'''w \}Kpu_B, W_B, \text{ Firm}] \rightarrow \text{Distributor} \tag{40}$$

$$\text{Distributor} \rightarrow [\{ V'''w \}Kpu_B, W_B, \text{ Firm}] \rightarrow \text{Buyer} \tag{41}$$

This process shows the approach used to avoid ambiguity between documents belonging to the Buyer and the Distributor, and the ambiguity between documents belonging to the Distributor and the Author. By following this method, the only entity that could have the three-watermarked document (V'''w) is the Buyer, and the TTP. Hence, if anyone different form the legal Buyer is found to have it, it will be considered the offender, supposing the TTP free of suspicions.

5   Conclusions and Future Work

In this paper we introduce the idea of applying 'attack trees' methodology to analyse the security provided by conventional commerce scenarios with watermarked digital contents.

This method has several advantages in this type of studies, such as its simplicity and practicality, as well as the possibility to analyse systems in a global way. Because of that, we have considered interesting the application of this approach to identify the weakness and vulnerabilities of the most demanding scenarios developed to provide property rights protection in electronic commerce with digital contents using watermarking.

As a result of this work we have found important issues to be solved in the analysed models. One of these is the ambiguity, which exists when two agents in any system have the same unencrypted content and it is impossible to resolve the identity of authors of fraudulent actions. This issue has been solved by introducing a new model for transactions, starting from the basics of the ECMS protocol, which we have called SecDP (Secure Distribution Protocol). This new scheme enhances the security of previous models, by removing the ambiguity problem and solving the MITM problem, while keeping things simple on the user's side, a basic requirement for these systems to success. This way, the feasibility of digital watermarking for copyright protection is strengthened.

However, this type of schemes still has some limitations, mostly due to the amount of operations involved in transactions. They use asymmetrical cryptography and depend completely in TTP services, hence making systems slow. Our approach uses symmetric encryption of the content together with asymmetric ciphering of a session key in the sendings (instead of asymmetric ciphering of the content) to improve the throughput of the transactions. The use of a TTP is necessary to resolve ambiguity, however. Besides, in this model the buyer's anonymity is not possible, being that a problem for cheap contents. Because of this, we consider the use of different business models for distinct categories of contents.

And finally it is also essential that copyright laws be developed in parallel with technical advances, before these systems can be applied in real-world environments.

References

1. Furon, T., Venturini, I., Duhamel, P. Unified approach of asymmetric watermarking schemes. Security and watermarking of multimedia contents III, P.W. WONG and E. Delp, eds., Proc. SPIE, col 4314 (2001), pp 269-279.

2. Piva, A., Bartolini, F., Barni, M.: Managing Copyright in Open Networks, IEEE Internet Computing, (2002).
3. Qian, L. Nahrstedt, K. Watermarking Schemes and Protocols for Protecting Rightful Ownership and Customer's Rights. Journal of Visual Communications and Image Representation, Vol. 9. No. 3 (1998) 194-202.
4. Memon, N., Wong, P. A Buyer- Seller Watermarking Protocol, IEEE Signal Processing Society. Electronic Proceedings. Los Angeles, USA (1998).
5. Amoroso, E.: Fundamentals of Computer Security Technology. Prentice Hall PTR (1994).
6. Schneier, B. Modeling security threats, Dr. Dobb's Journal (1999).
    http://www.schneier.com/paper-attacktrees-ddj-ft.html
7. Tomsich, P., Katzenbeisser, S.: Copyright protection protocols for multimedia distribution based on trusted hardware. PROMS 2000. Cracow, Poland (2000). Pag. 249-25
8. Piva, A., Bartolini, F., Barni, M.: Managing Copyright in Open Networks, IEEE Internet Computing, (2002).
9. Cox, I J., Miller, M.L., Bloom, J.A.: Digital Watermarking. Morgan Kauffman (2002).
10. Goirizelaia. I., Unzilla, J.J.: A new watermarking method using high frequency components to guide the insertion process in the spatial domain. Lectures Notes in Computer Science. Springer Verlang, Ljubljana, Slovenia (1999).
11. Katzenbesisser, S. Peticolas, F.A.P.: Information Hiding: techniques for steganography and digital watermarking. Artech House (2000) .
12. Johnson, N., Duric, Z., Jajodia, S.: Information Hiding: steganography and watermarking – attacks and countermeasures. Kluwer Academic Publishers (2001).
13. Peticolas, F.A.P., Anderson, R.J., Jun, M.G.: Attacks on copyright marking systems. Lecture Notes in Computer Science, Vol. 1525. Springer-Verlag, Portland Oregon, USA (1998) 415–438
14. Craver, S. Memon, B., Yeo, L. Yeung, M.M.: Resolving Rightful Ownerships with Invisible Watermarking Techniques: Limitations, Attacks, and Implications. IEEE Journal of Selected Areas in Communication, 16 (1998). 573-586.
15. Voloshynovskiy, S., Pereira, S., Pun, T.: Attacks on digital watermarks: classification, estimation-based attacks and benchmarks. IEEE Communications Magazine, Vol 39 (2001).
16. Amenaza Technologies Limited. http://www.amenaza.com

Digital Image Authentication Based on Turbo Codes[1]

Qing Yang and Kefei Chen

Department of Computer Science and Engineering, Shanghai Jiao Tong University
1954 Huashan Road, Shanghai 200030, China
yangqing@sjtu.edu.cn

Abstract. Image ownership authentication is an important part of copyright protection, and digital watermark can be used to implement this task. In this paper, we propose a new image authentication plan concentrating on its security performance. Secret information used as the copyright owner's signature is first turbo coded, encrypted, scaled and then processed in wavelet domain. The original image is also needed in signature extraction. Simulation results are finally given to draw our conclusions.

1 Introduction

Nowadays, more and more digital data can be accessed and processed easily. Almost anyone can download, store, and retransmit digital media, such as image, video and audio, without offering any rewards to the copyright owner. This has heavily interfered with the normal development of digital media industry and e-commerce. Digital watermark can provide a solution to this problem by embedding ownership information in the content of digital media. By labeling digital media, usually image, with secret information, such as binary sequence, plain text or other small image, the copyright owner can protect the embedded secret information from being removed or decoded by any attackers without the required and preset authentication keys. And in e-commerce, the potential application for digital watermark is to prove the ownership for copyright protection or to hide images for security reasons.

Research in digital watermark has experienced two major phases: at the beginning, various kinds of methods were proposed relying on psycho-visual techniques, but the researchers seldom paid much attention to the effective ways of detecting/extracting the watermark information, and there were not many effective methods for assessing these algorithms' performance. In the following period, started in late 1990's, people found the similarities between digital watermark and digital communication systems. Hence, a set of analysis and synthesis tools, generally based on statistical theories, is brought out.

Researchers have invented lots of watermarking plans these years, and some are designed for image authentication exclusively. For those plans, how to increase the security level has become one important and even crucial issue for final practical use.

---

[1] This work was supported by National Natural Science Foundation of China (NSFC) under the grant 90104005 and National Hi-Tech Research and Development Program of China (863 Program) under the grant 2001AA144060.

V. Roca and F. Rousseau (Eds.): MIPS 2004, LNCS 3311, pp. 276–285, 2004.

And the coding efficiency has become a bottleneck in some certain environment. But unfortunately, these problems are still unresolved and there is so much work to be done. In this paper, we try to overcome this problem and propose an experimental watermarking plan for image authentication based on public key cryptosystem [2] [7] and turbo code [1].

In section 2, we first explain some concepts about Discrete Wavelet Transform (DWT) and turbo code, present the public-key signature embedding and extracting plan based on turbo code in wavelet domain, and explain the signature extracting method. Here, the word "signature" refers to the secret information embedded in host image to realize ownership authentication, and the plan in this paper is not digital signature. The performance of turbo coding technique in watermarking is tested and analyzed through various distortions in the next parts. Finally, conclusions and further discussions are given in the last section.

2 Watermark Embedding and Extracting Plan

2.1 Wavelet Transform of Images

DWT is identical to a leveled subband system, while the subbands are logarithmically located in frequency.

For a 1-D digital image, a 1-D wavelet function Ψ and a 1-D scaling function Φ are chosen to iteratively decompose the signal into different-scaled high-frequency subbands. 1-D DWT can be implemented by Mallat's Direct Pyramid Algorithm [3].

For a 2-D digital image, the same wavelet function Ψ and scaling function Φ are used in both vertical and horizontal directions to decompose the image. For example, the scaling function $\Phi_{LL}(x, y)$ of low-low subband in a 2-D wavelet transform is implemented by $\Phi(x)\Phi(y)$. Three other 2-D wavelet functions are obtained by the wavelet function $\Psi(x)$ as follows:

$$\Psi_{LH}(x, y){=}\Phi(x)\Psi(y)$$

$$\Psi_{HL}(x, y){=}\Psi(x)\Phi(y) \tag{1}$$

$$\Psi_{HH}(x, y){=}\Psi(x)\Psi(y)$$

In the above equations, H means a high-pass filter and L is a low-pass filter.

The basic idea of 2-D DWT of images is described as follows. An image is first decomposed into four frequency subbands (i.e., LL_1, HL_1, LH_1 and HH_1) by corresponding filters. The subbands labeled with HL_1, LH_1 and HH_1 represent the finest scale wavelet coefficients. To obtain coarser-scaled wavelet coefficients, the subband LL_1 should be further decomposed and critically subsampled. This process is repeated for an arbitrary number of times, which is determined by the application needs. A layout of DWT subbands with three-level decomposition of Lena image is shown in Fig. 1.

In Fig. 1, Lena image is decomposed into ten subbands with three scaled levels. Each level has various band-information such as low-low, low-high, high-low, and high-high frequency subbands. Furthermore, the original image can be reconstructed from these DWT coefficients.

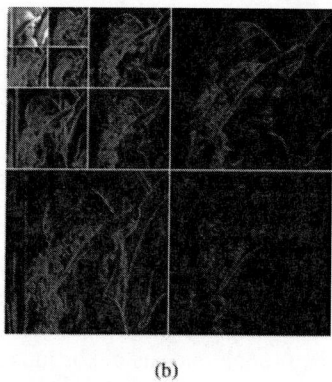

(a)   (b)

Fig. 1. Example of Wavelet Decomposition. (a) Original Lena image; (b) DWT decomposition of Lena image

2.2   Turbo Code in Watermarking

Turbo code is a new class of Error Correcting Codes (ECC) that was introduced along with a practical decoding algorithm. Parallel concatenation of convolutional codes is used to give the code structure so that it can be decoded easily. Pseudorandom interleaving is used to give the code performance which approaches that of random coding in Shannon's theory. The original turbo code is a parallel concatenation of two recursive systematic convolutional (RSC) encoders, while the turbo decoder consists of two concatenated decoder of the component codes separated by the same interleaver. The component decoders are based on a maximum a posteriori (MAP) probability algorithm or a soft output Viterbi algorithm (SOVA) generating a weighted soft estimate of the input sequence. The iterative process performs information exchange between the two component decoders [1].

2.2.1   Encoder

Turbo encoder is basically composed of two RSC encoders, and a rate R=1/3 turbo encoder is shown in Fig. 2.

In Fig. 2, each RSC encoder has two registers (M=2), and the same transfer function (7, 5) is used. One encoder receives the original information sequence and the other receives the interleaved version of the original one. The information sequence d_k is made up of independent bits taking values 0 and 1 with equal probability. Given an input d_k, the encoder outputs at time k are information bit d_k and parity bits Y_{ik}. The output, d_k and Y_{ik}, are modulated with Binary Phase Shift Keying (BPSK) and noise n_{ik}. The above process can be expressed as follows:

$$x_k = 2(d_k - 1) + n_{1k} \tag{2}$$
$$y_k = 2(Y_{ik} - 1) + n_{2k}$$

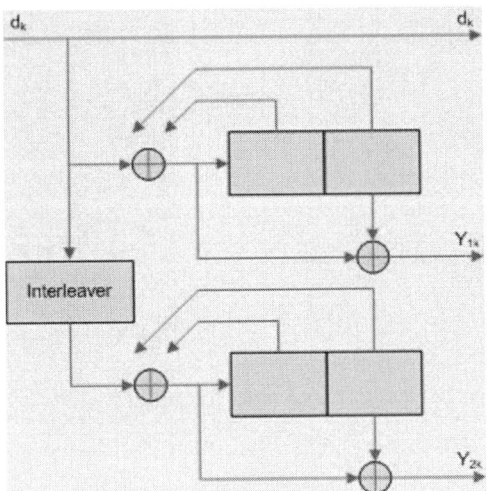

Fig. 2. Turbo Encoder

2.2.2 MAP Decoder

The MAP algorithm makes a decision on the information bit d_k based on Log-Likelihood Ratio (LLR), which is the logarithm of the ratio of the A Posteriori Probability (APP) of each information bit d_k. Let the state of the encoder at time k be S_k. The LLR of bit d_k is given by:

$$\Lambda(d_k) = \ln \frac{\sum_{S_k=0}^{2^M-1}\sum_{S_k=1}^{2^M-1} \gamma_1(R_k, S_{k-1}, S_k)\alpha_{k-1}(S_{k-1})\beta_k(S_k)}{\sum_{S_k=0}^{2^M-1}\sum_{S_k=1}^{2^M-1} \gamma_0(R_k, S_{k-1}, S_k)\alpha_{k-1}(S_{k-1})\beta_k(S_k)} \tag{3}$$

where the numerator and denominator of (3) represent the APP of each information bit 1 and 0, $R_k = (x_k, y_k)$ is the received channel output data at time k, $\gamma_i(R_k, S_{k-1}, S_k)$ denotes Branch Transition Probability (BTP) when the information bit is i, and $\alpha_k(S_k)$ and $\beta_k(S_k)$ are Forward Recursion (FR) and Backward Recursion (BR) respectively. They are given by:

$$\alpha(S_k) = \frac{\sum_{S_{k-1}=0}^{2^M-1}\sum_{i=0}^{1} \gamma_i(R_k, S_{k-1}, S_k)\alpha_{k-1}(S_{k-1})}{\sum_{S_k=0}^{2^M-1}\sum_{S_{k-1}=0}^{2^M-1}\sum_{i=0}^{1} \gamma_i(R_k, S_{k-1}, S_k)\alpha_{k-1}(S_{k-1})} \tag{4}$$

$$\alpha_0(S_0) = 1 \quad \text{for } S_0 = 0$$
$$\alpha_0(S_0) = 0 \quad \text{else}$$

$$\beta(S_k) = \frac{\displaystyle\sum_{S_{k+1}=0}^{2^{M-1}} \sum_{i=0}^{1} \gamma_i(R_{k+1}, S_k, S_{k+1})\beta_{k+1}(S_{k+1})}{\displaystyle\sum_{S_k=0}^{2^M-1} \sum_{S_{k+1}=0}^{2^M-1} \sum_{i=0}^{1} \gamma_i(R_{k+1}, S_k, S_{k+1})\alpha_k(S_k)} \tag{5}$$

$$\beta_N(S_N) = 1 \quad \text{for } S_N = 0$$
$$\beta_N(S_N) = 0 \quad \text{else}$$

The BTP in FR and BR is given by:

$$\gamma_i(R_k, S_{k-1}, S_k) = P(R_k / d_k = i, S_{k-1}, S_k)P(d_k = i / S_{k-1}, S_k)P(S_k / S_{k-1}) \tag{6}$$

where the first part of right side is channel transition probability (CTP), the second one is either 0 or 1 depending on whether the bit i is associated with the transition from S_{k-1} to S_k or not, and the third one is state transition probability (STP) that uses a priori probability of information bit d_k.

The turbo decoder is shown in Fig. 3.

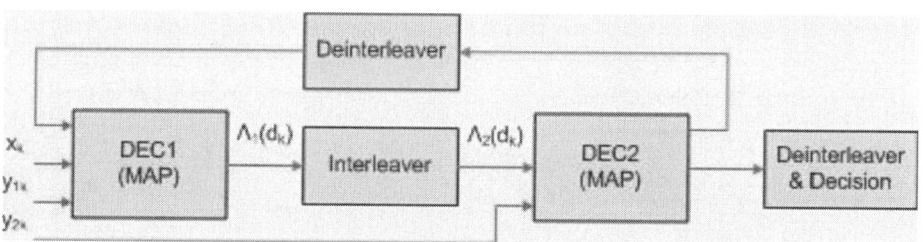

Fig. 3. Turbo Decoder

2.3 Signature Embedding and Extracting

In our plan, the signature for ownership authentication is embedded in the image as digital watermarks, and the watermarks are embedded in wavelet domain. The reason for adopt wavelet based watermarking plan is the fact that most network based images and video are in compressed form, and wavelets have played an important role in the new compression standards such as JPEG 2000 and MPEG-4.

The watermark embedding and extracting process is shown in Fig. 4.

In Fig. 4, DWT stands for Discrete Wavelet Transform and IDWT stands for the inverse process. KR is private key and KU is public one in public key cryptosystem.

The signature may be text or image. For image signature, it should be first compressed using vector quantization before the following procedure to be suitable for embedding into image. The quantized signature data are turbo coded [4] and encrypted by the copyright owner's private key KR.

According the perceptual model, the encrypted data are scaled to guarantee watermarking invisibility and added to the DWT coefficients of the original image. IDWT is then performed and we'll get the watermarked image.

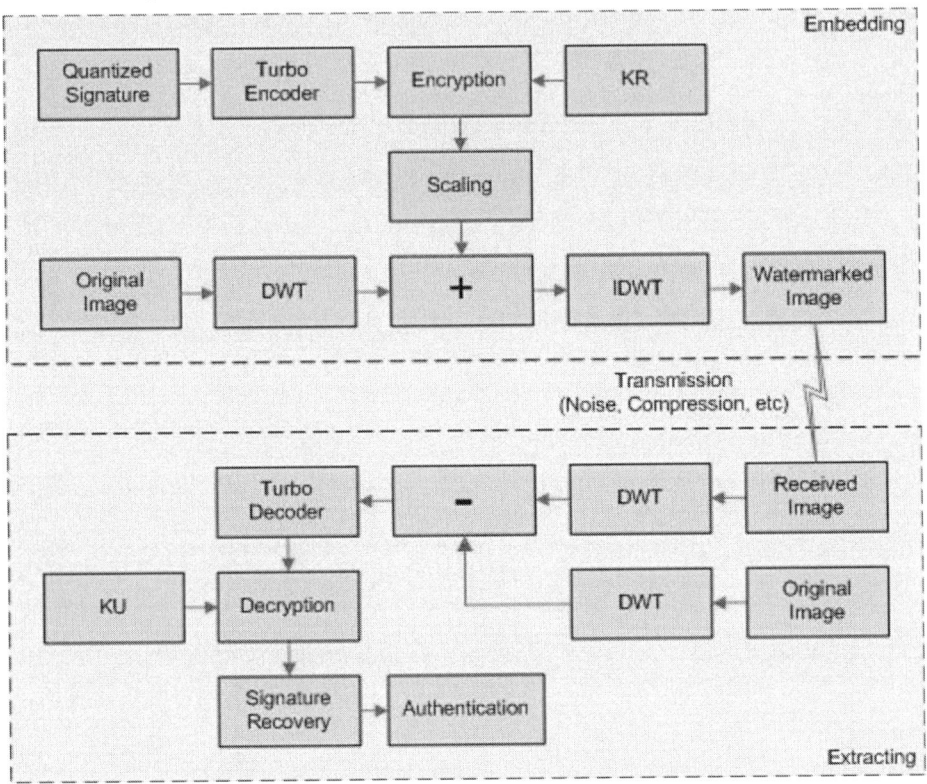

Fig. 4. Watermark Embedding and Extracting Process

During transmission or distribution, noise and compression will interfere with the quality of the watermarked image. And attackers are always trying to break the secret hidden in the image. But without the key and appropriate turbo decoder [5], attackers will not get anything useful and the image will be totally destroyed. In these conditions, there is no need for us to carry out watermark extracting and signature authentication in the following procedures [6].

For watermark extracting, original image is also needed. The DWT coefficients of the received watermarked image are subtracted with those of the original image, and we can get turbo coded data. After being sent to the turbo decoder, these data are decrypted using the public key KU, the signature is then recovered and may be used for authentication by the copyright owners and the authorized users.

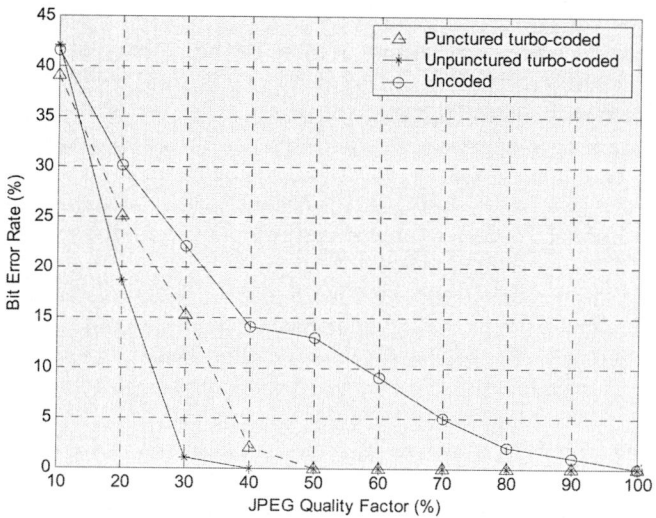

Fig. 5. Bit Error Rate vs. JPEG compression for turbo coded and uncoded messages

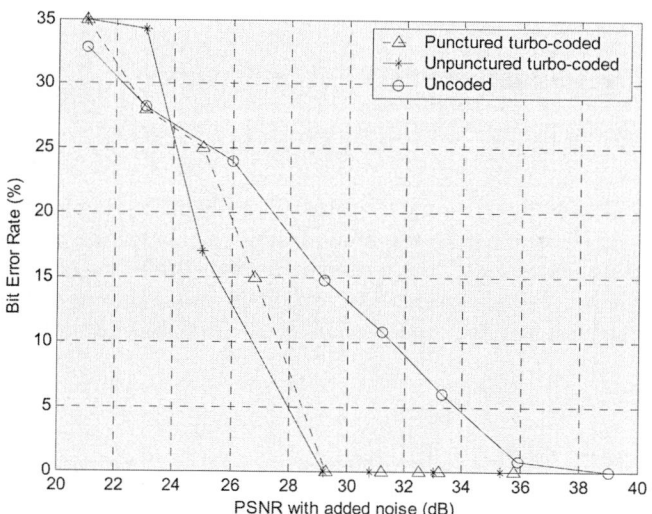

Fig. 6. Bit Error Rate vs. noise addition for turbo coded and uncoded messages

2.4   Performance of Turbo Codes in Watermarking

In order to demonstrate the effectiveness of turbo coding in our plan, we watermarked Lena image using turbo codes. We use the Haar wavelet transform for all simulations, and the scale parameter equals 10.

For our tests, we demonstrate the performance of turbo codes in watermark embedding and extracting when the watermarked image undergoes distortions. Here we introduce the two most common distortions: JPEG compression and noise addition. The watermark is then extracted and compared with the original watermark to measure the robustness and capability of our plan.

For the use of turbo codes in encoding the signature, the DWT coefficients of Lena image is modified to enable the hiding of encoded data. In our test, a random message of length 1000 bits is turbo coded using two parallel convolutional encoders of rate 1/2 with the transfer function (7, 5). Punctured encoding brings about a rate of 1/2 for the turbo codes while unpunctured encoding gives a rate of 1/3. The decoder uses logmap decoding with 5 iterations. Related results are shown in Figure 5 and 6 for the bit error rate vs. JPEG compression and noise addition respectively. The difference in Bit Error Rate (BER) between the coded and uncoded messages is quite significant especially for lossless recovery (BER=0). However, little difference exists between punctured and unpunctured codes. And the number of bits hidden in the case of punctured code will be smaller than that of unpunctured code.

3   The Coding Technique in Digital Watermark

Digital watermark system can be viewed as one digital communication system. To guarantee the survivability of the watermark message, i.e. the signature in this paper, reliable transmission of the watermark symbols through this particular communication channel is a must. The signal-to-noise ratio in this special channel is very low due to the imperceptibility constraints, so it's natural to adopt Error Correcting Code (ECC) in digital watermark. Among various ECC schemes, turbo code is a new comer and "whiz kid" compared with other "older" ECC codes. It's also the starting point of this paper.

3.1   Convolutional Coding

As we've stated previously, turbo code is the concatenation of convolutional codes. For convolutional code, it's been extensively used in digital communication systems because of its good performance and low computational complexity. Convolutional code is more powerful when it's combined with soft-decision Viterbi decoding, which is actually the optimum Maximum Likelihood (ML) decoding method for Additive White Gaussian Noise (AWGN) channel. For the convolutional codes, the classical decoding approach is to use the hard-decision decoder. This kind of decoder consists of two steps: first, a binary decision is made for each of the outputs of the AWGN channel, just comparing them with a preset threshold; then, the resulting bits, also called hard information, are fed into a binary decoder.

However, hard-decision decoding is not the optimum decoding approach since there will be some loss of valuable information in the hard decision process. Better performance can be achieved if the decoder takes directly the real-valued outputs from the AWGN channel and provides the decoded bit sequence as output results. The outputs of the AWGN channel are also called soft information, so this decoding strategy is called soft-decision decoding.

Soft-decision decoders in practice with block codes are proved to be computational complex, while convolutional codes allow low-cost implementations of soft-decision decoders employing the well-known Viterbi algorithm. This is the reason why this kind of codes can be superior to good block codes such as BCH.

Now, consider the AWGN channel model applied to the digital watermark system, we can use the channel codes commonly used in communications. In addition, the real valued outputs r_i, $1 \quad i \quad n$, of the equivalent channel are indeed soft information which can be used by a soft-decision decoder. For this reason, convolutional codes combined with a soft-decision Viterbi decoder will form a viable alternative to block codes.

3.2 Turbo Coding

As stated in section 2.2, the original term "turbo code" refers to parallel concatenation of two or more systematic convolutional encoders that are decoded iteratively, with the advantage of a complexity similar to that of the constituent codes. The basic part of turbo code is an interleaver. In the case of parallel concatenated codes, the interleaver acts upon the information bits that enter the parallel encoders. Before the practical ML decoding, the interleaver will work for several iterations and this will bring about some computational cost. This iterating process will be halted as soon as the decoding result reaches the preset threshold, and near-optimum performance can be achieved.

The decoding algorithm of turbo code is based on Maximum a Posteriori (MAP) probability, gives soft information along with hard decision and provides an effective way of exchanging information between decoders.

For better understanding of turbo code, concatenated code and interleaver, please refer to [8] and [9].

4   Conclusions and Discussions

This paper proposes an image authentication plan using public key and private key, in which turbo coding and DWT enhance the efficiency and data capability of digital watermarks. In our plan, public key cryptosystem and appropriate coding technique can guarantee the security of data embedding and extracting.

In these years, people have studied digital watermark in spatial, DWT, DCT, DFT and other domains, and have found that each research aspect has its own advantages and disadvantages. After detailed comparison, we think wavelet based watermark is more efficient and easer to handle. So we choose DWT as our research method.

The combination of turbo code and public key cryptosystem into digital watermark is a new idea in digital watermark research circle. There are different kinds of public key cryptosystems that can be used in this watermarking plan and we just propose such an idea. In the future, we will continue to work on this point and put forward more practical plans.

Though original image is needed in watermark extracting process, we don't think it's a weak point, for our plan is used for image ownership authentication, and not for secret communication. The copyright owner wants to know whether the copies he assigned to the corresponding users previously are illegally distributed or duplicated. And the end user wants to know whether he gets the genuine and legal copy of the copyright-protected image product or check the alleged owner's identification. That's to say, we just want to extract the related signature information embedded in the watermarked image to check the user's identification, the image copy's validity or serial number. In such conditions, the original image will be surely possessed by the legal copyright owner or the trusted third party who is responsible for key distribution and management. So, it's not difficult for the verifier to get the original image and extract related secret.

Furthermore, without the original image and corresponding key, no meaningful information can be extracted, and attackers can only get random-like noise. This will ensure the validity of copyright owner and copy user.

To the coding technique, better error correction performance can be acquired by adopting longer turbo code, but this will bring about longer decoding delay. So, there is some tradeoff between the performance and related efficiency in our plan.

An experimental plan is proposed in this paper, and much work has to be done to perfect it. Digital watermark needs continuous development and enrichment, so is our plan. In the following report, we'll try to convert this plan into a blind one, in which the original image is no longer needed to extract the watermark. We'll provide more testing results to support our conclusion and make our plan more practical.

References

1. Berrou, C., Glavieux, A.: Near Optimum Error Correcting Coding and Decoding: Turbo-codes. IEEE Transactions on Communications, Vol. 44, Issue 10 (1996) 1261-1271
2. Diffie, D., Hellman, M.: New Directions in Cryptography. IEEE Transactions on Information Theory, Vol. 22, No. 6 (1976) 644-654
3. Mallat, S.G.: Multifrequency Channel Decompositions of Images and Wavelet Models. IEEE Transactions on Acoustics, Speech, and Signal Processing, Vol. 37, No. 12 (1989) 2091-2110
4. Chou, J., Ramchandran, K.: Robust Turbo-based Data Hiding for Image and Video Sources. ICME (2002) 565-568
5. Cvejic, N., Tujkovic, D., Seppanen, T.: Increasing Robustness of an Audio Watermark Using Turbo Codes. ICME (2003) 217-220
6. Cox, I.J., Kilian, J., Leighton, F.T., Shamoon, T.: Secure Spread Spectrum Watermarking for Multimedia. IEEE Transactions on Image Processing, Vol. 6, No. 12 (1997) 1673-1687
7. Kuroda, K., Nishigaki, M., Soga, M., Takubo, A., Nakamura, I.: A Digital Watermark Using Public-key Cryptography for Open Algorithm. http://charybdis.mit.csu.edu.au/~mantolov/CD/ICITA2002/papers/131-21.pdf, 2003
8. Benedetto, S., Divsalar, D., Montorsi, G., Pollara, F.: Analysis, Design and Iterative Decoding of Double Serially Concatenated Codes with Interleavers. IEEE Journal on Selected Areas in Communications, Vol. 16, No. 2 (1998) 231-244
9. Hagenauer, J., Offer, E., Papke, L.: Iterative Decoding of Binary Block and Convolutional Codes. IEEE Transactions on Information Theory, Vol. 42, NO. 2 (1996) 429-445

Author Index